Practical Machine Learning Illustrated with KNIME

Yu Geng • Qin Li • Geng Yang • Wan Qiu

Practical Machine Learning Illustrated with KNIME

 Springer

Yu Geng
School of Software
Shenzhen Institute of Information Techn
Shenzhen, China

Qin Li
School of Software
Shenzhen Institute of Information Technology
Shenzhen, China

Geng Yang
School of Software
Shenzhen Institute of Information Techn
Shenzhen, China

Wan Qiu
School of Software
Sehnzhen Zhaoyang Institute of Information
Technology
Shenzhen, China

ISBN 978-981-97-3953-0   ISBN 978-981-97-3954-7   (eBook)
https://doi.org/10.1007/978-981-97-3954-7

Jointly published with Publishing House of Electronics Industry

This Springer imprint is published by the registered company Springer Nature Singapore Pte Ltd.
The registered company address is: 152 Beach Road, #21-01/04 Gateway East, Singapore 189721,
Singapore

If disposing of this product, please recycle the paper.

Preface

At the Shenzhen Institute of Information Technology (SZIIT), our faculty, many with PhD degrees, faced a big challenge: How should we break down complex machine learning (ML) and artificial intelligence (AI) concepts for vocational students? At first, we were concerned that our advanced qualifications might not be put to good use, but we quickly realized the key was to make these subjects simpler for our students. Their straightforward questions demanded clear, practical answers, influencing how we crafted this textbook.

Recognizing the scarcity of entry-level resources in AI and ML, we, the authors, have drawn from our experiences at SZIIT to develop an instructional methodology that revolves around visually illustrating concepts, adopting a top-down approach, steering clear of unnecessary excess, and promoting hands-on mastery. The distinctive feature of this book lies in its emphasis on graphical explanations, using visual representations to make complex ML principles more intuitive. By incorporating KNIME, a graphical ML tool, we make learning more accessible for beginners, eliminating the daunting prerequisite of programming.

Our "top-down" approach structures the content with a clear narrative on ML models, supported by two threads that align with our educational philosophy. The first thread involves mastering KNIME from an overall workflow to detailed operations, while the second transitions from a macro understanding of ML to the mastery of specific skills. The textbook focuses on graphical understanding of algorithms and application-oriented operations using KNIME, without delving into complex theoretical formulas. Hands-on mastery, another key feature, is demonstrated through the participation in Kaggle competitions, ensuring learners master ML processes, model configurations, data cleaning, and handling imbalanced datasets.

This textbook aims to provide a comprehensive and accessible introduction to AI and ML. It is structured into three parts: Introduction to AI Technology, Traditional Machine Learning, and Deep Learning. The first part lays the groundwork of essential knowledge, the second part explores various ML models and techniques, and the concluding part introduces the basics of deep learning.

Primarily crafted for beginners, this textbook is also a valuable resource for anyone interested in the practical applications of AI and ML. Our aspiration is that readers will, in the spirit of Caesar's renowned quote, "Veni Vidi Vici," truly come, see, and conquer the basics of ML.

My sincere appreciation extends to all collaborators, including co-authors, editors, reviewers, and educational institutions, for their invaluable contributions to this publication. As you engage with this content, we welcome your feedback, which is crucial for our continuous improvement.

In conclusion, whether you are a novice or an enthusiast, this textbook offers a holistic learning journey in the realms of AI and ML. "A visual representation can hold thousands of expressions," and we hope this book effectively portrays these groundbreaking technologies.

Shenzhen, China   Yu Geng
Shenzhen, China   Qin Li
Shenzhen, China   Geng Yang
Shenzhen, China   Wan Qiu

Contents

About the Authors

Yu Geng Yu Geng is an Associate Professor at the Shenzhen Institute of Information Technology. He holds a PhD from the Hong Kong University of Science and Technology and specializes in liquid crystals, semiconductor optoelectronic devices such as detectors and lasers, and AI application development. He has published over 30 papers, including several SCI papers as the first author, holds four invention patents, and has led and participated in numerous research projects.

Qin Li is an accomplished researcher with a Ph.D. from Hong Kong Polytechnic University, which he obtained in 2010. Following his postdoctoral work at Shenzhen University in 2013, he embarked on his journey as an associate professor at the Shenzhen Institute of Information Technology. His primary research focus is on the fundamental theories of pattern recognition.

Since 2001, he has made significant contributions to his field, publishing over 40 articles that have been recognized and cited in prestigious journals indexed by SCI/EI, with more than 10 of them being featured in SCI Zone 1. He has played a leadership role in guiding two Natural Science Foundation projects of Guangdong Province and holds five patents, including one in the United States. liqin@sziit.edu.cn.
Shenzhen Institute of Information Technology, Shenzhen, China

Geng Yang is a distinguished professional with an EngD degree earned from Hong Kong Polytechnic University in 2018. He currently holds the title of Senior Engineer and serves as an esteemed member of the Shenzhen Emergency Management Technology Informatization Consulting Expert Committee. His research primarily revolves around the innovative applications of artificial intelligence and blockchain technologies. His groundbreaking work has led to the successful implementation of numerous related achievements that have left a lasting impact in the field. yangg@sziit.edu.cn.
Shenzhen Institute of Information Technology, Shenzhen, China

Wan Qiu is the Chairman and Director of Shenzhen Zhaoyang Institute of Information Technology, an EMBA graduate from Tsinghua University, and holds a bachelor's degree in electronic information engineering from Zhejiang University.

Formerly, he served as the Director of the Training Department of Shenzhen High-tech Association, with a primary focus on the application of artificial intelligence in intelligent manufacturing, smart management, and data mining. qwvivienqiu@hotmail.com.

Shenzhen Zhaoyang Institute of Information Technology, Shenzhen, China

Chapter 1
Overview of Artificial Intelligence and Machine Learning

Artificial intelligence (AI) would be the ultimate version of Google. The ultimate search engine that would understand everything on the web. It would understand exactly what you wanted, and it would give you the right thing. We're nowhere near doing that now. However, we can get incrementally closer to that, and that is basically what we work on.

—Larry Page, Interview, Oct. 28, 2000

Abstract This chapter offers a comprehensive overview of artificial intelligence (AI) and machine learning (ML), essential for understanding these transformative technologies in the modern era. Beginning with a historical perspective, it charts the evolution of AI from its conceptual origins in the late 1930s to its present-day advancements. The discussion encompasses key milestones such as the Dartmouth Conference, the rise and fall of expert systems, and recent breakthroughs in deep learning and neural networks.

The chapter also delves into the essence of machine learning, highlighting its three primary types: supervised, unsupervised, and reinforcement learning. It illustrates how these methods are pivotal in pattern recognition and predictive analysis, forming the backbone of current AI applications. Additionally, the chapter explores the intersection of machine learning with traditional statistical methods, shedding light on their similarities, differences, and complementary roles in data analysis.

Through this chapter, readers gain a clear understanding of AI's capabilities and limitations, its historical context, and its potential future trajectory. This sets the stage for a deeper exploration of AI's practical applications and ethical implications in subsequent chapters.

Keywords Artificial intelligence · Machine learning · Supervised learning · Unsupervised learning · Reinforcement learning · Deep learning applications

This chapter covers

- **History of artificial intelligence.**
- **Definition of artificial intelligence.**

- **Capability of artificial intelligence.**
- **Definition of machine learning.**
- **Statistics of machine learning.**

In this chapter, we'll review the history of artificial intelligence (AI) and get a basic understanding of machine learning and deep learning. We'll also explore how machine learning and statistics are interconnected.

1.1 Overview of Artificial Intelligence

Artificial intelligence (AI) has become incredibly popular in recent years, sparking various opinions. Let's explore its history and uncover what AI is all about, as well as the exciting possibilities for its future.

Points for Discussion

- What knowledge have you acquired that can be applied in the realm of AI?
- What AI technologies have you encountered during your professional or academic pursuits?

1.1.1 A Brief History of Artificial Intelligence

By looking at both the past and the present, we aim to develop a clear understanding of the core of AI.

1.1.1.1 Starting up: Late 1930s to Early 1950s

The origins of AI research can be traced back to a series of scientific breakthroughs from the late 1930s to the early 1950s. Neurological studies have shown that the brain functions like an electronic network of neurons, operating in binary states of "on" and "off." Norbert Wiener's cybernetics explains the control and stability of electronic networks, while Claude E. Shannon's information theory explores digital signals using binary numbers of 0 and 1. Alan Turing's theory of computation establishes that digital signals can describe any type of computation. These interconnected ideas suggest the possibility of building an electronic brain.

This phase of work includes the development of some robots, such as the "Johns Hopkins Beast" and William Grey Walter's Turtle Robots. These machines do not use computers, digital circuits, or symbolic reasoning. They are controlled by purely analog circuits. Walter Pitts (logician) and Warren McCulloch (neuroscientist) analyzed idealized artificial neuron networks and pointed out their mechanisms for performing simple logical operations. They were the earliest scholars to describe the so-called neural networks. Their student Marvin Lee Minsky was a 24-year-old graduate student at the time. In 1951, Minsky and other researchers built the first neural network machine, known as SNARC. In the next 50 years, Minsky was one of the most important leaders and innovators in the field of AI.

In 1951, using the Ferranti Mark 1 machine of the University of Manchester, Christopher Strachey wrote a checkers program and Dietrich Prinz wrote one for chess. In the mid-1950s and early 1960s, Arthur Samuel developed a chess program that became decent enough to challenge an amateur. AI in games has always been considered as a standard for evaluating the progress of AI.

The birth of AI is marked by the Dartmouth Summer Research Project held in 1956, during which the name and goals of AI were coined. One of the key assertions made during this gathering was that "every aspect of learning or any other feature of intelligence can in principle be so precisely described that a machine can be made to simulate it." The participants comprised a distinguished group of scientists who would go on to make substantial contributions to the field of AI in its formative decade. At this workshop, Allen Newell and Herbert Simon discussed "logic theorist," while John McCarthy convinced participants to pick the name "artificial intelligence" for the new field. The 1956 Dartmouth Summer Research Project stands as a significant milestone in the history of AI, as it not only established the name and purpose of this field but also witnessed the emergence of pioneering achievements and a cohort of early researchers. Consequently, this event is widely acknowledged as the birth of AI.

1.1.1.2   Golden Age: 1956–1974

The Dartmouth Summer Research Project played a crucial role in sparking the first global wave of AI. During this time, there was a sense of optimism in the academic community, leading to numerous groundbreaking advancements in algorithms. One notable milestone was the development of the Bellman equation, which laid the foundation for reinforcement learning and the training of ChatGPT by OpenAI. Deep learning models, based on artificial neural networks, also originated in this period, with the perceptron being a pioneering example. Additionally, alongside algorithmic and methodological progress, the first generation of AI scientists created intelligent machines. Notable examples include STUDENT (1964), capable of solving practical problems, and ELIZA (1966), which engaged in basic human-machine dialog.

During this era, computers demonstrated their prowess in solving algebraic application problems, proving geometric theorems, and even acquiring and utilizing the English language. In private discussions and academic writings, researchers expressed a cautious yet hopeful belief that machines with advanced intelligence would become a reality within just two decades.

1.1.1.3 The First Setback: 1974–1980

However, pride goes before a fall. Due to a multitude of objective constraints during that period, it became evident that logical prover, perceptron, and reinforcement learning were only capable of executing rudimentary and specialized tasks, proving inadequate for even slightly broader endeavors. Two limitations emerged: firstly, the mathematical models and methods upon which AI is founded were discovered to possess inherent flaws; secondly, numerous computational tasks exhibited exponential complexity, rendering them insurmountable. Even the most exceptional AI programs could merely tackle a fraction of the challenges they aimed to address, thus relegating all such programs to mere "playthings."

Inherent flaws have led to AI encountering significant obstacles during its initial stages of development, marking the advent of its first setback. As a result, financial backing for AI has been slashed or altogether withdrawn in response.

1.1.1.4 Rejuvenation: 1980–1987

During the 1980s, a remarkable wave of advancements in AI mathematical modeling unfolded, giving birth to renowned innovations like multilayer neural networks and backpropagation algorithms, among others. By utilizing these algorithms, the automated recognition of postal codes on envelopes became possible with an impressive accuracy of over 99%, surpassing the abilities of an average person. As a result, society once again became captivated by the wonders of AI.

In the year 1980, Carnegie Mellon University pioneered the development of an expert system for Digital Equipment Corporation (DEC), yielding remarkable benefits. This expert system proved instrumental in enabling DEC to achieve annual savings of approximately 40 million dollars, particularly through its invaluable contributions to decision-making processes. As a result, in 1982, numerous countries, including Japan and the United States, started to make substantial investments toward the advancement of what was then referred to as the "fifth generation computer," commonly known as the AI-empowered computer.

Expert systems have the ability to simulate the expertise of human professionals in specific fields, which can lead to new trends in finance. The Japanese government, in particular, has been actively involved in spearheading the development of the fifth generation of computers. This has indirectly prompted the United States and the United Kingdom to renew their support for AI research. However, it's important to note that expert systems rely on a vast knowledge base. While this knowledge is comprehensive, it remains static and consists of conditional judgments. Humans meticulously input every piece of knowledge, requiring significant manual effort. Therefore, expanding and updating this knowledge base presents significant challenges. When faced with complex problems like chess, the number of decision points becomes overwhelming. These inherent complexities limit the practicality of expert systems.

1.1.1.5   The Second Setback: 1988–1992

In 1987, the advent of desktop computers manufactured by Apple and IBM had surpassed the AI computers created by companies such as Symbolics, leading to a natural decline in the prominence of expert systems.

Toward the end of the 1980s, the new leadership at Defense Advanced Research Projects Agency (DARPA) came to the realization that AI was not the anticipated "next emerging trend." Subsequently, in 1991, it became evident that Japan's ambitious "Fifth Generation Computer Project" had also failed to materialize. These undeniable truths gradually eroded the fervent pursuit of "expert systems," leaving disappointment in its wake. As a result, the field of AI research finds itself once again grappling with a funding crisis.

1.1.1.6   Return: 1993–2012

During this stage, AI technology has incorporated statistical methods. However, it is common for individuals to employ new terminology during this period, seeking to veil the somewhat tarnished reputation of "artificial intelligence." Terms such as informatics, knowledge systems, cognitive systems, or computational intelligence are frequently employed as substitutes.

However, in the year 1997, IBM's Deep Blue triumphed over the esteemed chess player Kasparov, resurfacing AI prominently in the public eye.

1.1.1.7   Explosive Development: 2012–Present

In the year 2012, the convolutional neural network known as AlexNet emerged victorious in the ImageNet image recognition competition, delivering a resounding blow to the second-place contender (support vector machine, SVM) in classification. It was this very competition that sparked the keen interest of numerous researchers toward convolutional neural networks.

In the year 2016, Lee Sedol lost the Go match to AlphaGo by the score of 1 to 4. This captivating event captured much attention, propelling AI to the forefront of both research and public consciousness.

Toward the end of 2022, OpenAI unveiled a remarkable conversational tool named ChatGPT. Boasting an extraordinary prowess in language comprehension, it possesses the capacity to simulate human cognition and linguistic expression. This groundbreaking innovation not only represents a major advancement in AI but also prompts deep reflection on the future impact of this technology. AI is set to play a crucial role in the upcoming global technological revolution and industrial transformation.

1.1.2   What Is Artificial Intelligence?

The basic idea of artificial intelligence (AI) is to create intelligence using artificial methods, where machines try to imitate human cognitive abilities. However, there is no unanimous agreement within the academic community on the exact scientific definition of AI.

According to the introduction in *Artificial Intelligence: A Modern Approach*, several definitions of artificial intelligence have been proposed: systems that think like humans, think rationally, act like humans, and act rationally (Russell 2010).

1.1.2.1   Think Rationally

The concept of "think rationally" can be understood as the act of thinking in a sound and logical manner. Its origins can be traced back to the rational thought processes of the ancient Greek philosopher Aristotle. What are the principles that govern such cogent thinking? If we possess an understanding of these principles, we can engage in accurate and rational thought. These rules can be encoded, enabling machines to engage in rational thinking and attain intelligence. However, it is evident that this approach faces challenges in scalability (it's actually an expert system).

1.1.2.2   Think Like Humans

How can machines think like humans? This inquiry delves into the realm of cognition. To attain humanlike intelligence in computers, we must study our own cognitive functions. It is akin to reverse engineering the complexities of the human brain, a formidable task indeed. Must we patiently await the advancement of cognitive science to a more advanced stage for AI to make significant strides? If machines cannot think like humans, does that mean a lack of intelligence on their part?

1.1.2.3   Act Like Humans

We redefine the very essence of the concept and approach problems from a fresh perspective. Rather than focusing on the mechanics of thinking, our focus lies in understanding the outcomes that such thinking produces. In simpler terms, it entails studying human behavior and actions. This approach can be traced back to the Turing test.

However, here's the catch: can you figure out the square of 756,946,124? Or do you know what "bonjour" means? Probably not. Similarly, AI would also struggle with this kind of knowledge, which would limit its ability to pass the Turing test. As a result, it falls short when compared to simple calculators or electronic dictionaries.

1.1.2.4   Act Rationally

"Act rationally" focuses on how to make decisions, and we are committed to studying a system of rational action. Here, "rationally" can be understood as "optimally."

A modern AI solution should integrate optimization, statistics, and various mathematical techniques. The essence of AI lies in its ability to optimize our intended objectives, for example, getting a room cleaned. Rather than dwelling on the definition of "cleaning" or consulting a manual, AI determines the initial course of action and strategically plans subsequent steps to progressively approach the desired state of "a clean room."

1.1.3   The Ability of AI

The potential of AI is not a question; it can be seen through careful observation. We can witness its widespread presence in our daily lives, from facial recognition systems at train stations to voice assistants on our phones. The case studies in this book mainly focus on how AI is applied in industries and businesses. By harnessing this technology, companies can improve manufacturing processes, personalize products, and achieve targeted marketing. This transformative approach allows for the conversion of products into services and vice versa. Traditional manufacturing companies can use these innovative technologies to enhance efficiency, reduce failure rates, and deliver products accurately and effectively, essentially turning the product into a service. Similarly, the sales industry can utilize these new technologies to analyze individual customer needs and standardize and customize services, ultimately turning their service into a product.

Taking the manufacturing industry as a prime example, ensuring the smooth operation of production equipment holds paramount significance. The ability to address issues proactively, even before they arise, can immensely benefit the company's production process. By analyzing machine operation data, it becomes feasible to assess and predict the real-time running conditions of each individual machine. Through the implementation of remote status monitoring and predictive fault diagnostics, intelligent maintenance of production equipment can be achieved, leading to a substantial reduction in maintenance costs. Moreover, companies can directly offer their products to end consumers in the form of services, facilitated by cutting-edge technologies like AI.

If we look at the retail industry, businesses exhibit great concern regarding the distinction between new and returning customers, the potential risk of customer attrition, and individual customer preferences. Assuming the merchant possesses an intelligent platform, upon a customer's entry into the store, facial recognition technology can be employed to ascertain whether they are a returning patron. Subsequently, the platform queries their purchase history to gauge their inclination to make a purchase on this occasion. In the case of a new customer, consumption habits are determined based on comprehensive insights drawn upon big data and fundamental information derived from facial recognition, thereby enabling more precise and standardized marketing strategies. Furthermore, should the system detect that a particular long-standing customer is at risk of attrition, it can automatically send targeted discounts and other pertinent information to reengage their interest.

AI has already changed the present and will definitely shape the future.

1.2 Overview of Machine Learning

Machine learning, an integral part of AI, stands as a pivotal and prevailing approach in contemporary times. In this section, you will learn the basics about the machine learning.

1.2.1 What Is Machine Learning?

Herbert Simon, a renowned scholar in AI, believes that learning entails the augmentation or refinement of one's own capabilities through the iterative repetition of tasks, thereby enabling the system to outperform its current state and exhibit heightened efficiency in subsequent executions of similar or identical tasks.

As shown in Fig. 1.1, machine learning serves as the very essence of modern AI, with deep learning emerging as a pioneering technique within this domain. Notably, machine learning can be divided into three fundamental categories: supervised learning, unsupervised learning, and reinforcement learning.

- Supervised learning. In the realm of machine learning, supervised learning involves inferring a function from labeled training data. This process entails

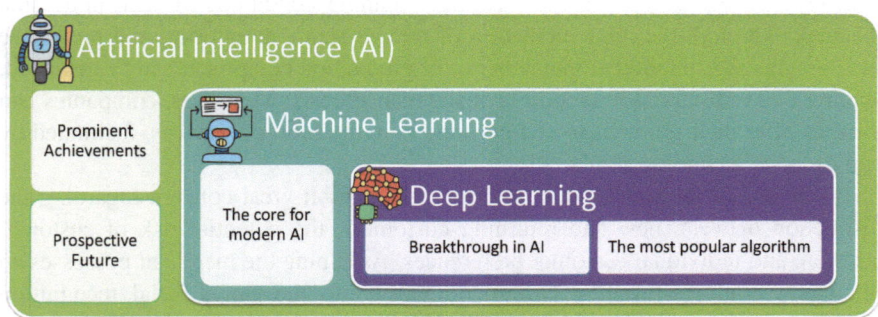

Fig. 1.1 The relationship between AI and machine learning (This figure has been designed using images from Flaticon.com)

adjusting the parameters of a classifier using a set of known samples with assigned categories, aiming to achieve the desired performance.

- Unsupervised learning. Unsupervised learning revolves around the exploration of unlabeled data in search of concealed structures. Since the instances provided to learners lack annotations, there are no error signals available to evaluate potential solutions.
- Reinforcement learning. Within the realm of AI, reinforcement learning entails the agent's acquisition of knowledge through a "trial and error" approach. Guided by rewards obtained through interactions with the environment, the agent strives to maximize its cumulative reward.

This book will focus solely on supervised learning, which is a fundamental aspect of machine learning and widely used in this field.

1.2.2 The Task of Machine Learning

Let's take supervised learning as an example to see what a machine learning task looks like.

Simply put, the task of machine learning is summarizing experiences, discovering patterns, mastering rules, and predicting the future.

As human beings, we possess the remarkable ability to identify patterns based on our past experience. When confronted with novel situations, we draw upon the knowledge acquired from our previous encounters to forecast unforeseen future events, as depicted in Fig. 1.2.

Machine learning systems, on the other hand, acquire knowledge by learning from historical data to construct a model. When confronted with novel challenges, these systems leverage the acquired model to forecast future inputs, as illustrated in Fig. 1.3.

1.2.3 Learning Task

In simple terms, the task of machine learning (supervised learning) is to discover the correspondence between input data x and output data y, $f(x)$, such that $y = f(x)$). Assuming $y = f(x)$, the objective is to determine the values of w and b given a substantial set of x and y.

1.2.4 The Fundamental Problem That Machine Learning Aims to Solve

Machine learning (supervised learning) tackles two distinct problems: regression and classification.

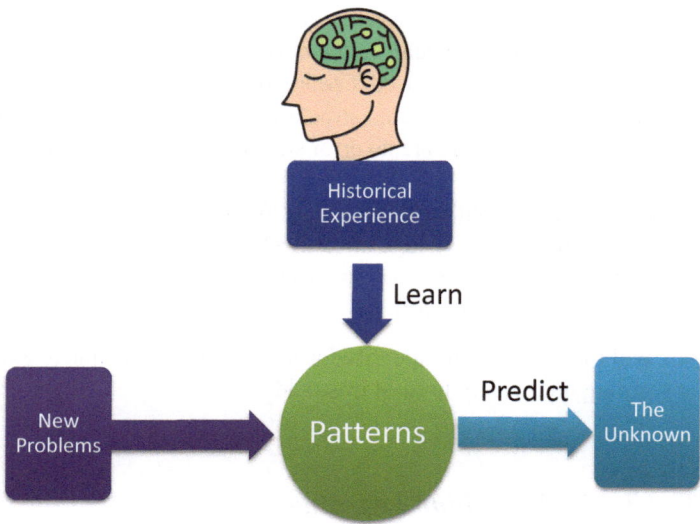

Fig. 1.2 Task of human learning (This figure has been designed using images from Flaticon.com)

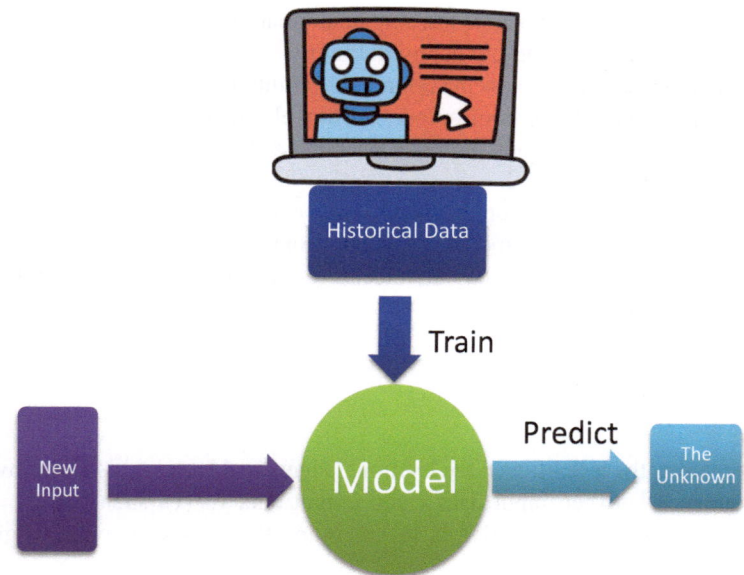

Fig. 1.3 Task of machine learning (This figure has been designed using images from Flaticon.com)

- Regression: identifying the quantitative relationship of interdependence between two or more variables, for instance, predicting an individual's height based on factors such as age and gender

- Classification: mapping input data to a specific category within a given set of classes, for example, determining whether a person is male or female based on their appearance and other characteristics

1.2.5 How Does Machine Learning Optimize Models

One crucial aspect of machine learning is the optimization of a model's parameters.

We can seek a scoring criterion for the model to optimize it, aiming to maximize or minimize this score. The question then arises: what is the evaluation criterion and how can we optimize it? This evaluation criterion is known as the **loss function**. By continuously reducing the value of this loss function, we iteratively optimize the model's parameters. In the following chapters, we will delve deeper into understanding the concept of a loss function and its application.

1.2.6 Machine Learning Workflow

The process of machine learning can be summarized as follows: data acquisition, data preprocessing, model training, model testing, model optimization.

This process is actually quite similar to the traditional software development workflow. We start by acquiring various known datasets and then preprocess them to enhance their computational suitability. Next, we input the data into an initialized model to train its parameters [assuming $f(x) = wx + b$, we aim to determine the values of w and b given a large set of x and y]. Finally, we evaluate the performance of the trained model through testing. And we may keep optimizing it as the model continues to run.

1.2.7 Prior Knowledge Needed

Machine learning is an intricate field that cannot be mastered overnight. It requires a strong grasp of mathematics, programming, data processing, and computer fundamentals. However, if you're looking to get started with machine learning applications, a basic understanding of computer fundamentals will be enough.

1.3 Overview of Deep Learning

Deep learning, a subset of machine learning, uses algorithms with multiple layers of processing to create complex structures and perform numerous nonlinear transformations. This allows for the creation of advanced data abstractions. You may find the structure of a deep learning network in Fig. 1.4.

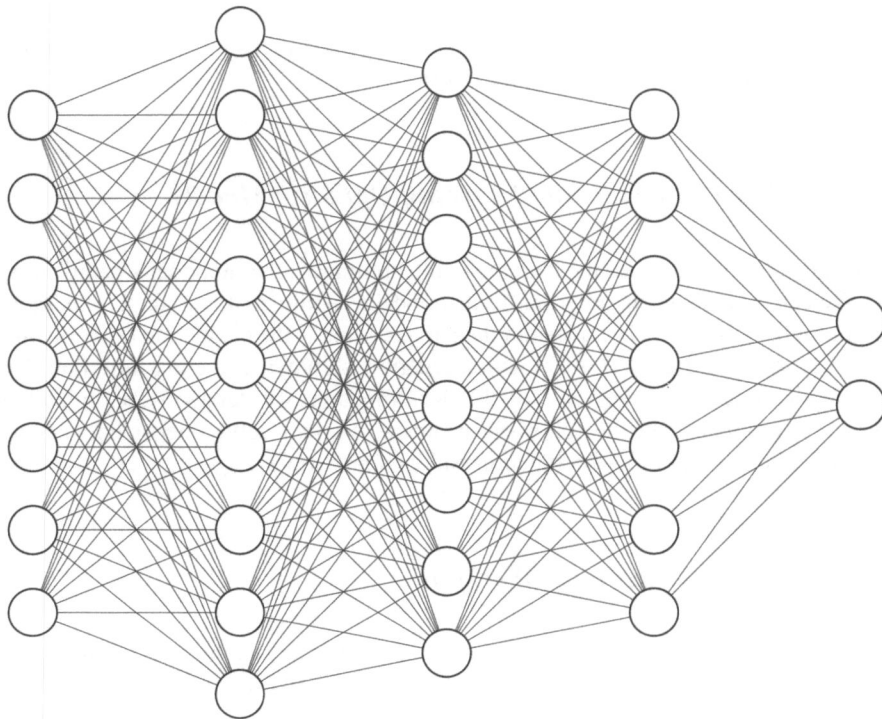

Fig. 1.4 Schematic diagram of deep learning networks

Deep learning is mainly used in specific areas such as machine vision, natural language processing, and recommendation systems. It is often used in systems that focus on achieving accurate results rather than analyzing the underlying causes extensively. For common applications like image recognition, there are already many APIs available. So, for beginners, once we determine the right deep learning model for a particular situation, finding the corresponding API to use is easy.

1.4 Machine Learning and Statistics

Machine learning and statistics have close connections, as they often use similar methods and address similar tasks. However, their analytical approaches are not exactly the same, and they also produce slightly different results. In essence, they have both differences and interdependencies. Here is the comparison of the two:

- Similarities. The goal of statistical methods and machine learning methods is to discover the relevant information from data. They are all trying to find a function $f(x) = wx + b$ to explain and predict y with x.
- Differences. The process of characterizing $f(x)$.

- Statisticians employ a methodology that involves making certain assumptions about the shape of $\tilde{f}(x)$ and distribution of y. For instance, they may assume that $f(x)$ follows a linear model or that y conforms to a normal distribution. Subsequently, they seek the optimal solution under specific criteria. Afterward, by considering the assumed data distribution or the law of large numbers, they can determine the uncertainty of parameter estimation or the standard deviation, thereby constructing a confidence interval.

    Advantages. It allows for the measurement of uncertainty. Simple models possess strong interpretability. When the assumptions are met, the model is scientific, rigorous, and accurate.
    Disadvantages. In complex scenarios, it becomes challenging to validate the assumptions.

- Machine learning experts tend to refrain from making excessive assumptions about the distribution of y. Instead, they rely on cross-validation to directly assess the efficacy of models, determining their quality.

    Disadvantage: lack of scientific rigor.
    Advantages: straightforward and direct.

People often say that "statisticians focus on making models interpretable, while machine learning experts prioritize their predictive power." Generally, models that are highly interpretable may sacrifice their predictive ability, while models with strong predictive ability can be difficult to explain. If you're interested in delving deeper into this topic, you can find numerous online discussions about it.

In our practical work, it is imperative to carefully choose the suitable model and methodology based on specific scenarios. Blindly resorting to deep learning is not advisable. Not only does deep learning lack interpretability, but it is also unnecessary to employ such a powerful "sledgehammer" to crack a "tiny nut."

Points for Discussion

- AI has achieved many significant breakthroughs, but it also faces some challenges and controversies. Can you raise some ethical or social issues that need to be addressed in the development of AI technology?
- The development of AI not only requires powerful algorithms and technologies but also relies heavily on a large amount of data. In reality, data privacy and security is an important issue. How do you think we can promote the development of AI while protecting data privacy?

1.5   Practice

1. Please observe the diverse range of intelligent products around you, and elaborate on the AI techniques they employ. In your opinion, what distinctive advantages does AI offer in these domains?
2. Please provide examples to explain the distinction between regression problems and classification problems.

Reference

Russell, Stuart J. (Stuart Jonathan), 1962-. Artificial Intelligence : a Modern Approach. Upper Saddle River, N.J.: Prentice Hall, 2010.

Chapter 2
Basics of Machine Learning

Veni, vidi, vici.

—*Gaius Julius Caesar*

Abstract This chapter serves as an introductory guide to machine learning (ML), offering a straightforward approach for beginners. Starting with the basics, it categorizes data into categorical and numerical types, elucidating their role in ML. The chapter then delves into fundamental statistical concepts crucial for ML, such as mean, median, standard deviation, and normal distribution, making these concepts accessible even to those with limited statistical background.

A significant portion of the chapter is dedicated to understanding linear regression, a cornerstone of ML. It covers simple and multiple linear regression, explaining their practical applications and geometric interpretations. The chapter also introduces the least-squares method for optimizing regression models and discusses how to assess the quality of these models using measures like the coefficient of determination and adjusted R^2.

To make ML more approachable for novices, the chapter introduces KNIME, a user-friendly, graphical tool for ML. It provides insights into downloading, installing, and basic usage of KNIME, highlighting its advantages in simplifying and visualizing ML tasks.

Overall, the chapter aims to demystify ML for beginners, blending theoretical knowledge with practical tools, making it a comprehensive starting point for anyone aspiring to delve into the world of machine learning.

Keywords Machine learning · Linear regression · KNIME · Classification · Regression

This chapter covers

- Classification of data.
- Linear regression.
- Least-squares method.

© The Author(s), under exclusive license to Springer Nature Singapore Pte Ltd. 2024   15
Y. Geng et al., *Practical Machine Learning Illustrated with KNIME*,
https://doi.org/10.1007/978-981-97-3954-7_2

- Determining the quality of the fit.
- Introduction to KNIME.

For machine learning to truly work its magic, it needs to be used by a wider range of people. However, machine learning is never an easy task. It involves a range of knowledge, especially statistics, which can be a real headache for many learners. This often leaves aspiring learners hesitant to dive into the world of machine learning. But here's the thing: getting started in any subject doesn't always require too much prior knowledge, and machine learning is no exception. That's why we have this book. It aims to help you take the first step into machine learning. It doesn't bombard you with endless formulas, but it does provide a solid understanding of the underlying principles.

2.1   Mathematical Foundations

When it comes to getting started with machine learning, the math requirements are actually quite manageable. You don't need to be a math genius. In fact, the mathematical knowledge needed for the initial stages of machine learning mainly revolves around the basic math concepts you learned in secondary school, along with some basic statistical knowledge. The beauty of it is that you don't have to tackle complex mathematical problems on your own, as computers have already taken care of that for us (https://www.deeplearning.ai/courses/machine-learning-specialization/).

2.1.1   Types of Data

Data can be simply divided into categorical data and numerical data, as shown in Fig. 2.1.

- Categorical data. Categorical data represents different categories or groups. For example, when classifying people based on gender, we have two categories: male and female. Categorical data can be further divided into two subtypes: nominal data and ordinal data. Nominal data is also called unordered data. As the name suggests, this type of data does not have a specific sequence or order, such as the categories of male and female. On the other hand, ordered data refers to categories that have a clear size or order relationship, like small, medium, and large.
- Numerical data. Numerical data comprises information expressed in numerical form, such as the heights and weights of a group of teenagers or someone's salary. Numerical data can be further categorized into two types: discrete data and continuous data.

In real machine learning projects, numerical data is commonly encountered, but there is also a presence of categorical data. By referring to the content covered in

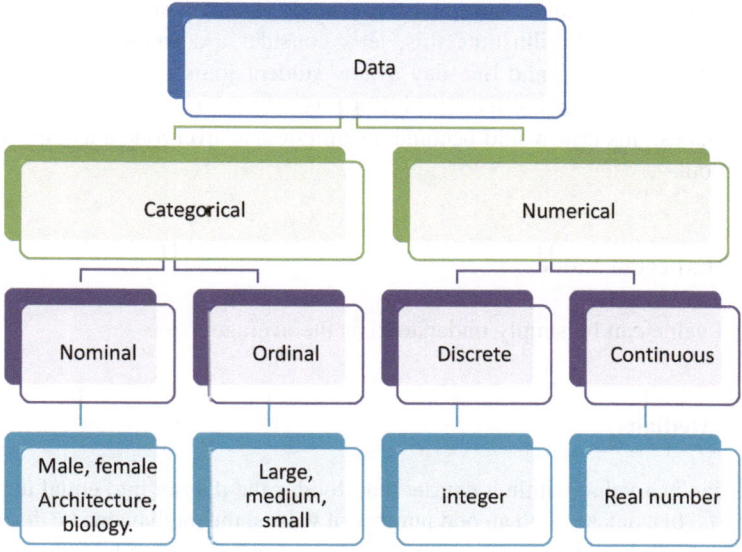

Fig. 2.1 Types of data

Chap. 1, we should be able to envision how numerical data is utilized. However, incorporating categorical data into machine learning models is not as straightforward. It requires imagination and creative thinking to explore how categorical data can be effectively integrated. In the upcoming sections, we will delve into the utilization of categorical data and its application in machine learning.

Points for Discussion

- What data in life is categorical? How about numerical?
- Assuming there are a large number of buildings around, the number of floors in these buildings ranges from 1 to 5. Should the number of floors be represented using categorical data or numerical data? If the number of floors ranges from 1 to 100 floors, which type of data should we adopt?

2.1.2 Basic Statistical Terms

2.1.2.1 Mean

The concept of average (mean) holds significant importance in statistics as it serves as the most commonly employed measure of central tendency. Its purpose is to determine the balance point of a dataset. The average value is a statistic that we are often familiar with, as we frequently encounter average wages, average prices, and similar figures in news reports. Due to its intuitive and concise nature, the average value is commonly used in our daily lives.

However, it is crucial to note that the average is also a statistical value that is frequently misused. To illustrate this, let's consider a scenario where our class consists of 49 students, and one day a new student joins us who happens to be from the wealthiest family in the world. If we were to calculate the average assets of our classmates, this task would become meaningless as everyone's assets would be averaged out.

2.1.2.2   Expected Value

Expected value can be simply understood as the average value.

2.1.2.3   Median

The median is a value within a dataset that divides the dataset into equal upper and lower parts. In a dataset with an odd number of values, the median can be determined by arranging all the values in ascending order and selecting the middle one. If there is an even number of values, there will be more than one. And typically, the average of the two middle values will be considered as the median. For a finite set of numbers, the median is a type of value such that half of the data in the group is greater than it, while the other half is smaller.

Let's consider the scenario of a child from the world's richest family joining our class. By using the median, we may notice that his presence or absence in our class has minimal impact on the median. Therefore, the median effectively avoids the issue of data being skewed by extreme values.

Points for Discussion

- Which scenes in life are better suited for the use of the median? Which situations are better suited for using the average value?

2.1.2.4   Standard Deviation

The standard deviation, denoted by the mathematical symbol σ (sigma), is widely employed in probability and statistics to quantify the dispersion or variability within a set of values.

2.1.2.5   Normal Distribution

Normal distribution, also known as the Gaussian distribution, is a widely encountered continuous probability distribution. It holds significant importance in statistics and is frequently employed in the natural and social sciences to model unknown

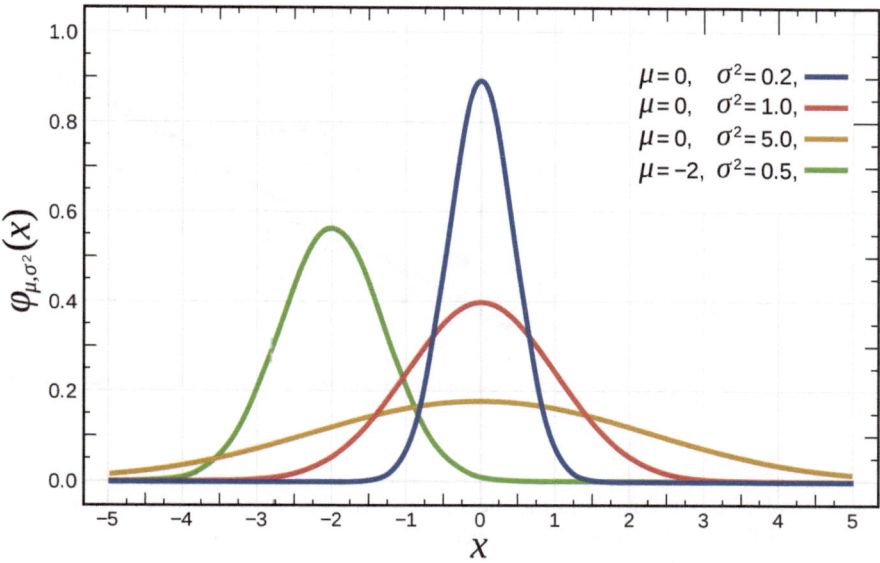

Fig. 2.2 Normal distribution (By Inductiveload—self-made, Mathematica, Inkscape, Public Domain, https://commons.wikimedia.org/w/index.php?curid=3817954)

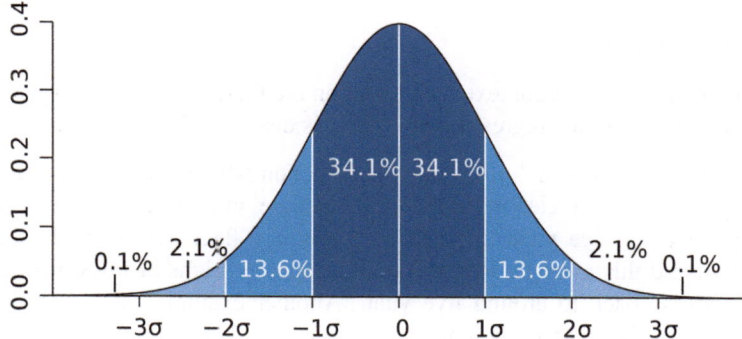

Fig. 2.3 Normal distribution and standard deviation (By M. W. Toews—Own work, based (in concept) on figure by Jeremy Kemp, on 2005-02-09, CC BY 2.5, https://commons.wikimedia.org/w/index.php?curid=1903871)

random variables. In Fig. 2 2, μ represents the mean, σ^2 denotes the variance, and the graph illustrates the normal distribution under various mean and variance settings. When the expected value is 0 and the variance is 1, it is referred to as the standard normal distribution (https://en.wikipedia.org/wiki/Normal_distribution).

Figure 2.3 reveals the relationship between the standard deviation and the normal distribution. It demonstrates that roughly 68% of the values fall within one standard deviation from the mean, approximately 95% of the values lie within two standard deviations from the mean, and about 99.7% of the values are encompassed within

Fig. 2.4 Regression is to regress to the mean value

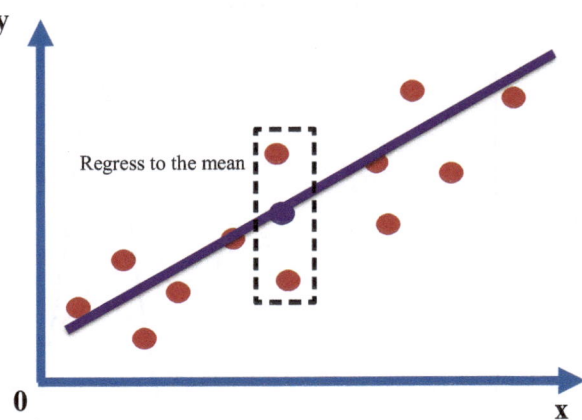

three standard deviations from the mean. This principle is commonly referred to as the "68-95-99.7 rule" (https://en.wikipedia.org/wiki/Standard_deviation).

2.1.3   Regression

Points for Discussion

- Do not read the following text. Just based on the literal meaning of "regression," what do you think the regression algorithm is designed for?

The concept of "regression" refers to a phenomenon where values tend to return to an average point, as depicted in Fig. 2.4. To illustrate, in everyday life, if an apple is typically priced at five yuan, the market price may fluctuate but will ultimately stabilize around this average. Regardless of temporary highs or lows, it will eventually "regress" back to around five yuan. Another example can be observed in biology, where the average height of a family is 1.7 m. If one family member happens to be exceptionally tall, let's say 2.3 m, it is unlikely for his or her children to inherit such a height. They will "regress" toward the family's average height of approximately 1.7 m.

In statistics, regression is a statistical analysis method employed to establish the quantitative relationship between two or more variables that are interdependent.

In the realm of machine learning, regression entails discovering a model that ensures its predicted values gravitate toward the true values, minimizing any significant deviations.

Regression, in a broader sense, can be categorized into two fundamental types: linear regression and logistic regression. Linear regression is commonly referred to as "regression," whereas logistic regression deals with classification.

To better comprehend the essence of regression, let's examine a simple linear regression example.

2.1.3.1 Simple Linear Regression

Simple linear regression can be represented by the "$y = b + wx$" equation. In the context of machine learning, "y" represents the label, "x" denotes the feature, "w" signifies the weight, and "b" represents the bias. The objective of simple linear regression is to determine the unknown values of "w" and "b" based on a set of known "x" and "y" values.

Using the example of the relationship between annual salary and years of work experience as shown in Fig. 2.5, let's examine the significance of the simple linear regression equation.

The bias corresponds to the point where the straight line intersects the vertical axis, while the weight represents the slope. The objective of simple linear regression is to determine suitable weights and biases that minimize the distance between this line and the distribution of data points.

Points for Discussion

- Assuming you are facing a distribution of points similar to Fig. 2.5, can you approximately draw a regression line? Why did you draw such a regression line? You do not need to consider how to calculate. Just draw it.

2.1.3.2 Multiple Linear Regression

Similar to simple linear regression, the objective of multiple linear regression is to solve the equation $y = b + w_1x_1 + w_2x_2 + \ldots + w_nx_n$, determining unknown variables w_i and b through known variables x_i and y. In scenarios where factors like education level and work experience, in addition to seniority (YearsExperience), contribute to an individual's annual salary, we may use multiple linear regression.

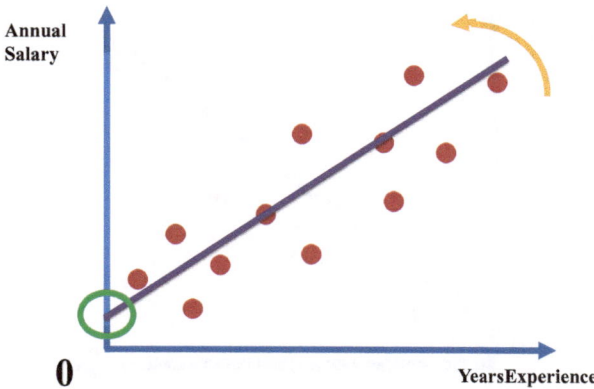

Fig. 2.5 Example of simple linear regression

2.1.4   Least-Squares Method

We are aware that the goal of regression is to minimize the distance between the regression line and the distribution of data points. However, how can we ensure that this distance is minimized? Are there any measurement methods available? The least-squares method can be employed to address this issue.

As depicted in Fig. 2.6, we can establish a measure of closeness between the observed actual data points and the fitted line. To achieve this, we draw a perpendicular line from each data point y_i and determine the ordinate of the intersection point with the fitted line, denoted as $\widehat{y}_i$. A good fit implies that we desire each pair of y_i and $\widehat{y}_i$ to be as close as possible. However, if we simply adopt the difference $\widehat{y}_i - y_i$ and calculate the sum of these differences, $\sum(\widehat{y}_i - y_i)$, the positive and negative values may cancel each other out, failing to truly reflect the desired proximity. To address this, we employ the squared difference $(\widehat{y}_i - y_i)^2$ as a measure of the closeness between each observation point and the fitted value. The overall degree of closeness is determined by the sum of these squared differences, $\sum(\widehat{y}_i - y_i)^2$. It is evident that a smaller value indicates a better fit. The least-squares method aims to minimize this value. Since $\sum(\widehat{y}_i - y_i)^2$ represents the deviation between the actual data and the fitted data, it can be viewed as a form of loss or error reflected by the fitted data in relation to the actual data. Hence, it serves as a loss function. While there are various types of loss functions, their common purpose is to measure the deviation between actual data and fitted data.

The concept of mean squared error (MSE) can be frequently seen in various contexts. Essentially, MSE is calculated using the formula $\frac{1}{m}\sum(\widehat{y}_i - y_i)^2$, where m represents the number of samples. It is worth noting that dividing the loss function by m or not does not affect the comparison of the quality of fit for a dataset.

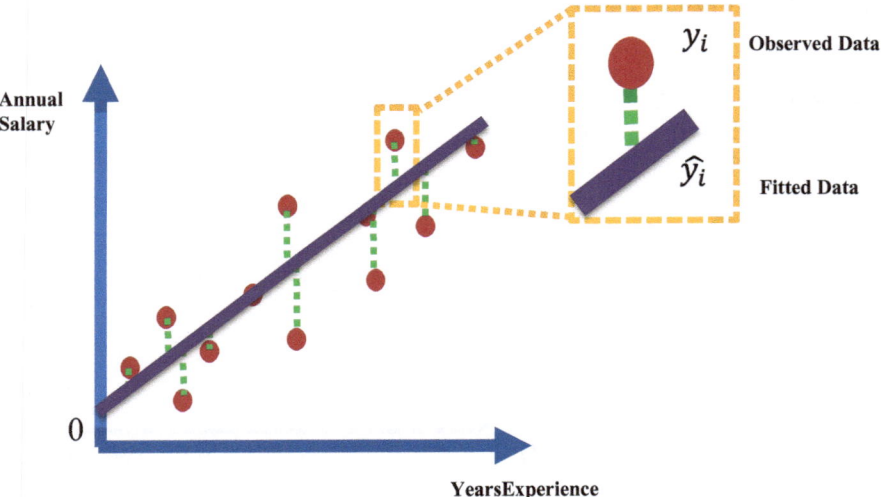

Fig. 2.6 Data observed and your data

2.1.5 Determining the Quality of the Fit

If you want to determine the interpretability of the model, which measures the proportion of the variation in the response that can be explained by the independent variables, relying solely on the loss function is not enough (https://scikit-learn.org).

2.1.5.1 Coefficient of Determination

Points for Discussion

- Suppose you are given a set of data $X_1, X_2, \ldots, X_n$, with all the numerical values except X_n provided. Your task is to make an educated guess about the value of X_n. Which is more reasonable: choosing the maximum value, the minimum value, the mean value, or simply making a random guess? Why?

Most people would probably feel that the average (mean) is more appropriate. As depicted in Fig. 2.7, if you plot this average value and measure the distance from each actual observation point to this average line, you can calculate the sum of the squares of these distances, denoted as $\sum (y_i - \bar{y})^2$, where $\bar{y}$ represents the mean value.

Below are two definitions:

$$SS_{res} = \sum (y_i - \hat{y}_i)^2$$
$$SS_{tot} = \sum (y_i - \bar{y})^2$$

The first formula above represents the residual sum of squares, which is utilized to assess the proximity between observation points and the regression line. On the other

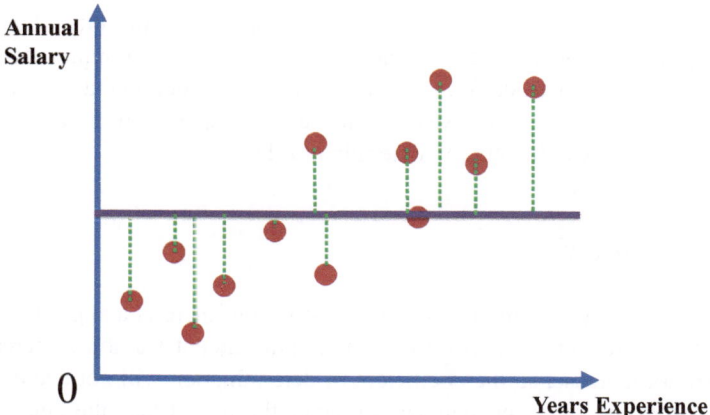

Fig. 2.7 Distance between each point and the mean value

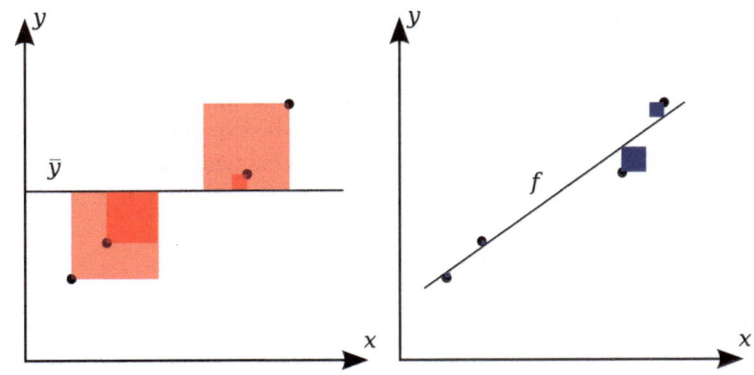

Fig. 2.8 The total sum of squares (left) and the sum of residual squares (right) (By Orzetto—Own work, CC BY-SA 3.0, https://commons.wikimedia.org/w/index.php?curid=11398293)

hand, the second formula represents the total sum of squares, which helps determine the proximity between individual measurement points and the mean value. Now, let's define the coefficient of determination as

$$R^2 = 1 - \frac{SS_{res}}{SS_{tol}}$$

Looking at the formulas and referring to Fig. 2.8, we can observe that as the linear regression (right) outperforms the mean (left), the coefficient of determination tends to approach 1 more closely.

Since we are dealing with the calculation of sums of squares, it can be visualized as the area of a square. When calculating the total sum of squares, it can be understood as the residual sum of squares between the data points and the mean value. Each squared residual represents the area of a square with the distance from the data point to the mean value as its side length. That's to say, the total sum of squares is the sum of the areas of all these squares. On the other hand, when calculating the sum of squared residuals, each squared residual represents the area of a square with the distance from the data point to the fitted line as its side length. The sum of all these areas gives us the sum of squared residuals (https://en.wikipedia.org/wiki/Coefficient_of_determination1).

2.1.5.2 Adjusted R^2

R^2 provides a better measure of the quality of fit, but there is a slight issue: as the number of features, or in statistics known as independent variables, increases, the SS_{res} does not increase, so the coefficient of determination will not decrease. Does this imply that the more independent variables the model has, the better it is? In reality, the more complex the model becomes, the more likely it is to encounter problems. Simpler models are preferred (Occam's razor). To compare models with different numbers of independent variables more fairly, we introduce the adjusted R^2 (adjusted R^2):

$$\text{Ajd } R^2 = 1 - \left(1 - R^2\right) \frac{n-1}{n-p-1}$$

Among them, p represents the number of independent variables, and n represents the number of samples. In essence, it imposes a penalty on p, whereby the larger p becomes, the greater the penalty. This serves to avoid the misunderstanding that a more complex model is inherently better.

2.1.6 Summary

Now, we understand what linear regression is and know how to determine the quality of fit of a regression line, so we already have the basic knowledge needed to further study machine learning.

2.2 Reading Charts

Understanding the data is essential before taking on any machine learning project. One of the most intuitive approaches to comprehend the data is by analyzing different types of data charts. For machine learning engineers, it is not only important to interpret the charts accurately but also to determine which type of chart is most suitable for displaying specific data. Here, we present several commonly utilized charts that can greatly facilitate data comprehension in the future.

2.2.1 Distribution of Numerical Data

To quickly visualize the distribution of a dataset, we can employ a histogram. As shown in Fig. 2.9, the horizontal axis represents the data values, while the vertical axis represents the quantity of data. Each bar in the histogram corresponds to a specific range of values, with taller bars indicating a greater amount of data falling within that range. By examining the histogram, we can easily see which ranges of data are more prevalent and which are less so, thereby gaining insights into the overall distribution. This serves as a fundamental step for subsequent analysis.

One-dimensional data can be examined using a histogram, while two-dimensional data can be visualized through a joint distribution, as shown in Fig. 2.10. In this diagram, the graph in the center is a scatter plot, where each point represents a distinct data point, while the histograms positioned on the top and right

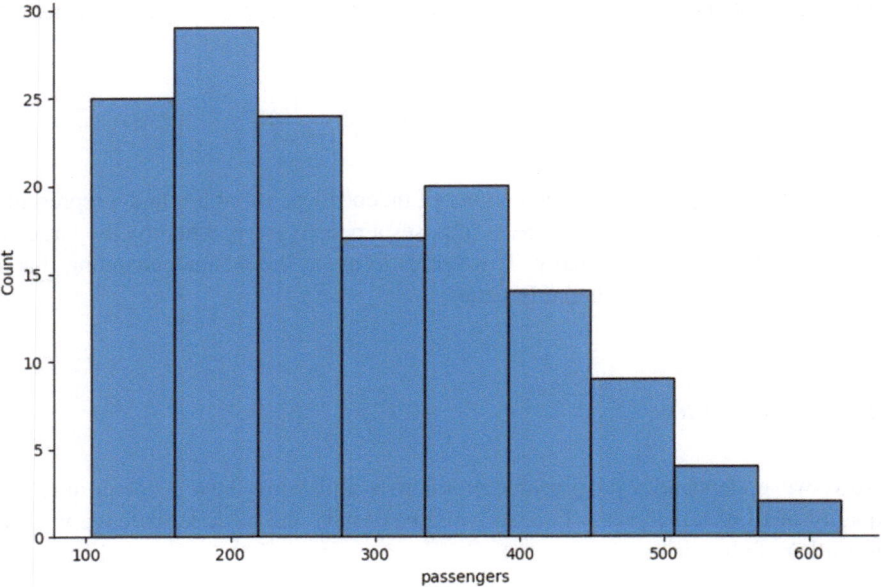

Fig. 2.9 Histogram

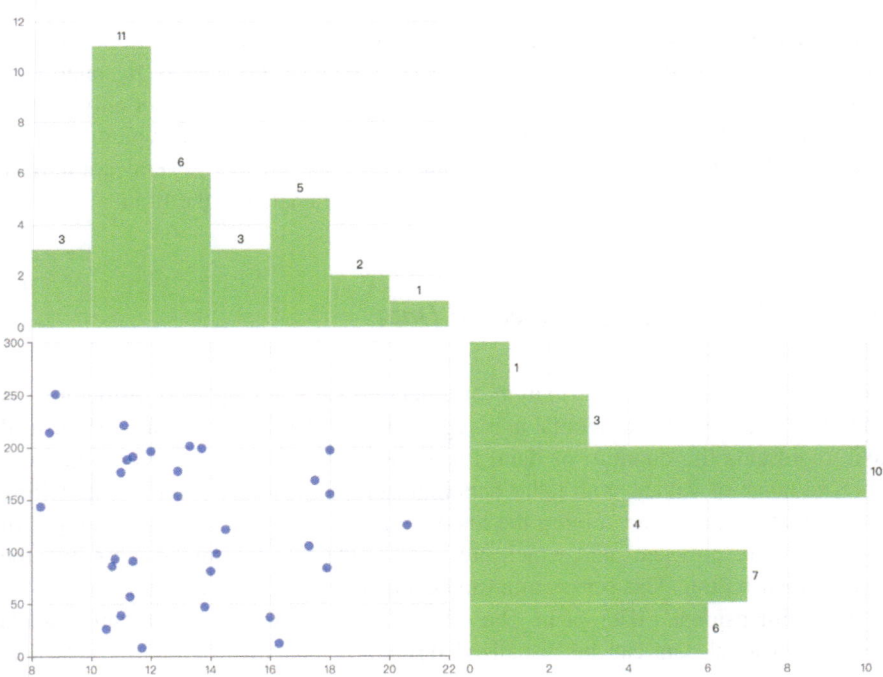

Fig. 2.10 Joint distribution of 2D data

sides are the corresponding marginal distribution. Let's take the histogram in Fig. 2.9 as an example: the length of each bar signifies the cumulative data amount, encompassing all y-values, within a specific x-range. This graphical representation not only unveils the distribution of individual data points but also unravels the intricate relationships within the data, bringing many insights to the viewers. In many instances, we only look at the scatter plot.

2.2.2 The Distribution of Categorical Data

2.2.2.1 Quartile

In statistics, the quartile is a specific type of quantile. It divides a set of values, arranged in ascending order, into four equal parts. The values at the three dividing points are known as quartiles, as shown in Fig. 2.11.

- The 1st quartile (Q1), also referred to as the "lower quartile," represents the 25th percentile of the sorted values in the sample.
- The second quartile (Q2), commonly known as the "median," is the value that divides the sample into two equal halves when arranged in ascending order.
- The 3rd quartile (Q3), also known as the "upper quartile,", represents the 75th percentile of the sorted values in the sample.
- The difference between Q3 and the Q1 is also known as the interquartile range (IQR).

2.2.2.2 Box and Whisker Plot (Box Plot)

This is a type of statistical graph used to display a distribution of data. As shown in Fig. 2.12, it can display the maximum, minimum, median, first quartile, and third quartile of a set of data. The line in the middle of the box is called the median line, and the vertical lines above and below the box are called whiskers. In general, the

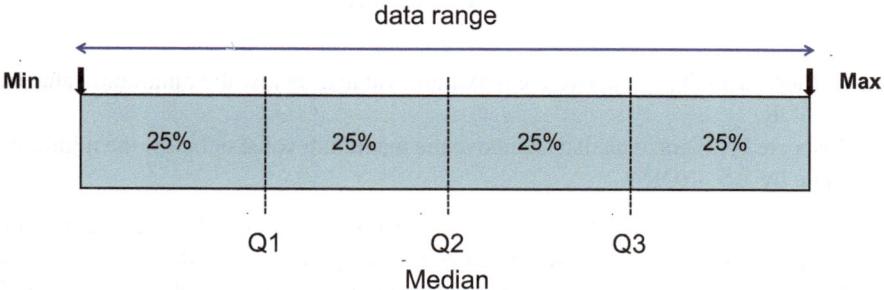

Fig. 2.11 Quartiles

Fig. 2.12 Box and
whisker plot

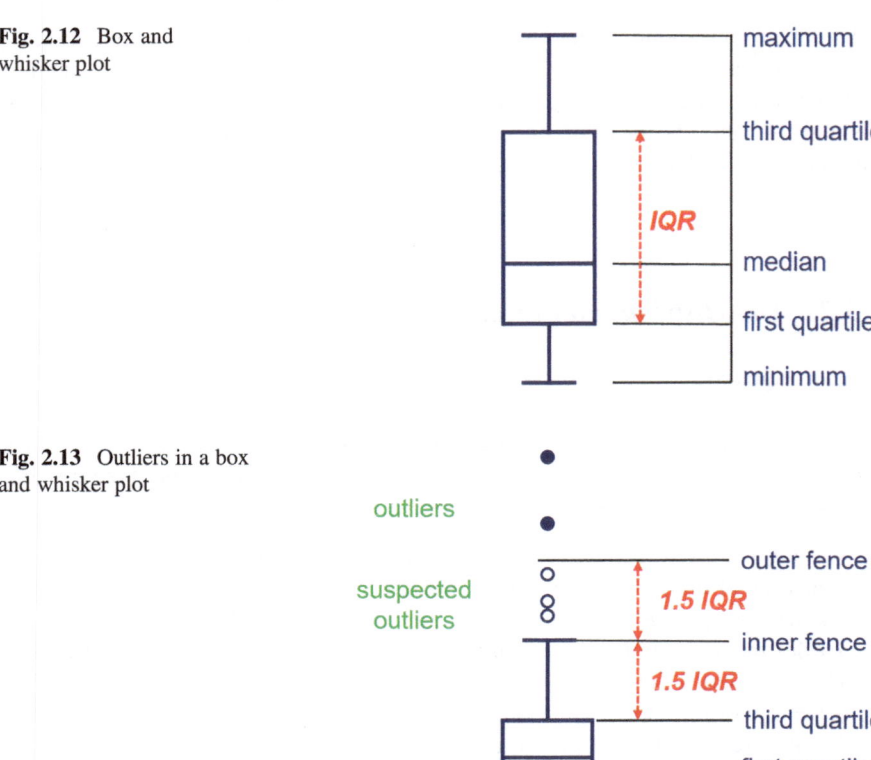

Fig. 2.13 Outliers in a box
and whisker plot

end of the whiskers represents the maximum or minimum value (https://en.
wikipedia.org/wiki/Box_plot).

The basic box and whisker plot provides a visual representation of the complete
range of data distribution, spanning from the minimum to the maximum values. It
also displays the potential interquartile range (IQR) and highlights the typical values,
such as the median (not the mean). Outlying values, which deviate significantly from
the norm, are depicted as outliers positioned outside the maximum and minimum
values. John Tukey, the creator of the box and whisker plot, offers two definitions for
outliers.

- Outliers: data that are above the maximum value or below the minimum value by
  3 × IQR
- Suspected outliers: data that are above the maximum value or below the minimum
  value by 1.5 × IQR

As shown in Fig. 2.13, if any of the aforementioned abnormal values are present,
the endpoint of the whisker in the box and whisker plot serves as the cutoff point for
outliers, known as the inner fence. This inner fence is positioned at Q3 + 1.5 × IQR
(and Q1 − 1.5 × IQR), and suspected outliers are denoted by hollow circles.

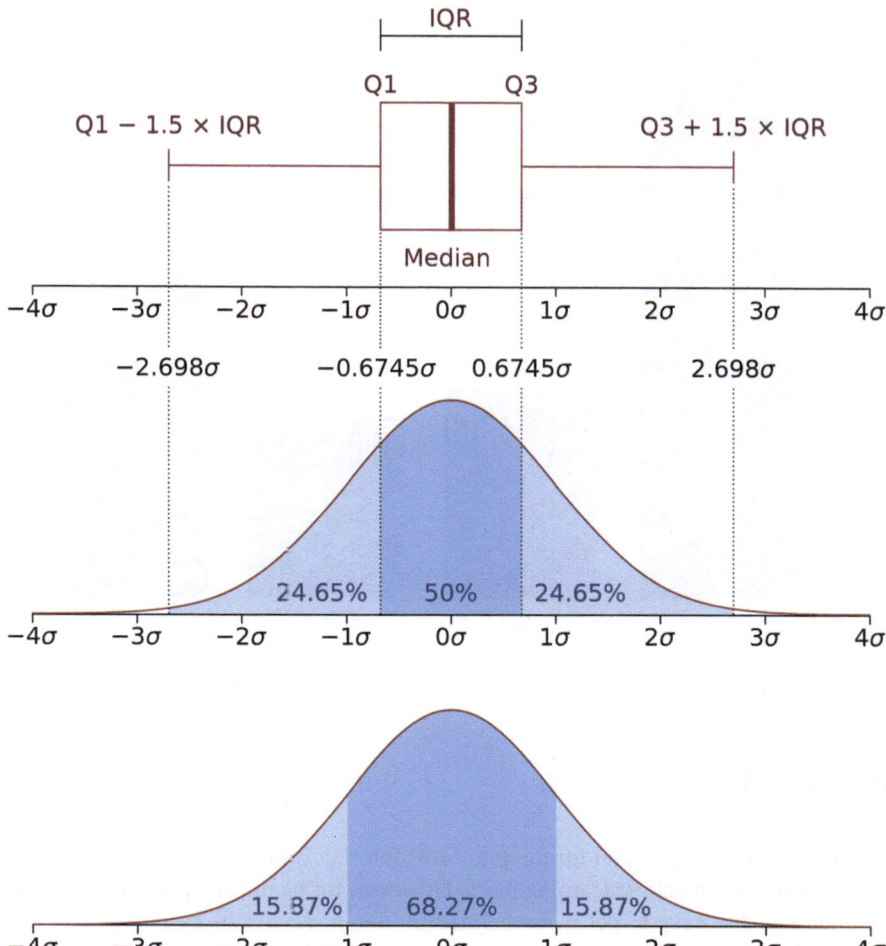

Fig. 2.14 Relationship between the box and whisker plot and quartiles (By Jhguch at en.wikipedia, CC BY-SA 2.5, https://commons.wikimedia.org/w/index.php?curid=14524285)

Similarly, the outer fence is located at 3 × IQR, and the corresponding outliers are represented by solid black dots.

Figure 2.14 provides us with a more intuitive representation of the relationship between quartiles and a box and whisker plot.

2.2.2.3   Bar Chart

Just like numerical data histograms, categorical data can be represented using bar charts. These bar charts allow us to visualize the distribution of each category. As shown in Fig. 2.15, the horizontal axis represents different categories, while the vertical axis indicates the quantity of data for each category.

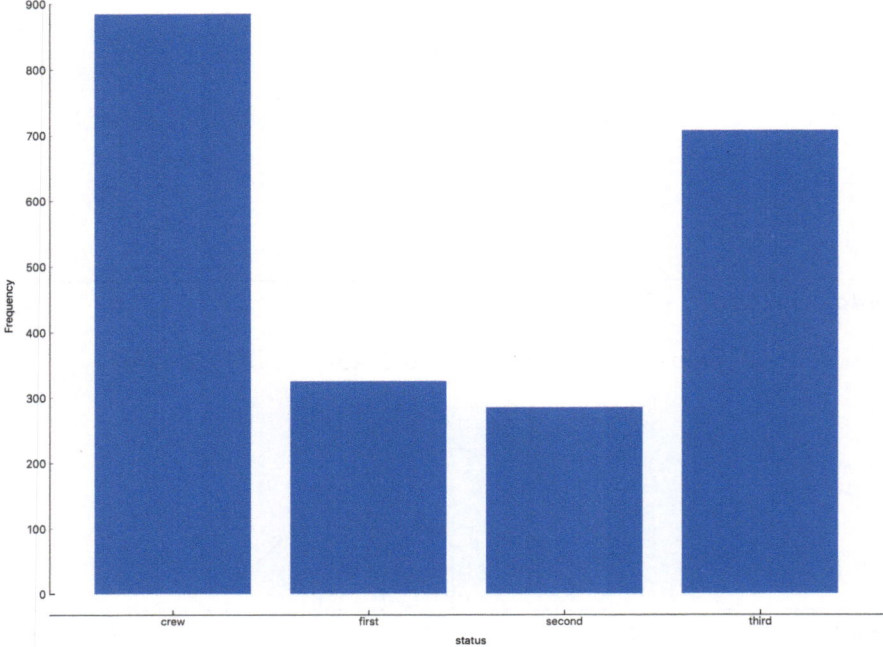

Fig. 2.15 Bar chart

2.3 KNIME

One must find the right tool for the job. Most tutorials and books rely on Python as the go-to tool for machine learning tasks. However, for beginners, programming can be daunting. It's easy to get overwhelmed by the intricacies of Python syntax and ignore the core concepts of machine learning itself. That's why we introduce KNIME, a user-friendly graphical tool for machine learning. Not only is KNIME simple and intuitive, but its standalone version is also completely free of charge, making it an ideal starting point for beginners venturing into the world of machine learning. Moreover, KNIME has the capability to handle enterprise-level machine learning projects, effectively supporting businesses in their digital transformation and growth (https://www.knime.com/getting-started-guide).

2.3.1 KNIME Introduction

KNIME ([naim]), short for Konstanz Information Miner, is a powerful machine learning tool developed by Michael Berthold's team at the University of Konstanz. It operates within the Eclipse development environment and follows a data flow approach to construct machine learning workflows. The beauty of KNIME lies in its seamless integration with other systems like Python, Java, and Tableau. Additionally, KNIME boasts an extensive collection of third-party extensions, making it a versatile choice for a wide range of machine learning tasks. Notably, KNIME has gained significant popularity among manufacturing companies.

2.3.2 Download and Install

Downloading and installing the software is very easy. Simply head over to the official website (https://www.knime.com/) and grab the latest version. KNIME recently underwent a major upgrade in July 2023, so make sure you get your hands on the latest version. In this book, we'll be using version 5.1, which has some exciting changes compared to the older versions. I highly recommend sticking with this updated version of KNIME for a smooth learning experience.

While installing KNIME, it's important to consider whether all users require permission to write to the installation directory. If you check the box of "Make installation folder writable for everyone" (Fig. 2.16), it grants all users the ability to write to the installation directory, which may pose certain risks to the system. However, if you leave it unchecked, you'll need to enable administrator privileges

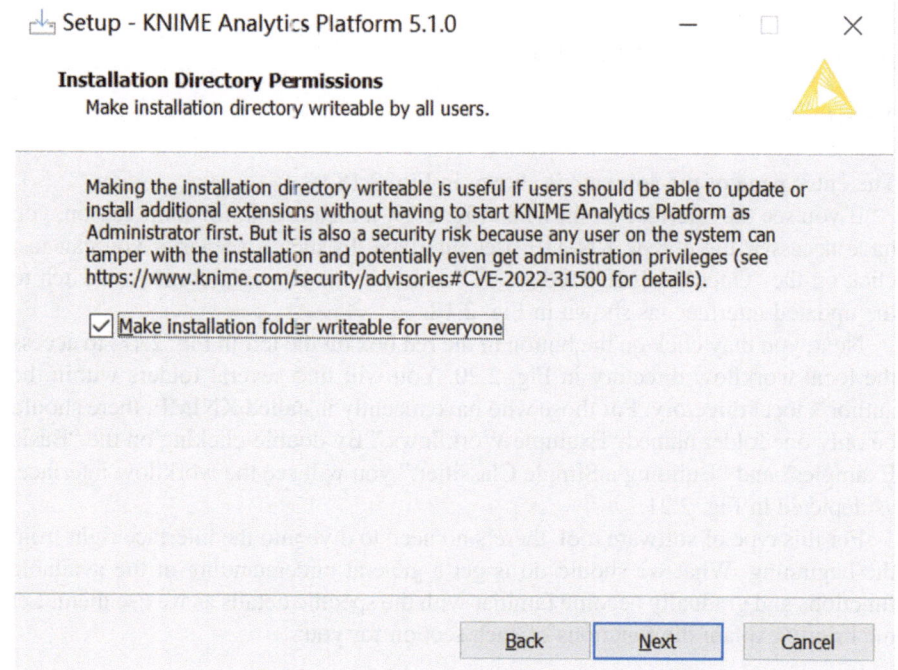

Fig. 2.16 Make installation folder writable for everyone

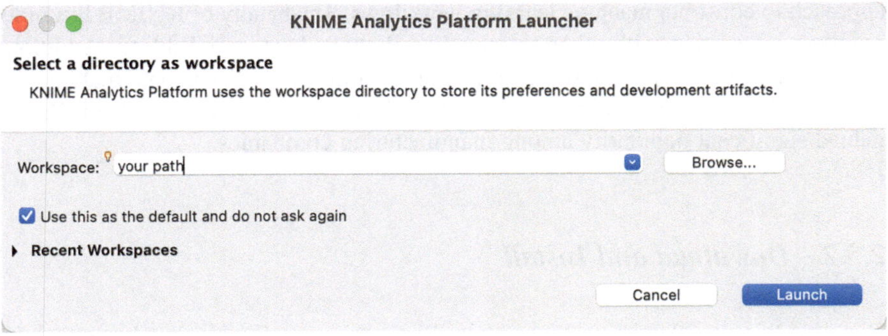

Fig. 2.17 Directory settings

each time you upgrade or install plug-ins. We have checked the box in Windows system to ensure convenient operations in the future.

Once the installation is complete, it's advisable for everyone to set up the appropriate working directory when launching the software as shown in Fig. 2.17. Don't forget to check the box of "Use this as the default and do not ask again." By doing so, you won't have to go through the same process every time you open the software in the future.

2.3.3　Basic Use of KNIME

2.3.3.1　Interface

The entry page of the software is shown in Fig. 2.18.

If you see the interface as shown in Fig. 2.19, it means that, for some reason, you have accessed the classic KNIME user interface. In such a scenario, you can just click on the "Open KNIME Modern UI" option in the top right corner to switch to the updated interface, as shown in Fig. 2.18.

Next, you may click on the button in the red box on the left in Fig. 2.18, to access the local workflow directory in Fig. 2.20. You will find several folders within the author's local directory. For those who have recently installed KNIME, there should be only one folder named "Example Workflows." By double-clicking on the "Basic Examples" and "Building a Simple Classifier," you will see the workflow interface, as depicted in Fig. 2.21.

For this type of software tool, there's no need to dive into the interface right from the beginning. What we should do is get a general understanding of the available functions and gradually become familiar with the specific details as we use them. Let me briefly explain the functions of each section for you.

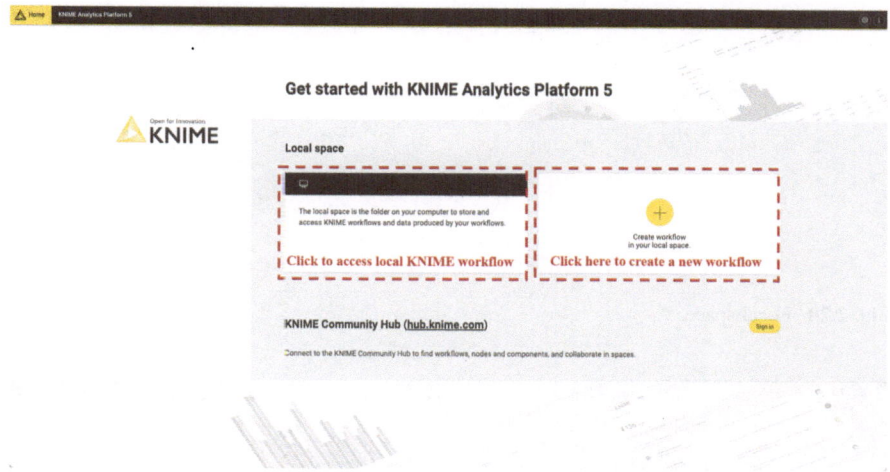

Fig. 2.18 KNIME entry page

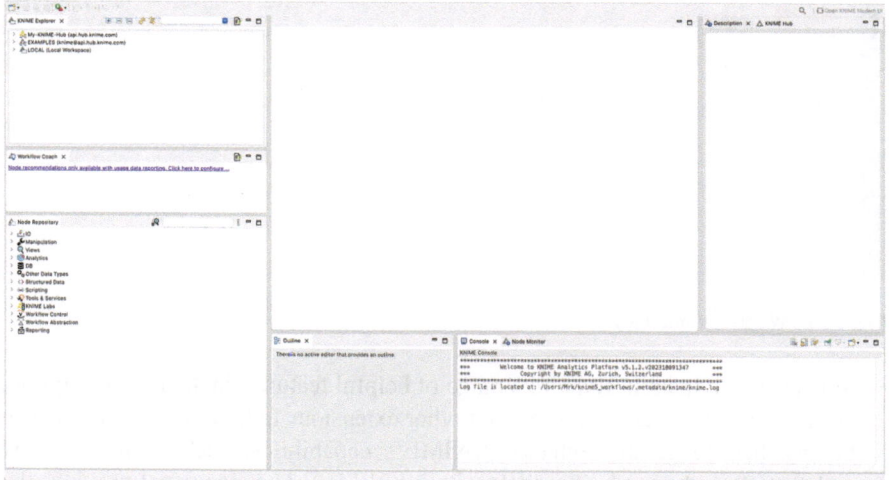

Fig. 2.19 Classic KNIME user interface

- Application tabs. Easily switch between the entry page and other opened workflows.
- Settings and info page. Access and manage KNIME functions, install additional extensions, and explore community resources.
- Toolbar. Conveniently interact within the whole workflow like save, undo, and redo or with actions that are applicable for the current selected nodes.
- Workflow editor. Canvas for editing the currently active workflow.
- Node monitor: Shows the output of the current selected node and also the flow variable values.

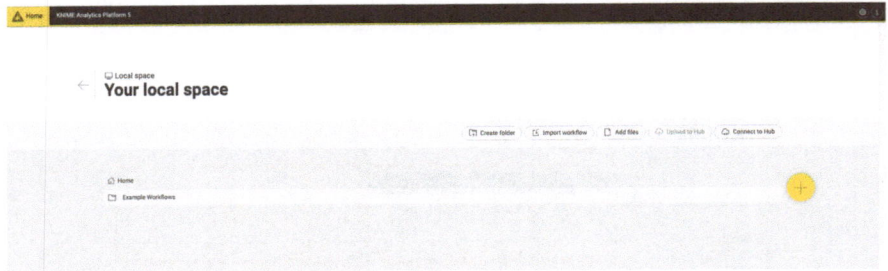

Fig. 2.20 Local space

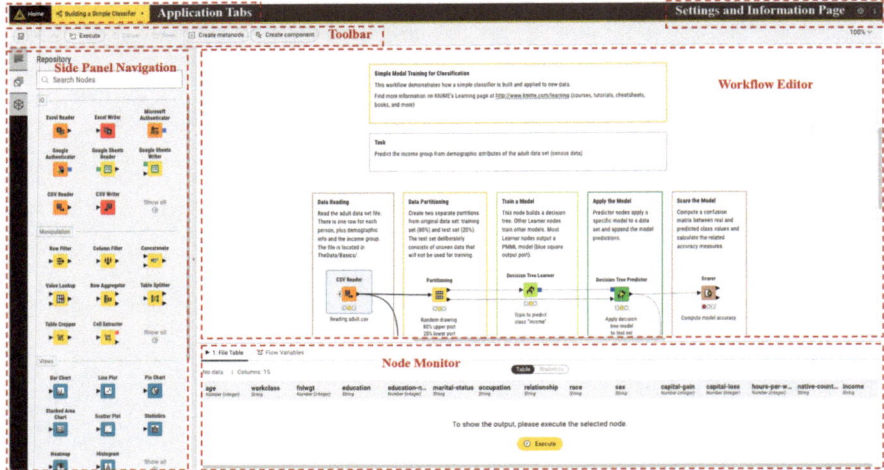

Fig. 2.21 Workflow interface

- Side panel navigation. Explore a range of helpful features, including description, node repository, space explorer, and other extensions (additional extensions can be installed to greatly enhance KNIME's capabilities, such as utilizing AI assistants for advanced AI models within KNIME). The side panel provides the following features:

  - Description ☰ . View the description of the selected node.
  - Node repository ⬚ . Browse all nodes available.
  - Space explorer ⊗ . Navigate local or cloud-based KNIME Hub workflow spaces.

2.3.3.2 Open Example Workflows

We can check example workflows to help us get started and become more familiar with KNIME's functionality. You may find many facts and notes in the

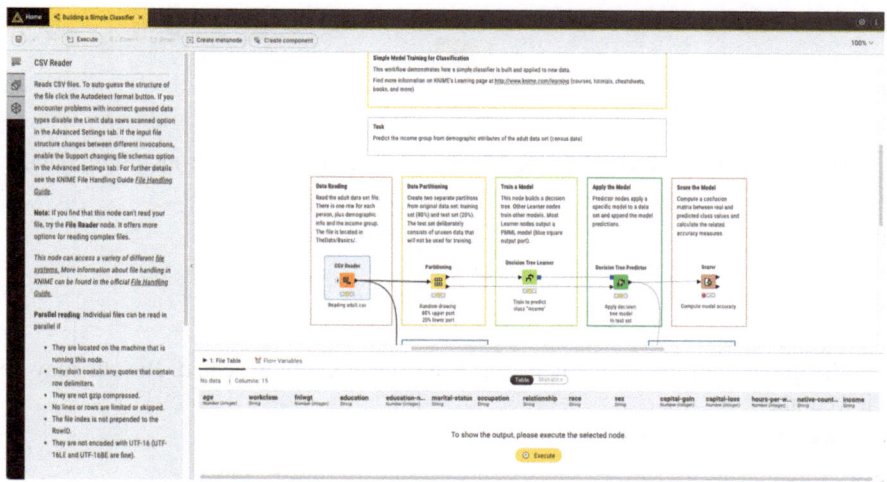

Fig. 2.22 Workflow description

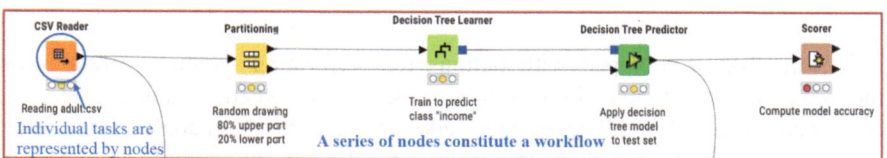

Fig. 2.23 Nodes and workflows

"Description" on the left side panel. You're encouraged to review them attentively. In Fig. 2.22, by clicking on the "CSV Reader" located in the upper-left corner of the workflow editor and opening the description, you can access all facts you need to know about this module.

By comparing Figs. 2.21 and 2.22, you can have a detailed understanding of the functionality of each part of the UI. The main part of the UI is the workflow editor. The workflow serves as a practical approach to implementing machine learning processes through KNIME's graphical interface.

2.3.3.3   Nodes and Workflows

As shown in Fig. 2.23, the nodes in KNIME are individual units that perform a specific task. A series of nodes work together to complete a task, forming a workflow.

Nodes have different working states, which are displayed in the form of traffic lights.

The yellow light signifies the node has been configured correctly and can be executed:

The red light indicates the node is waiting for configuration or incoming data:

The green light means the node has been successfully executed. Results may be viewed and
used in downstream nodes:

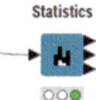

The toolbar is equipped with buttons for performing various operations on individual nodes and the entire workflow, depending on whether a node is selected. In the current workflow, all nodes have yellow or red lights. You don't need to select any nodes and click on a blank area of the workflow, as the toolbar will appear as shown in Fig. 2.24, where you can click on "Execute all" to run the entire workflow. By clicking "Reset all" as shown in Fig. 2.25, all nodes will be reset, turning either yellow or red. When you click on an unexecuted node, the toolbar will display, where you can click on "Execute" to run that specific node as shown in Fig. 2.26.

When you hover your mouse over a node, the node action bar and flow variable ports will appear. You can find a detailed explanation of the node in Fig. 2.27.

The node action bar often includes functions such as configure, execute, cancel, reset, and open view. Different functionalities can be performed by clicking on different icons.

Fig. 2.24   Clicking on the blank area of the workflow

Fig. 2.25   Clicking on the blank area of the workflow after clicking "Execute all"

Fig. 2.26 Clicking an unexecuted node

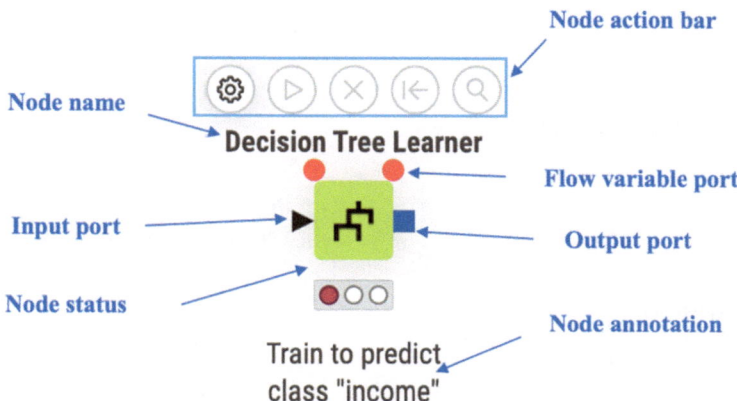

Fig. 2.27 Action bar of a node

Give It a Try

- Try different node setup approaches and run either a single node or the entire workflow.

2.3.3.4   Observing Data

Before we begin using the model to create a workflow, it's helpful to first observe our data. KNIME provides a range of data observers, allowing us to view the data in the node monitor. However, simply looking at numbers can make it challenging for us to develop an intuitive understanding of the data.

On one hand, we can gain an initial understanding of the data by examining statistical values. To do this, we can select the "CSV Reader" node as depicted in Fig. 2.28 and navigate to the "Statistics" tab in the node monitor to observe the statistical values of the data.

On the other hand, using diagrams can be more user-friendly, especially for beginners. We can opt for a histogram to visualize the distribution of ages. Here, we will provide a brief introduction to its usage, and more detailed instructions will be given in specific projects.

To begin, locate the "Histogram" node in the "Views" section of the node repository in the side panel (Fig. 2.29). Once selected, drag the node and drop it into the workflow editor and place it in a suitable position. Next, click on the triangle (output port) of the "CSV Reader" node on the right, and drag the mouse to create a connection line, linking it to the triangle (input port) of the "Histogram" node on the

▶ 1: File Table　　　♡ Flow Variables

Rows: 15 | Columns: 14　　　　　　　　　　　　　　　　　　　　Table　Statistics

Name	Type	# Missing values	# Unique values	Minimum	Maximum	25% Quantile	50% Quantile (Me..	75% Quantile	Mean	Mean Absolute D..	Standard Deviation	Sum	10 most common...
age	Number (integer)	0	73	17	90	28	37	48	38.582	11.189	13.64	1,256,257	36 (898; 2.76%), 31 ...
workclass	String	0	9	⊘	⊘	⊘	⊘	⊘	⊘	⊘	⊘	⊘	Private (22696; 69...
fnlwgt	Number (integer)	0	21648	12,285	1,484,705	117,821.5	178,356	237,054.5	189,778.367	77,608.219	105,549.978	6,179,373,392	123,011 (13; 0.04%...
education	String	0	16	⊘	⊘	⊘	⊘	⊘	⊘	⊘	⊘	⊘	HS-grad (10501; 32...
education-num	Number (integer)	0	16	1	16	9	10	12	10.081	1.903	2.573	328,237	9 (10501; 32.25%), ...
marital-status	String	0	7	⊘	⊘	⊘	⊘	⊘	⊘	⊘	⊘	⊘	Married-civ-spouse...
occupation	String	0	15	⊘	⊘	⊘	⊘	⊘	⊘	⊘	⊘	⊘	Prof-specialty (414...
relationship	String	0	6	⊘	⊘	⊘	⊘	⊘	⊘	⊘	⊘	⊘	Husband (13153; 4...
race	String	0	5	⊘	⊘	⊘	⊘	⊘	⊘	⊘	⊘	⊘	White (27816; 85.4...
sex	String	0	2	⊘	⊘	⊘	⊘	⊘	⊘	⊘	⊘	⊘	Male (21790; 66.92...
capital-gain	Number (integer)	0	118	0	99,999	0	0	0	1,077.649	1,877.273	7,385.292	35,089,324	0 (29849; 91.67%), ...
capital-loss	Number (integer)	0	92	0	4,356	0	0	0	87.304	166.462	402.96	2,842,700	0 (31042; 95.33%), ...
hours-per-week	Number (integer)	0	94	1	99	40	40	45	40.437	7.583	12.347	1,316,684	40 (15217; 46.73%)...
native-country	String	0	42	⊘	⊘	⊘	⊘	⊘	⊘	⊘	⊘	⊘	United States (291...
income	String	0	2	⊘	⊘	⊘	⊘	⊘	⊘	⊘	⊘	⊘	<=50K (24720; 75.9...

Fig. 2.28 View data statistics

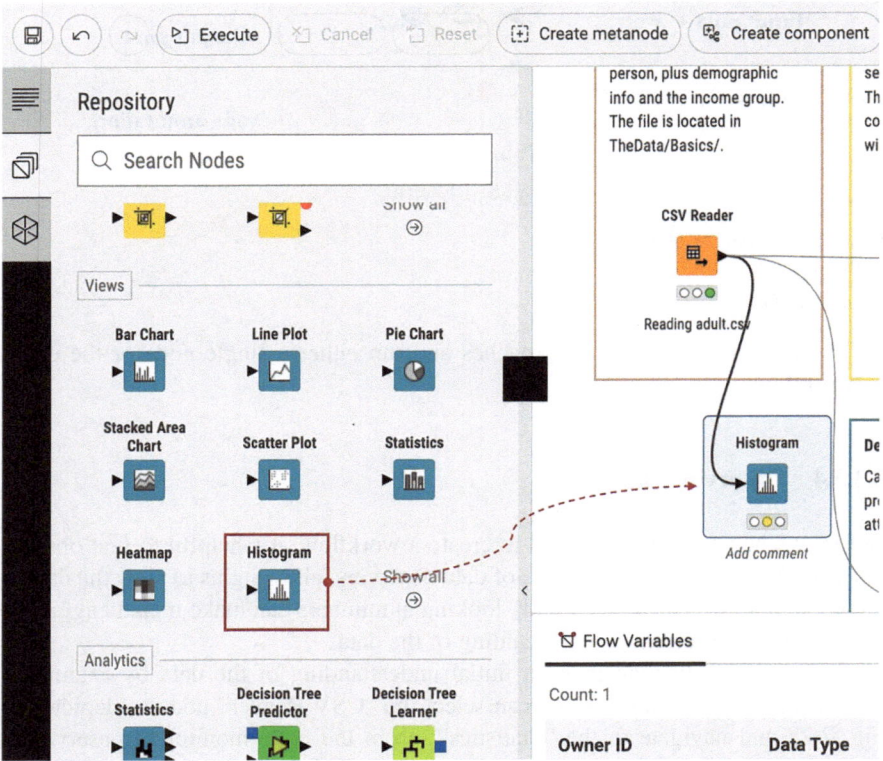

Fig. 2.29 Find the "Histogram" node and connect it to the "CSV Reader" node

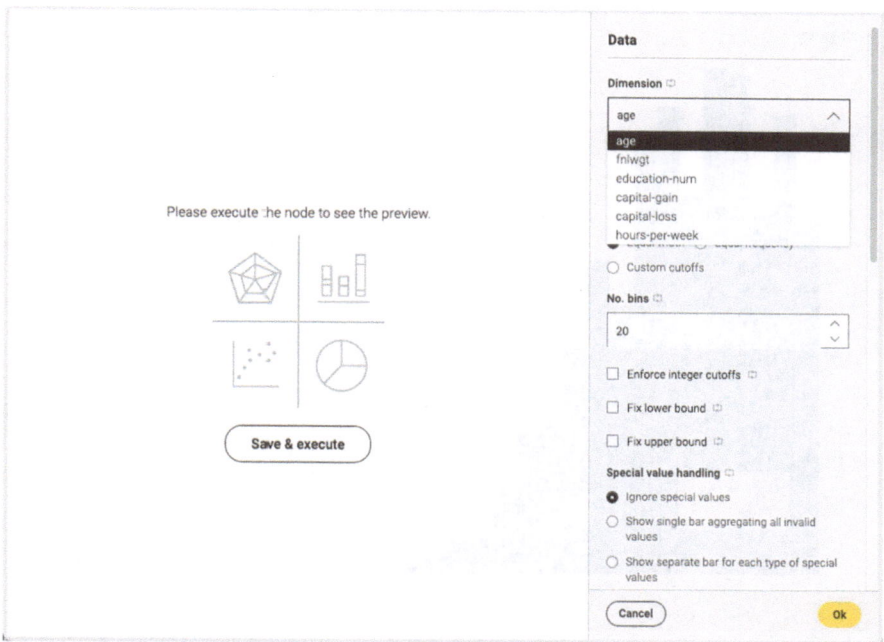

Fig. 2.30 Histogram plotting settings

left. After selecting the "Histogram" node, simply click the icon of "Execute" above the node or the "Execute" button up in the toolbar, and the node status will change to a green light.

Next, double-click the "Histogram" node, or click the gear-shaped icon of "Configure" of the node action bar to open the dialog window of plotting settings. As shown in Fig. 2.30, the basic settings have already been established here, so we can just have a general look at it. Select "age" in "Dimension" menu of the "Data" panel on the right side, and then click the "Save & execute" on the left side to preview the diagram, as shown in Fig. 2.31.

Click "OK" to save the settings and return to the workflow. Hover the mouse over the node, and click the ⓠ "Open View" above the node or right-click the "Histogram" node and select "Open View" in the pop-up shortcut menu as shown in Fig. 2.32, to observe the histogram plotted in Fig. 2.33.

At this point, it is sufficient to only have a basic understanding of KNIME operations. We don't need to delve into the details of how to use the software but rather focus on grasping its functions.

Points for Discussion

• Combining various methods of viewing data, talk about your understanding of the data.

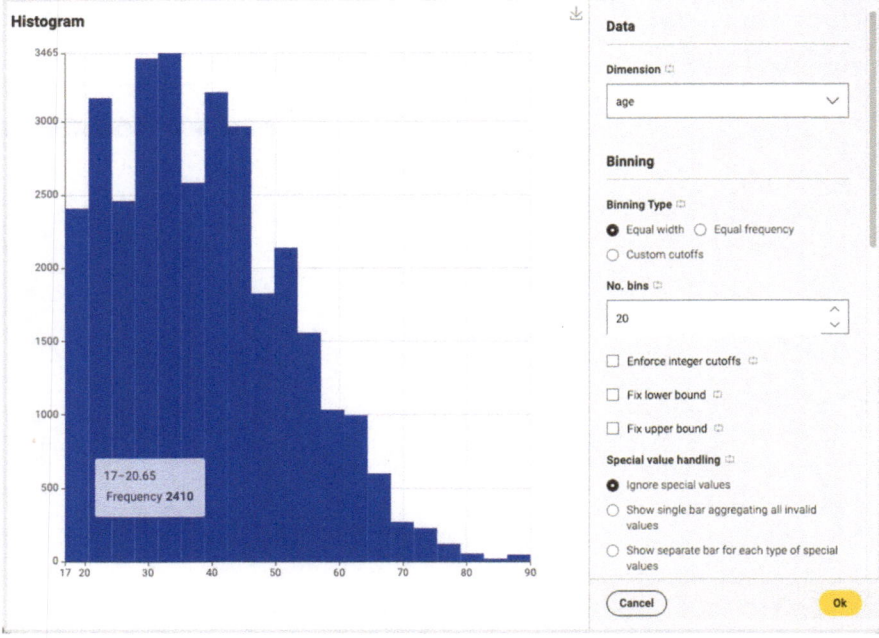

Fig. 2.31 Histogram preview

Fig. 2.32 Right-click the node to open the shortcut menu

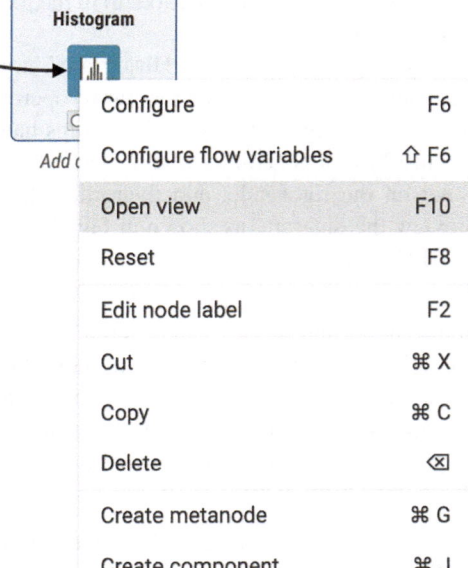

Histogram

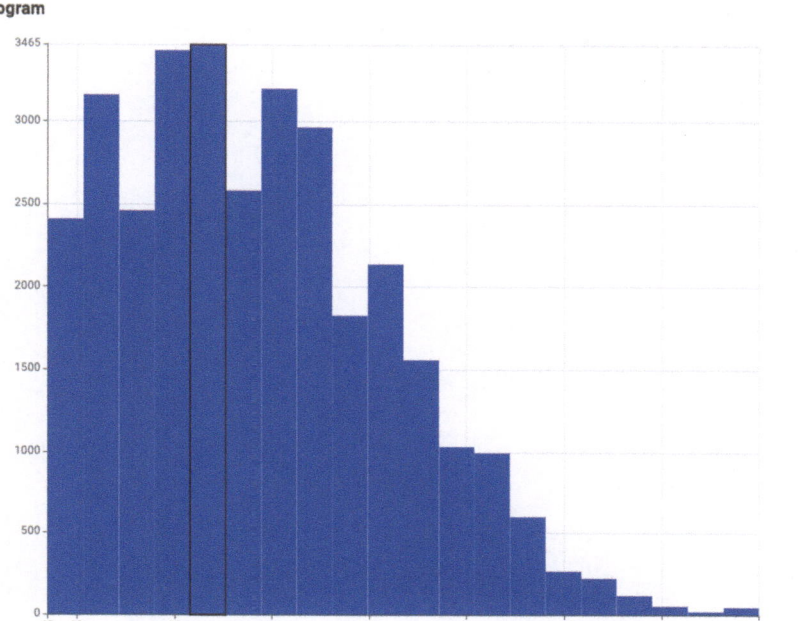

Fig. 2.33 Histogram plotted by node "Histogram"

Fig. 2.34 Scoring node

Scorer

Compute model accuracy

2.3.3.5 Observing Model Results

We can also observe the results of the model to determine whether it meets the requirements. Click the "Scorer" node in the workflow to observe the scoring (Fig. 2.34). Hover the mouse over this node, and select the "Open view" to see the scoring results of the model. Here is the scoring result of the confusion matrix, as shown in Fig. 2.35.

You can also view the confusion matrix and its scoring results separately in node monitor, as shown in Fig. 2.36.

In this case, there is no need to under the details of the confusion matrix or model evaluation. We simply need to be familiar with the basic operations.

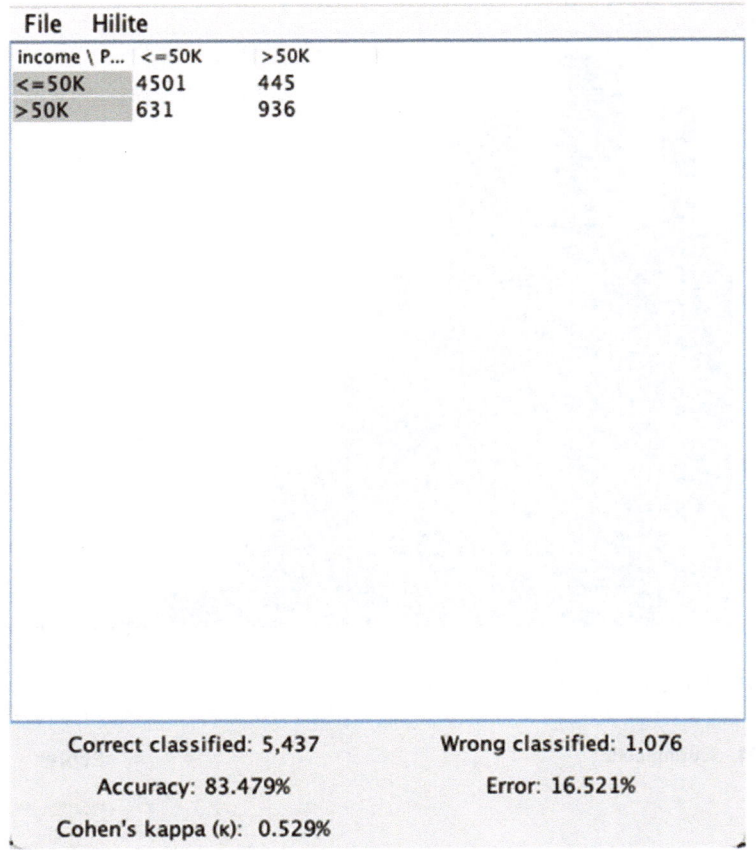

Fig. 2.35 Scoring of the confusion matrix

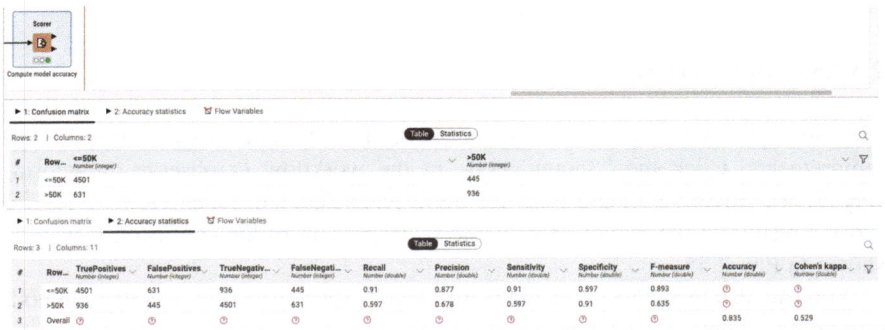

Fig. 2.36 View the results of the "Scorer" node in the node monitor

2.3.4 Summary

In the previous section, we gained a basic understanding of the functions and usage methods of KNIME. It's important to note that readers may not be familiar with using this software or understand some of the professional terms. However, that's perfectly fine. The purpose of this section is to grasp the software's functionality, not the specifics of its usage. Throughout this process, we discovered that KNIME offers the advantages of simplicity and visualization. We will utilize KNIME to tackle various machine learning tasks and participate in Kaggle competitions, which are highly regarded by machine learning enthusiasts worldwide. This will allow us to practice machine learning through real-world projects.

2.4 Practice

1. When is it not advisable to use the mean value?
2. What parameters determine the "thickness" or "thinness" of a normal distribution?
3. Can you briefly explain what linear regression is?
4. Using simple linear regression as an example, can you explain the geometric meaning of each parameter?
5. Can we determine the data distribution from a box and whisker plot? Do you think box and whisker plot can be used for detecting outliers?
6. Can we solely rely on the coefficient of determination to judge the quality of fit?
7. Take some time to explore other functionalities of KNIME by yourself.

References

https://scikit-learn.org
https://www.knime.com/getting-started-guide
https://www.deeplearning.ai/courses/machine-learning-specialization/
https://en.wikipedia.org/wiki/Normal_distribution
https://en.wikipedia.org/wiki/Standard_deviation
https://en.wikipedia.org/wiki/Coefficient_of_determination
https://en.wikipedia.org/wiki/Box_plot

Chapter 3
Linear Regression

On average, more towards the middle.

—*Francis Galton*

Abstract This chapter provides an insightful journey into linear regression, a fundamental technique in machine learning. It begins with an exploration of simple linear regression through a fabricated dataset, gradually transitioning to multiple linear regression (MLR) with a real-world Kaggle competition. The chapter emphasizes the importance of understanding and implementing key concepts like data visualization, training and testing sets, p-value, loss function, feature selection, and normalization in the context of MLR.

Practical application is a central theme, with the chapter offering hands-on guidance on using KNIME, a powerful tool for machine learning. Readers are introduced to workflow creation in KNIME, starting from data acquisition to model training and testing. Special attention is given to understanding and interpreting statistical measures like p-value and coefficient of determination (R^2) in the context of linear regression models.

The chapter delves into advanced topics such as feature engineering, overfitting, and model optimization techniques like forward selection. It also explores feature normalization methods, including Z-score and min-max scaling, to enhance model stability and performance. This comprehensive chapter aims to equip readers with both theoretical understanding and practical skills in applying linear regression in machine learning projects.

Keywords Linear regression · Machine learning · Data visualization · KNIME · Feature selection · Model optimization

This chapter covers

- **Data visualization.**
- **Training and testing.**
- ***P*-value.**
- **Loss function.**

- **Feature selection.**
- **Feature normalization.**

In this chapter, we'll start by giving an overview of the machine learning process and cover some key concepts like simple linear regression and *p*-value. We'll then dive into multiple linear regression (MLR), building upon these concepts to create a solid foundation for our model. As we progress, we'll enhance our model by incorporating modules such as data visualization and data processing, making it truly practical and feasible.

In simple linear regression, we'll work with a set of fabricated data to get acquainted with the machine learning process. Moving on to MLR, we'll take part in our first Kaggle competition and tackle a real-world project.

3.1   Simple Linear Regression

Having already grasped the fundamental principles of linear regression, we are now ready to put our theoretical knowledge into practice and delve deeper into machine learning. The overall process of a machine learning project is illustrated in Fig. 3.1. It begins with acquiring the necessary data, followed by data preprocessing, which involves observing and partitioning the dataset. Next, we proceed to train and test the model, continuously refining it based on evaluation methods like the loss function (https://www.deeplearning.ai/courses/machine-learning-specialization/).

3.1.1   Scene Description

In this example, we will analyze the correlation between years of work and salary using a table and make predictions about the salary for a certain number of years of work.

As we can see, this is a straightforward problem of using years of work to forecast salary via simple linear regression.

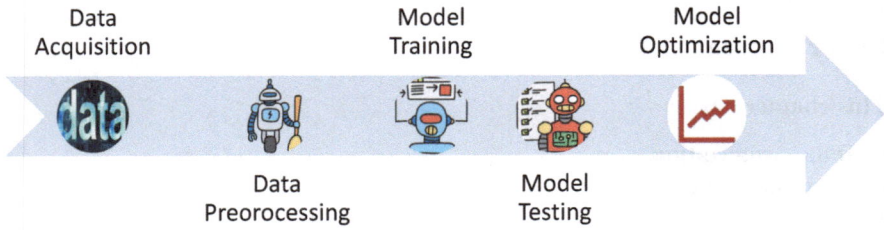

Fig. 3.1 Machine learning process (This figure has been designed using images from Flaticon. com)

3.1.2   KNIME Workflow Creation

Let's use KNIME to create a workflow of simple linear regression. The goal is to get familiar with KNIME's functionalities and the general steps of machine learning. This will serve as a stepping stone for more complicated tasks in the future.

We can save a workflow as a file in a folder known as a "workflow group" in KNIME. To do this, simply click on the icon of "Space Explorer" ⬡ located in the side panel, and then click on the icon of "More actions" (⋮) . In the pop-up menu, you'll see "Create Workflow" and "Create folder" as shown in Fig. 3.2.

Feel free to create workflows or folders based on your own needs. Additionally, you can click on the "Home" button to go back to the entry page as shown in Fig. 3.18 and create a new workflow. Now, we can really get started.

3.1.3   Data Acquisition

Any machine learning task requires data, so we need to import the data first.

KNIME provides a range of nodes for importing data. In our case here, we want to import a CSV file. To do this, we can select the "CSV Reader" node. In the "Node Repository" of the side panel, under the category of "IO," you can easily find the "CSV Reader" node. Or you may simply type "CSV" in the search field to find all nodes related to CSV (Fig. 3.3). Once you have selected the "CSV Reader" node and

Fig. 3.2 Create workflow

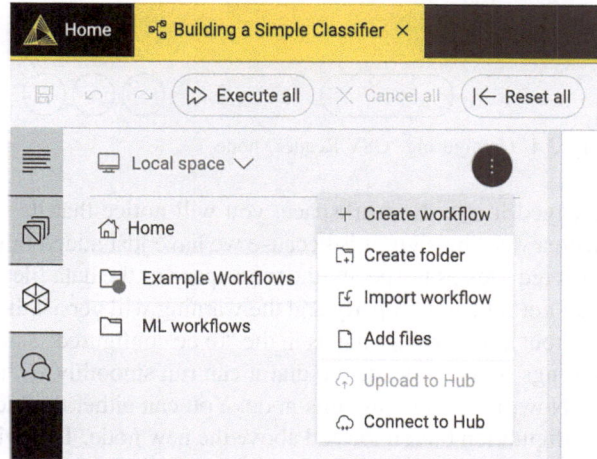

Fig. 3.3 Find the node you need

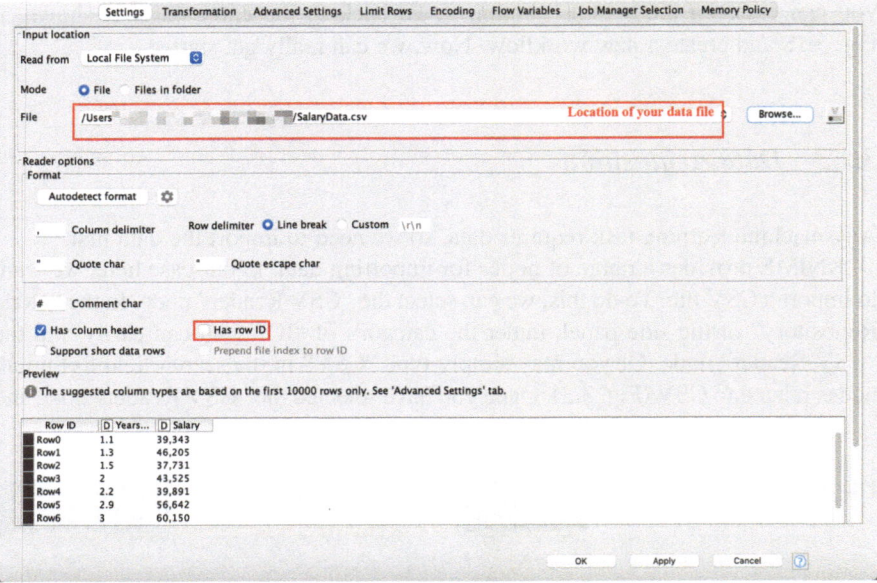

Fig. 3.4 Configuring "CSV Reader" node

dragged it into the workspace, you will notice that its status is "to be configured" with a warning sign. It is because we have just added this node to the workflow and thus requires us to specify the workspace of the data file. Without this step, the node will not function properly and the warning will persist there. If a new node is created without any warnings and is in the "to be configured" status, it means that the default settings of the node ensure that it can run smoothly without any errors.

Now, let's configure this node. You can either double-click on it or click on the configuration button located above the new node. This will open a dialog window as shown in Fig. 3.4, where you can specify the data and its location (SalaryData.csv). Depending on whether you require the first column data as row names, you can check or uncheck the "Has Row ID/Has Column Header" option. In the case of this

data file, it has column names but no row names, so we will uncheck "Has Row ID."
Once you have made the necessary selections, simply click on the "Confirm," and
you will notice that the node status changes to "Configured."
Give It a Try

- KNIME also provides an easier and faster way to create a "CSV Reader" node.
  Click on "Space explorer" in the side panel, and try dragging it into the KNIME
  workflow editor. What can you see?
- Try checking or unchecking the box of "Has Row ID/Has Column Header," and
  observe the changes in the output data.

3.1.4   View the Data

The initial step in data analysis and other machine learning processes is to observe
the data. This involves examining the statistical patterns and distributions of the data
generally. By viewing the data, we can develop an initial understanding, identify
relationships between different data points, and gain a rough understanding of the
connections between various features and data labels. However, it is crucial to avoid
having personal subjective judgments during this step.

3.1.4.1   View the Data Table

Click on the "CVS reader" node, and then click on the execute button above the
node. Then you will see the output content of "CVS reader" in the node monitor. By
default, in the "File Table" section, the "Table" tab is automatically selected. You
can use this table to browse the data (Fig. 3.5).

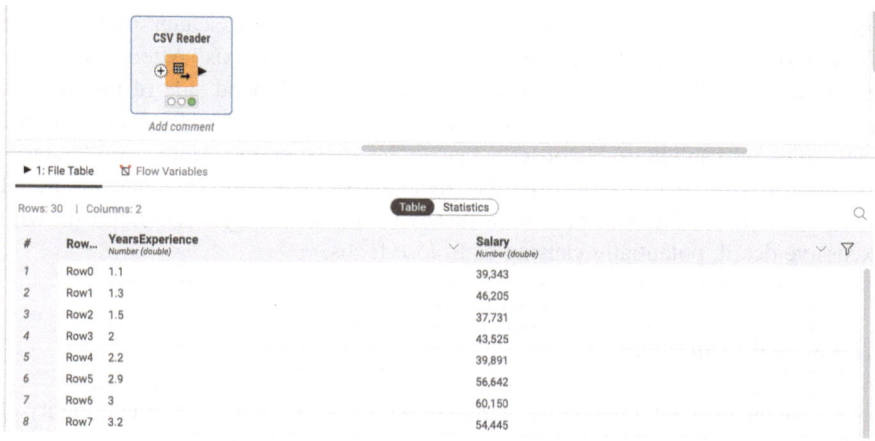

Fig. 3.5 View the data

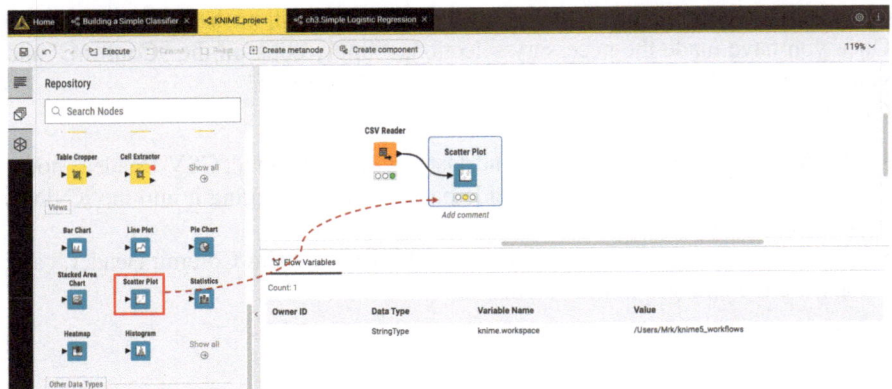

Fig. 3.6 Create a "Scatter Plot" node and connect it to the "CSV Reader" node

Give It a Try

• In the "CSV reader" node, select the "Statistics" tab in the "File Table" section to observe the statistical data. Can you find the mean, median, and other statistical information?

3.1.4.2 Data Visualization

Here, the scatter plot is used as an example to illustrate the general method of data visualization. We can create a new "Scatter Plot" node, as shown in Fig. 3.6, and drag it into the workflow editor. Then we can connect it to the "CSV Reader" node.

The "Scatter Plot" node takes the output of the "CSV reader" node as its input. To configure the "Scatter Plot" node, click on the configuration button. In the configuration area on the right, you will find "Horizontal dimension" and "Vertical dimension," which represent the X and Y variables of the scatter plot. Choose "YearsExperience" for the X-axis and "Salary" for the Y-axis. After making the selections, click the "Save & execute" button on the left-hand side of the plotting area to preview the scatter plot as described in Fig. 3.7. By visually examining the scatter plot, we can intuitively observe that as "YearsExperience" increases, "Salary" also tends to rise. Scatter plots offer the advantage of displaying each data point, but at the same time, this can also be a disadvantage as it may overwhelm us with excessive detail, potentially causing us to lose focus.

3.1.4.3 Add Comment

So far, we have added three nodes, and in order to easily identify the functionality of each node, annotations can be added to each node. In Fig. 3.8, we can double click

Fig. 3.7 Scatter plot preview

Fig. 3.8 Add comments for nodes

on "Add comment" at the bottom of the node to enter the content of your comment in the box, and click the "√" button to confirm.

3.1.5   Dataset Partition

Dataset partitioning is an extremely important step in the machine learning process and directly affects the effectiveness and reliability of the model.

3.1.5.1   Training Set and Test Set

We need to break down the imported data into two categories: training set and test set.

- Training set. Its role is to fit the model by optimizing the model's parameters during training.
- Test set. After obtaining the model from the training set, use the test set to evaluate the model.

For instance, let's consider the scenario where we aim to fit the equation $y = b + wx$, with the ultimate objective of estimating the value of y when new data x becomes available in the future. The purpose of a training set is to learn and determine appropriate w and b based on the known x and y values, ensuring that the predicted values closely align with the actual ones. However, if we employ all the known data for training, including both x and y, we won't be able to assess the proximity between the predicted y and the actual y when new x values are introduced. So, what should we do in such a situation? This is where the test set comes into play.

To address this, we can divide the known data into two parts: the larger portion (e.g., 75%) will serve as the training set, while the smaller portion (e.g., 25%) will be designated as the test set. The training set fulfills its intended purpose and usage as above mentioned, while the test set allows us to simulate a scenario where we are unaware of the actual y values. We input the x values, transform them using the equation $b + wx$, calculate $y^\wedge$, and then compare $y^\wedge$ with the y values in the dataset. This analysis helps us evaluate the effectiveness of the model and its predictive capabilities.

How well does $y^\wedge$ approximate y? Do you recall the loss function and R^2 that were discussed in the previous chapter? We can utilize these metrics to evaluate the performance.

Also note that the test set needs to meet the following two conditions:

- It should have a sufficiently large scale to yield statistically meaningful results.
- It should be representative of the entire dataset. In other words, the selected features in the test set should align with those in the training set.

Only when the test set meets these conditions can we obtain a model that effectively generalizes to new data.

And it is important to remember

It is absolutely prohibited to use test data for training purposes!

Fig. 3.9 Searching for
"Partitioning" node

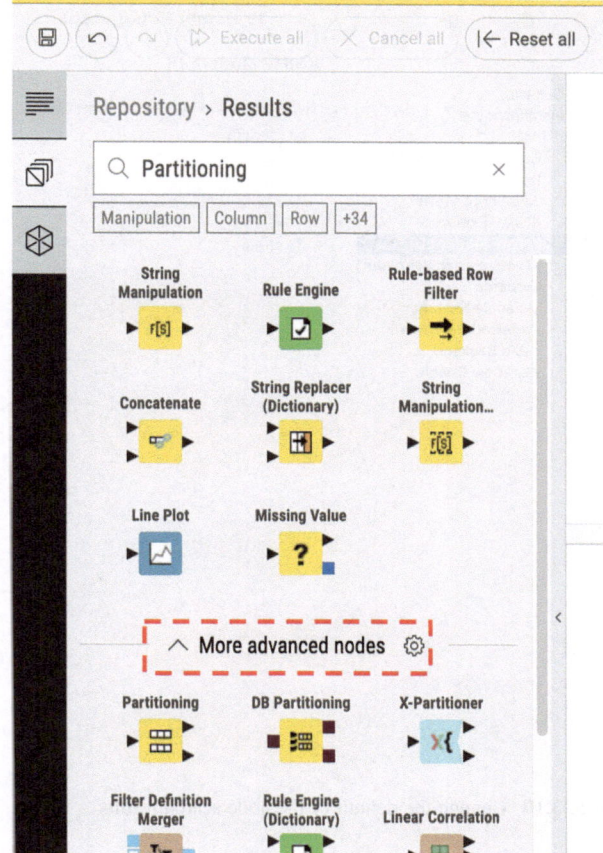

3.1.5.2   KNIME Operations

By utilizing the fundamental idea of dataset partitioning, we can employ the
"Partitioning" node to divide the dataset. As depicted in Fig. 3.9, you can find the
"Partitioning" node by searching in the node repository. Please note that this node
may not be readily visible in the default search field, so you will need to click on
"More advanced nodes" below to access additional nodes.

The reason behind this is that KNIME 5.1 has a limitation on the number of node
search results displayed, showing only basic functional nodes by default. However,
since machine learning often involves advanced nodes, we need to modify this
default setting. Click the cog icon 🔘 on the top right to go to the KNIME Modern

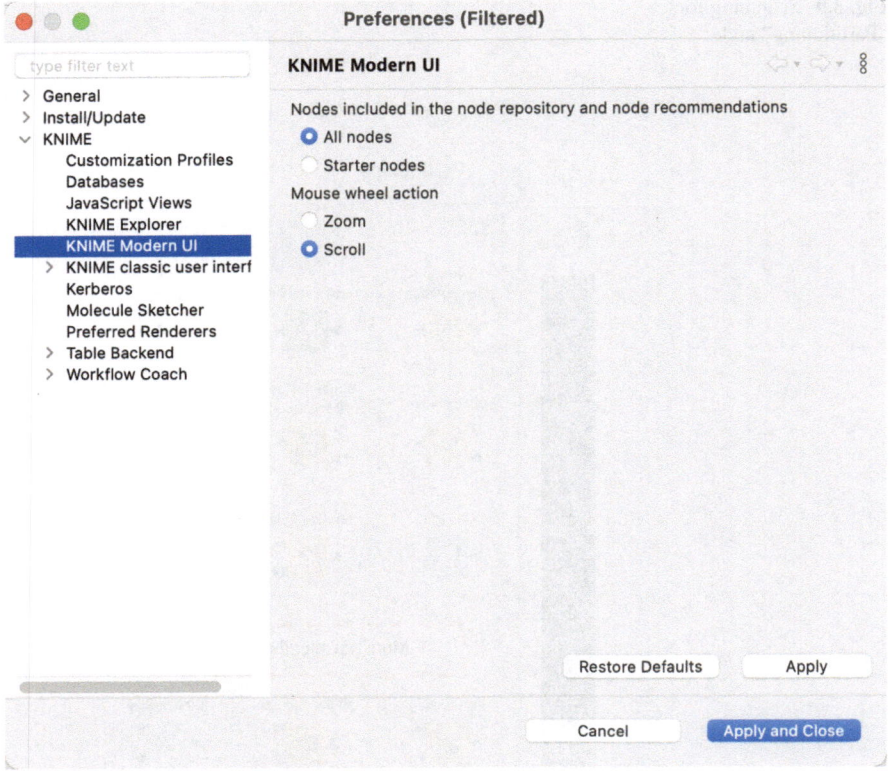

Fig. 3.10 Change the default of the node search results

UI as shown in Fig. 3.10. Within the "Nodes included in the node repository and node recommendations" option, we can select "All nodes" to change the default of the node search results.

Suppose we allocate 75% of the available data to the training set and 25% to the test set. This division can be accomplished by following the straightforward steps illustrated in Fig. 3.11. There are two important considerations to keep in mind during this partitioning process. Firstly, we have opted for a "Relative" percentage rather than an "Absolute" quantity. Secondly, we need to select "Draw randomly." As a result, this configuration randomly draws 75% of the data from the original dataset to form the training set.

Figure 3.12 displays the current workflow and the resulting data partitioning. In the node monitor, we may observe that the "Partitioning" node has generated

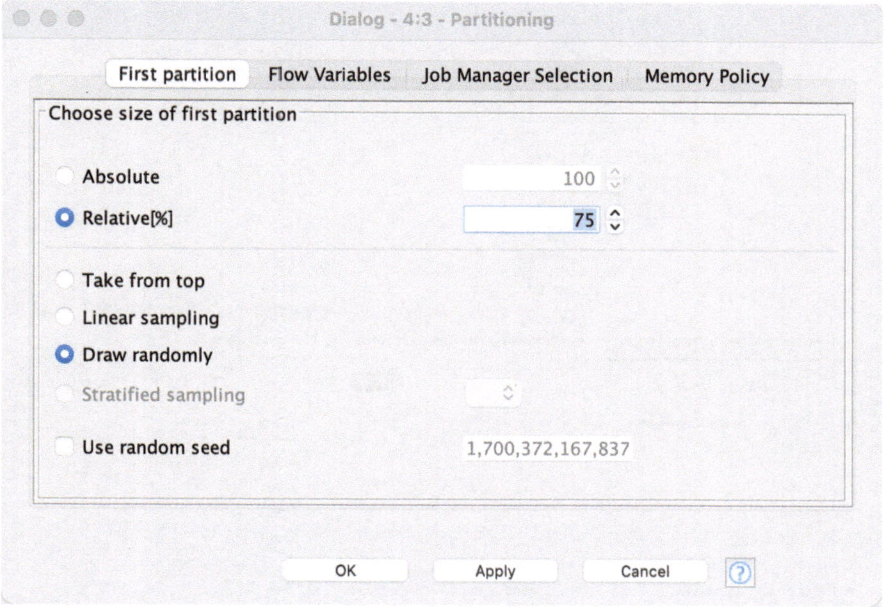

Fig. 3.11 Partitioning

22 rows for the first partition and eight rows for the second one. This indicates that the data has been divided by extracting 75% from the original dataset.

3.1.6   Train the Model

Once the data is ready, we can begin training the model. We'll be using the linear regression method, so we'll create a new "Linear Regression Learner" node. When we configure this node in the setting, we'll select "Salary" as the target and "YearsExperience" as the feature, just like in Fig. 3.13. In the feature selection, there are two columns: the left, marked in red, is for excluded features, and the right is for required features. Since we only have one feature, the left side is left empty and the right side is "YearsExperience."

Connect the input of this node to the first output of the "Partitioning" node (training set), as shown in Fig. 3.14.

Running this node allows you to check out some information about the trained model: when you click on "Open View" of the node or "Coefficients and Statistics"

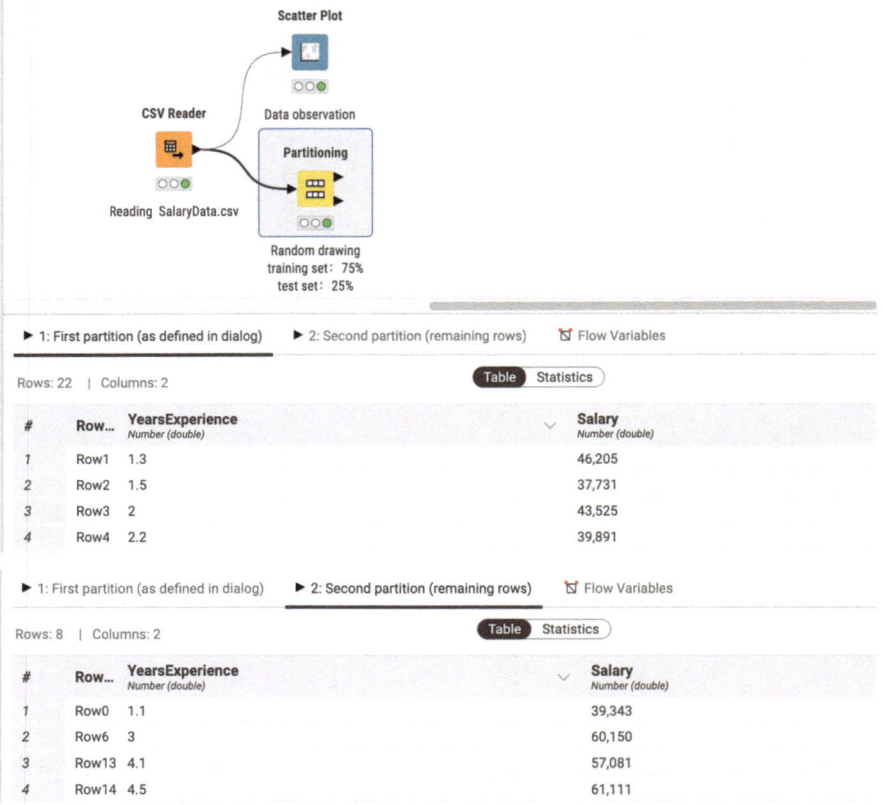

Fig. 3.12 Current workflow and data partitioning results

in the node monitor, you can see the fitting result (Fig. 3.15). If you switch back to the classic KNIME UI, you can also view the fitting line by right-clicking and selecting "View: Linear Regression Scatterplot View" from the shortcut menu (Fig. 3.16). It's important to note that this step is for the training set only. By visually inspecting the fitting line shown in Fig. 3.16, we can see that the fitting line passes through the various data points. From the perspective of the mean squared error (MSE) introduced in the previous chapter, this line aims to minimize the MSE as much as possible, and it can no longer reduce the MSE through upward or downward shifts or angle adjustments.

Give It a Try

• How to go back to the classic UI.

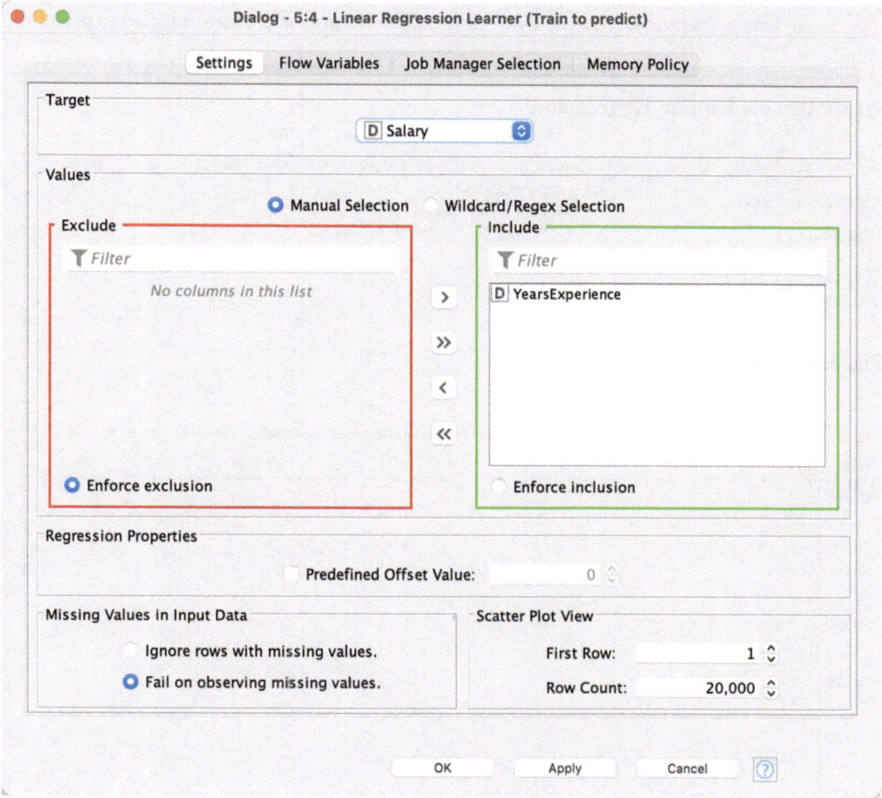

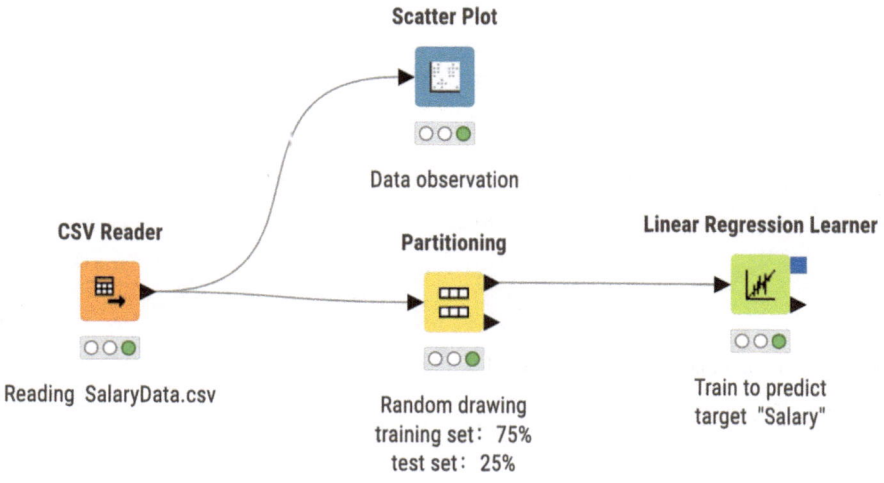

Fig. 3.13 Setting up a linear regression model

Fig. 3.14 Training model

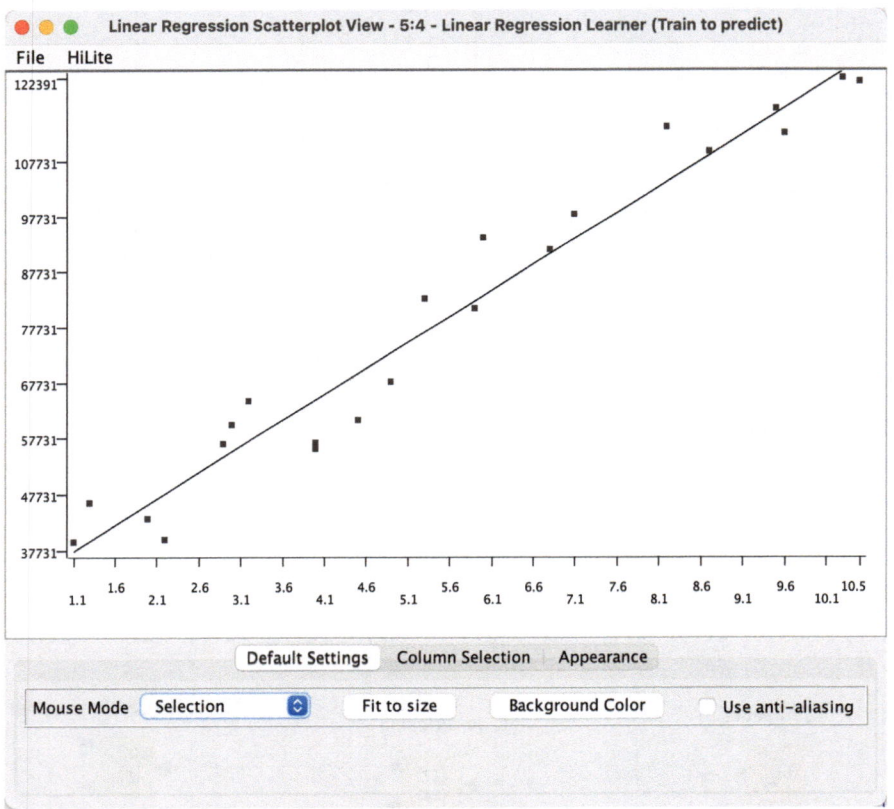

Statistics on Linear Regression

Variable	Coeff.	Std. Err.	t-value	P>|t|
YearsExperience	9,367.9459	443.3771	21.1286	3.77E-15
Intercept	27,454.1613	2,760.6154	9.9449	3.47E-9

R-Squared: 0.9571
Adjusted R-Squared: 0.955

Fig. 3.15 Fitting result

Fig. 3.16 Fitting line in classic KNIME UI

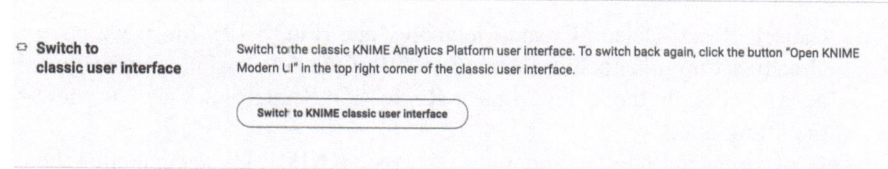

Fig. 3.17 Switch back to the classic UI

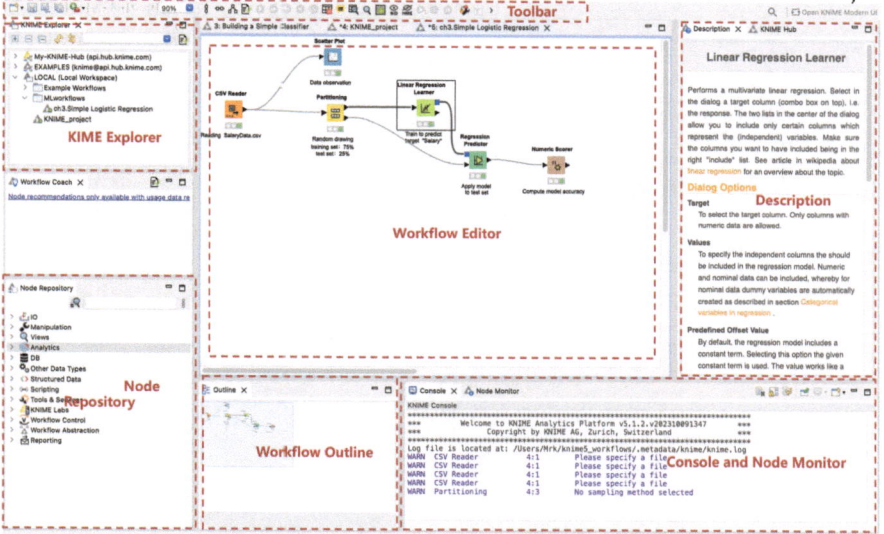

Fig. 3.18 Classic KNIME UI

Click on the "Go to info page" button [i] in the top right corner to enter the settings, as shown in Fig. 3.17. Find the "Switch to classic user interface" button, and click on it to switch back to the classic UI.

The classic UI differs quite significantly from version 5.1, but the node operation functions are similar. We mainly use the classic UI to operate on specific functions and nodes that are missing or phased out in the new version, so it is only necessary to have a brief understanding of it. As shown in Fig. 3.18, the classic UI mainly includes a toolbar, a workflow editor, a KNIME explorer, a node repository, description, etc. The workflow editor is consistent with the 5.1 version in principle, but there are significant differences in the style of node operations.

Taking the "Linear Regression Learner" node as an example, right-click on this node, and in the pop-up shortcut menu, you can choose the following functional operations: Configure, Execute, Execute and Open Views,

(continued)

Cancel, Reset, Delete, Create Metanode, etc. (Fig. 3.19). Since we have already set up this node in the modern UI of the 5.1 version, now we just need to click on the "View: Linear Regression Scatterplot View" to view the fitting line.

After experiencing the functionalities of classic KNIME UI, we can click the button in the upper right corner to return to the modern UI of 5.1 version and continue with the following learning.

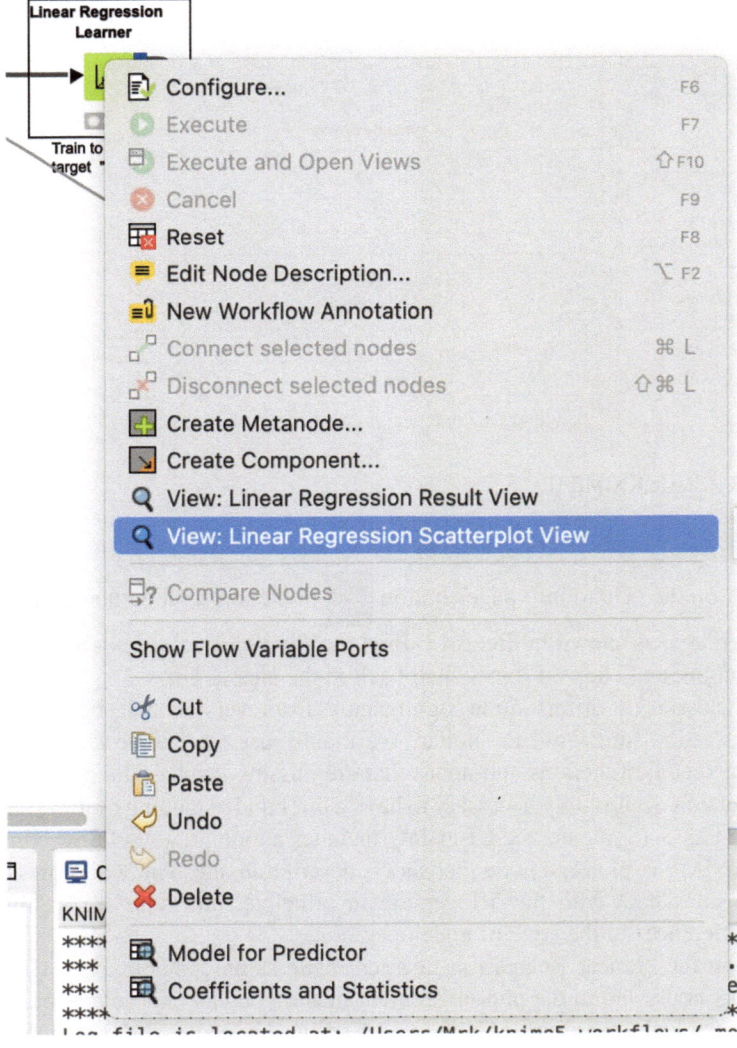

Fig. 3.19 Node operations in classic KNIME UI

Let's analyze some details of the model.

3.1.6.1 Slope and Intercept

In Fig. 3.15, the "Coeff" for "YearsExperience" represents the slope, while the "Coeff" for "Intercept" indicates the vertical intercept of the fitted curve. In simpler terms, the "Intercept" represents the base salary without any work experience, and the "Coeff" for "YearsExperience" shows how much the salary increases for incremental work experience on a yearly basis.

3.1.6.2 *P*-Value

First, we start with the null hypothesis, which suggests that this feature has no connection to the target. Then, we check if we can dismiss this null hypothesis, indicating that we are rejecting the idea that there is no link between the feature and the target.

Please note that rejecting the null hypothesis, which suggests no connection between features and target labels, does not automatically mean confirming a relationship between the features and the target. It's a bit like a court trial, where the presumption of innocence is upheld, and the burden of proof lies with the prosecuting lawyer to demonstrate the guilt of the criminal suspect. If they fail to do so, the suspect is released. However, in this case, it's more about being unable to prove the guilt of the suspect (unable to reject the null hypothesis) rather than proving their innocence.

The p-value represents the likelihood of observing the sample results or more extreme results if the null hypothesis is true. A very small p-value indicates a very low probability of the null hypothesis being true, giving us cause to reject it. The smaller the p-value, the stronger the basis for rejecting the null hypothesis. In essence, the smaller the p-value, the more significant the relationship between the target and features. Typically, a p-value less than 0.05 is considered small. In other words,

- If the p-value <0.05, then we can reject the null hypothesis.
- If p-value ≥ 0.05, the null hypothesis cannot be rejected.

3.1.6.3 Coefficient of Determination

The remaining two important indicators are coefficient of determination R^2 and *Adj R Minister*. Simply put, the closer these two values are to 1, the better.

3.1.7 Test the Model

Now, it's time to put the model to the test. Create a "Regression Predictor" node, and feed the trained model and test data into this node separately on the input side. You can refer to Fig. 3.20 for the workflow. The testing data comes from the second output port of the "Partitioning" node. One thing to note in KNIME is that the blue input or output port of a node is related to the model, while the black triangle is related to data.

Then create a "Numeric Scorer" node and connect it as shown in Fig. 3.21.

Double-click this node to set the "Reference column" and "Predicted column" (Fig. 3.22), which correspond to the actual value and predicted value, respectively. Right-clicking on the "Open View" button of the "Numeric Scorer" node allows you to observe the results, as shown in Fig. 3.23. Here we are essentially feeding test data into a trained model, comparing the predicted values with the actual values, and finally obtaining statistics for the model's performance. The ones that we have

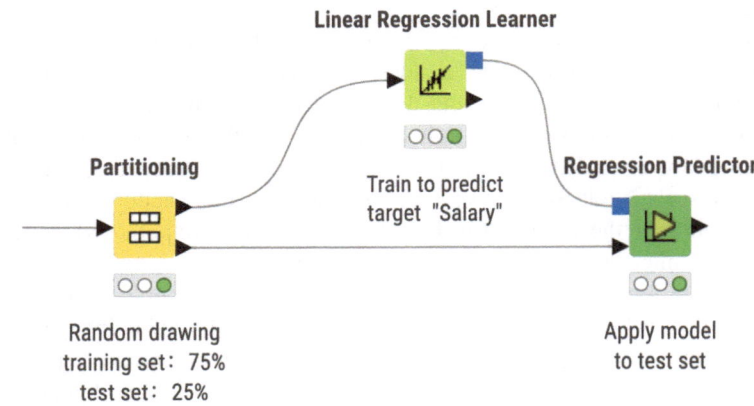

Fig. 3.20 Test the model

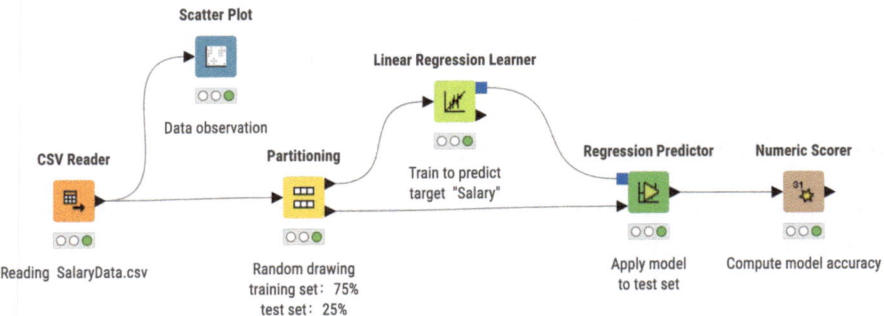

Fig. 3.21 Overall workflow

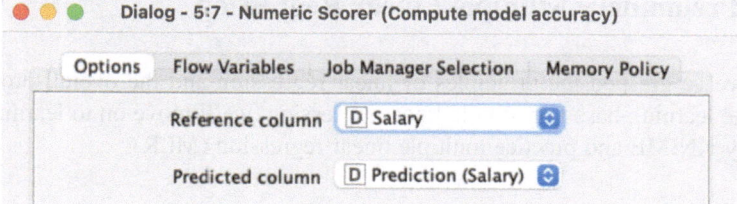

Fig. 3.22 Set reference column and prediction column

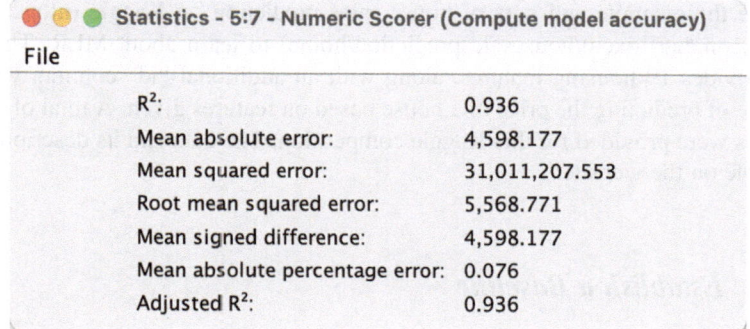

Fig. 3.23 Model scorer

already known include R^2 and mean squared error. In simple terms, for the former, the larger the better, while for the latter, the smaller the better.
Points for Discussion

- Why is it the larger the better for R^2, while the smaller the better for mean squared error?

3.1.8 Loss Function

The loss function is used to measure how much the predicted values of a model differ from the true values, and it is a nonnegative real-valued function. The smaller the loss function, the better the model. We will minimize the loss function using a method called the "gradient descent algorithm" to optimize the model. In Fig. 3.23, "mean squared error," "mean absolute error," "root mean squared error," and "mean signed difference" can all be used as loss functions for our model. We generally use "mean squared error" as the loss function, which is represented by $\frac{1}{m} \sum (\hat{y}_i - y_i)^2$, where m represents the number of samples. We have already introduced this function in Chap. 2, so we won't go into details here.

3.2   Preliminary Multiple Linear Regression

Once we've grasped the principles of linear regression and the overall process of machine learning based on simple linear regression, we'll move on to learning how to apply KNIME and practice multiple linear regression (MLR).

3.2.1   Task and Data Description

We use the example and data of house sales prediction on Kaggle (https://www.kaggle.com/harlfoxem/housesalesprediction/home) to learn about MLR. This project provides 19 housing features, along with an additional "id" column, with the purpose of predicting the price of a house based on features given. A total of 21,613 samples were provided for this Kaggle competition. The data and its description are available on the website.

3.2.2   Establish a Baseline

Establishing a baseline is the first stage of machine learning tasks and reflects the concept of "baseline thinking" in the field of machine learning.

Based on the simple linear regression introduced earlier, you can create a workflow as shown in Fig. 3.21 and correspondingly modify the features (shown as "Values" in KNIME) and "target" for training (see Fig. 3.24).

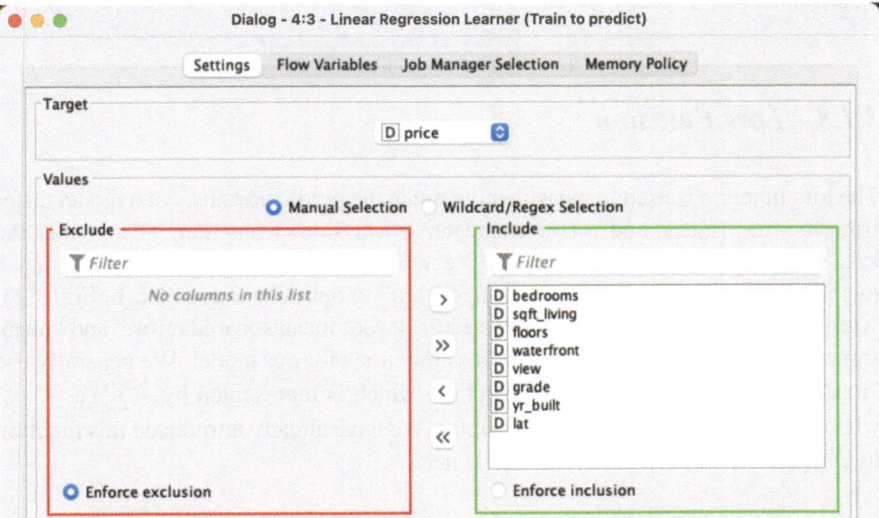

Fig. 3.24 Set the target and values

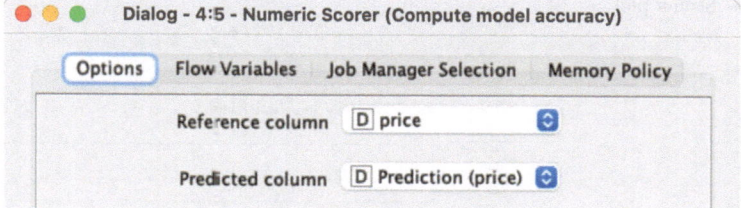

Fig. 3.25 Set reference column and prediction column

The corresponding scorer can be set as shown in Fig. 3.25.

We've managed to wrap up a MLR task swiftly with just a simple setup. This really shows off the straightforward and user-friendly features of KNIME. Of course, the model isn't perfect yet. We'll work on improving it step by step to create a better version. This model will set the baseline for our future work and give us a simple, quick explanation of the project's feasibility.

3.2.3 Read and Observe the Data

To better understand the problem, let's first observe the data. We use the "CSV Reader" node to read the data (kc_house_data.csv) and then prepare to observe the data.

3.2.3.1 Use Scatter Plots to Observe the Distribution of Data

The simplest and most intuitive way to observe data distribution is through a scatter plot. Enter "scatter" in the Node Repository search field and find the "Scatter Plot" node as shown in Fig. 3.26.

Click on the configuration button of this node, as shown in Fig. 3.27, and set the "X Column" as "sqft_living" (area) while keeping the "Y Column" as "price." This will allow us to observe how the price changes with the size of area. It's noticeable that the price generally goes up as the area increases, which is in line with our expectations. We can also see that lower-priced houses are more densely distributed and have a more concentrated price range. Take note of the warning at the bottom of the scatter plot that says "WARNING Only the first 2500 rows are displayed". Why do we often come across this warning when using KNIME? This is because displaying graphics consumes a significant amount of computer resources, so KNIME has taken the easy way out. In the settings for plotting, as shown in Fig. 3.27, you can choose the number of rows to be displayed in "Max rows," and the default setting is recommended in our case. Increasing the amount of data in scatter plots will only add burden to our brains.

You can try to discover more secrets behind these data.

Fig. 3.26 Scatter plot

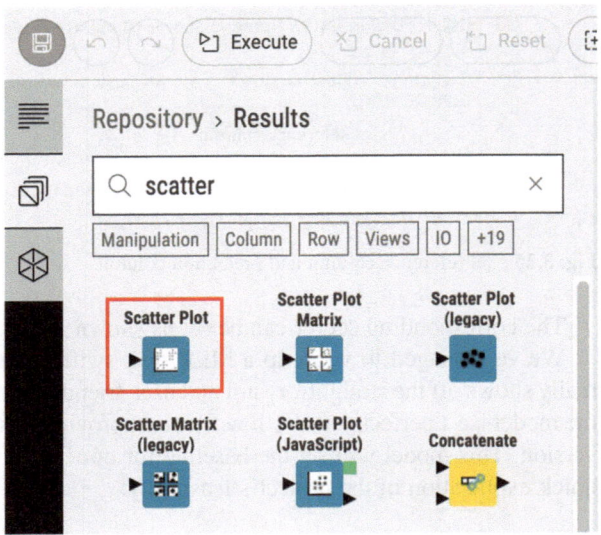

Fig. 3.27 Relationship between area and price

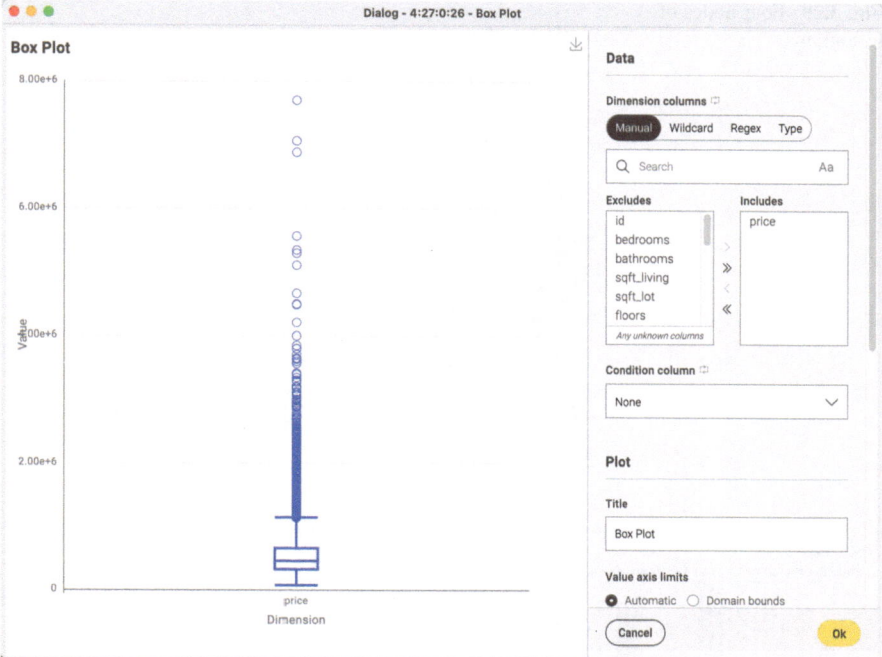

Fig. 3.28 Preview of box plot

3.2.3.2 Use a Box Plot to Observe the Distribution of Data

Next, let's use a box plot to check out how the data is spread out. Type "box" in the search field of the Node Repository and select "Box Plot" (*not* "Box Plot (JavaScript)," but if you may, have a try if you're interested). Once you've chosen this node, click on the node configuration button and head to the settings shown in Fig. 3.28. Now, we're only interested in the price, so we'll just select "price" in the "Includes" section. Click the "Save & execute" button to see the pot preview. The box plot reveals that most houses are priced on the lower side, but there are still some with higher prices, but very few with extremely high prices. Feel free to explore other features and see what you can find out.

3.2.3.3 Use Histogram to Observe the Distribution of Data

Next, let's take a look at the data using a histogram and a bar chart. Remember, the histogram is for numerical data, while the bar chart is for categorical data. However, in KNIME, both of these graphs are referred to as "histogram," but we still need to make a clear distinction between them.

When you type "hist" in the search field of the Node Repository, you'll see four nodes related to histograms, as shown in Fig. 3.29. Choose the last one, i.e., "Interactive Histogram (legacy)." "Histogram" and "Histogram (JavaScript)" might give you slightly prettier graphs, but their functionality is pretty average. "Histogram (legacy)"

Fig. 3.29 Four nodes of histograms

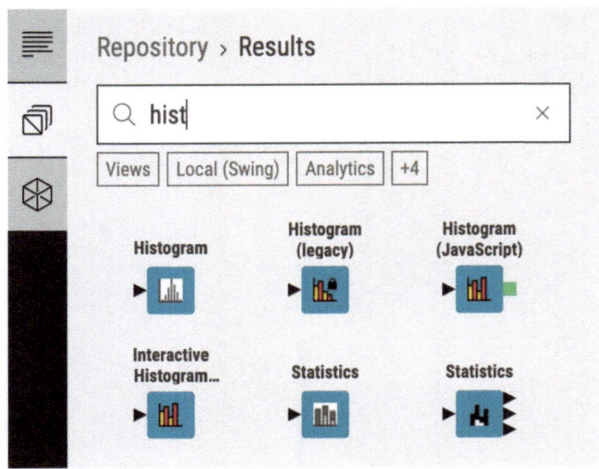

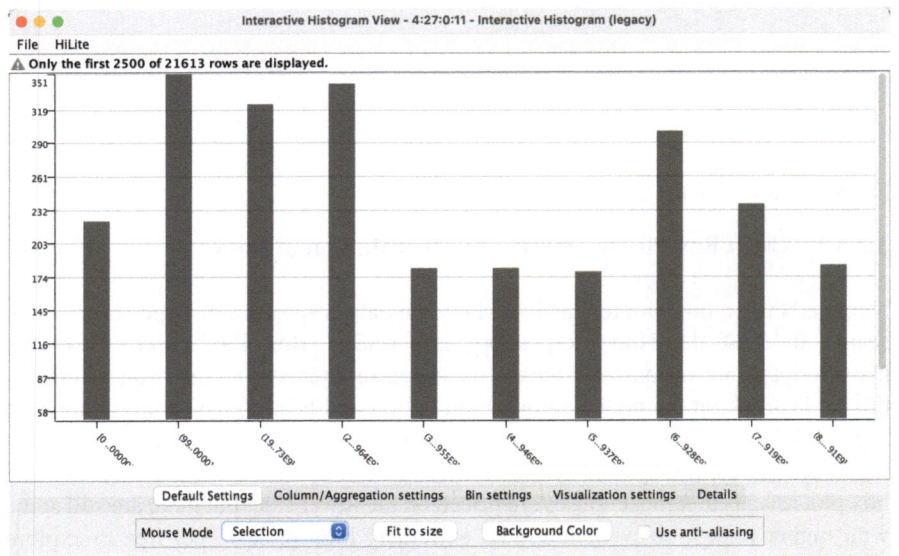

Fig. 3.30 Initial interface of interactive histogram (legacy)

is a bit of a hassle to use, so it's best to go with the last one, "Interactive Histogram (legacy)." (Note. Nodes with the label "legacy" are from the classic KNIME UI. They're still available in the modern UI but might be removed in the future.)

After connecting it with the "CSV Reader" node, you can run it directly without any additional settings, and then click the "Open View" command of the node to start exploring the data using the graphical approach.

In the initial interface, as shown in Fig. 3.30, there's a warning at the top that says "Only the first 2500 of 21,613 rows are displayed." As histograms don't need to plot

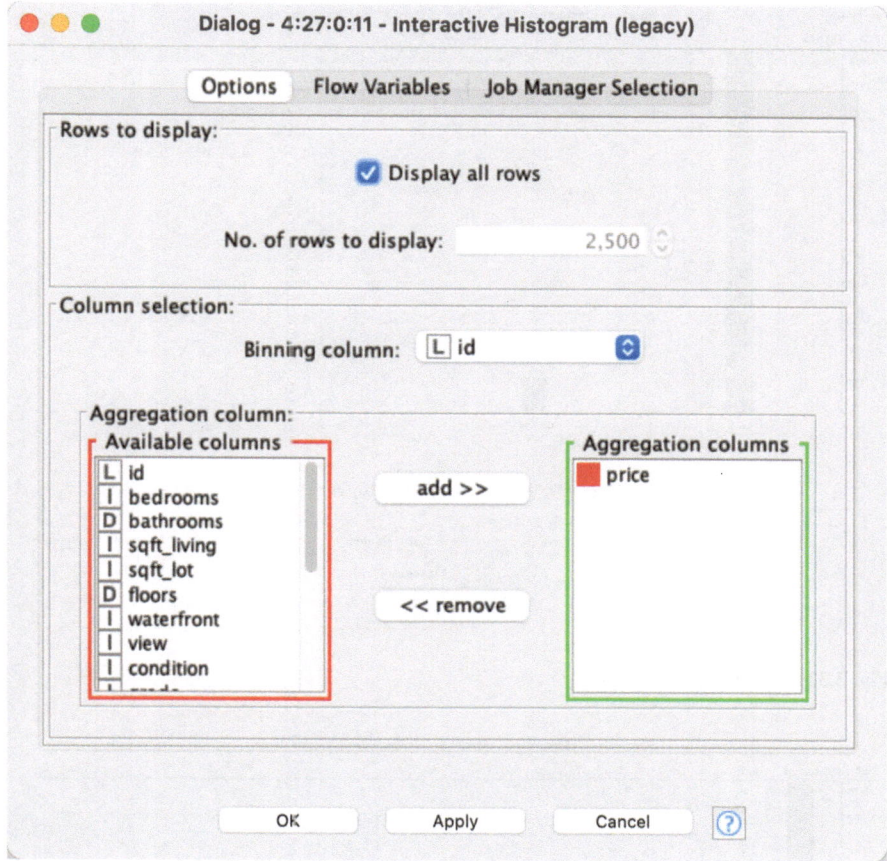

Fig. 3.31 Interactive histogram (legacy) node settings

all the data like scatter plots, the plotting process itself doesn't consume many resources; only the initial calculations do. Therefore, when we use histograms to observe data, we can see and observe all of them. Click the configuration button of this node and make the settings as shown in Fig. 3.31. Check the "Display all rows" option at the top to observe all the data. We don't need to set other options; they can be set later when viewing the data.

There will be no warning when clicking the "Open View" again for the node. Now let's explore the data. Select the second tab "Column/Aggregation settings" as shown in Fig. 3.32, and choose "bedrooms" in the "Binning column" area; you can find that the number of data corresponding to [0–4] is the largest.

To further divide the data, you can go to the "Bin settings" and adjust the number of bins, as shown in Fig. 3.33. The default is set to 10, but we can clearly see only four bars. Is the software unable to count? Obviously, this logic doesn't hold, so we carefully explore the options and find a checkbox labeled "Show empty bins" that, when selected, reveals other empty bins. By default, if a bin is empty, it won't be

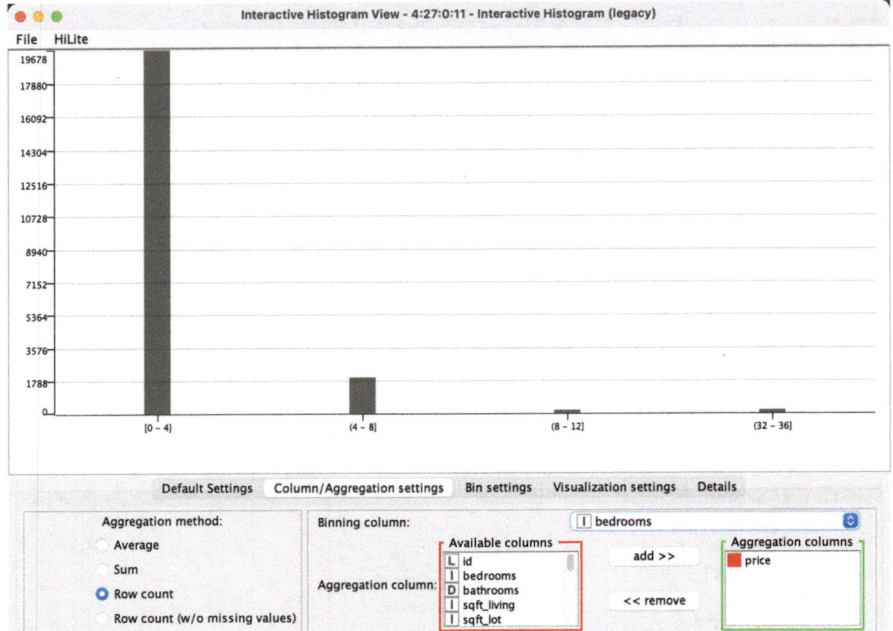

Fig. 3.32 Set data view

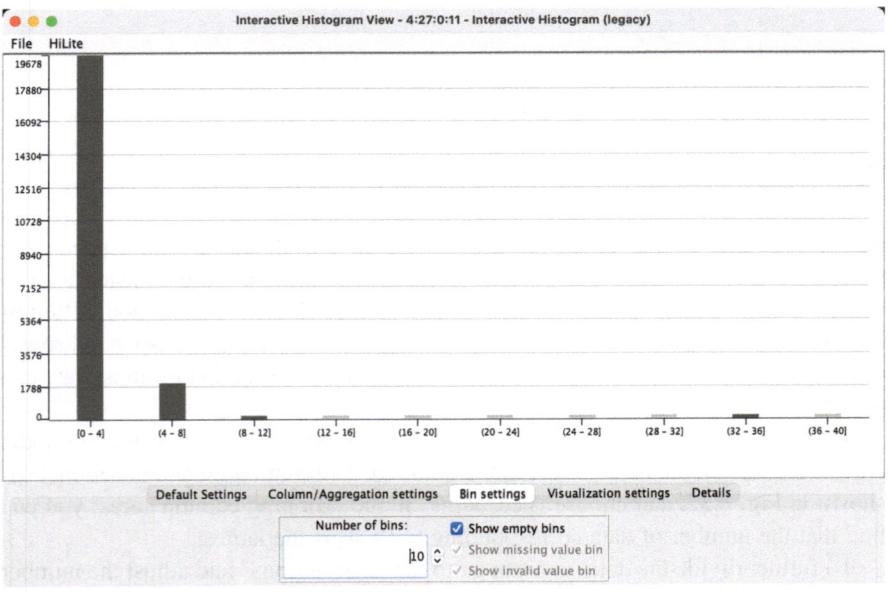

Fig. 3.33 Bin settings

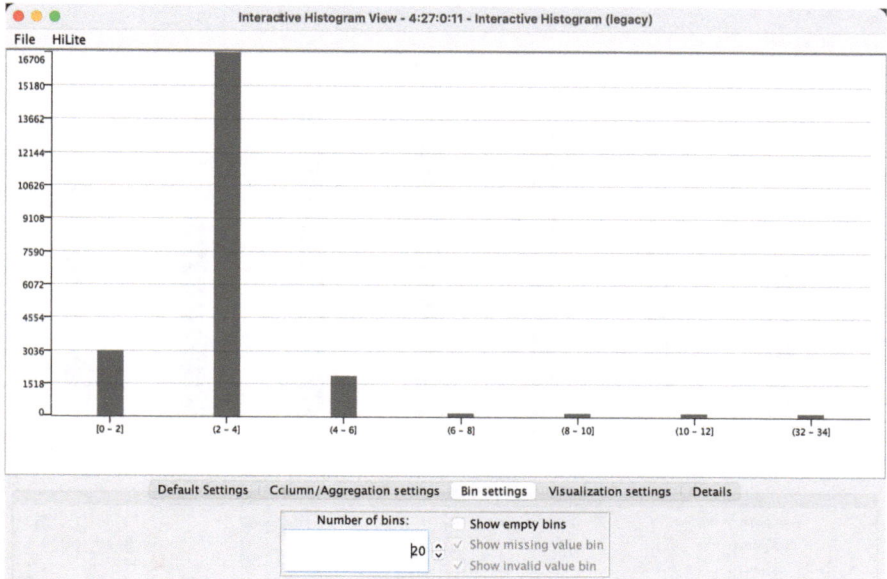

Fig. 3.34 Set the number of bins

displayed. Generally, we don't need to show these empty bins, so it's fine to stick with the default setting.

If you wish to have a more detailed data division, you only need to adjust the "Number of bins." You will notice that the bin ranges have changed to [0–2], (2–4], etc., as depicted in Fig. 3.34. Pay attention to the inclusivity of the data: "(" indicates exclusive; "]" indicates inclusive. Since the number of "bedrooms" is definitely an integer, the range (2–4] can only include 3 or 4. From this graph, it is evident that houses with three and four bedrooms are the most common. Considering the tendency for Americans to have larger houses, this data aligns with our expectations.

What if we want to explore other relationships within the data? For instance, if you want to examine the relationship between the number of bedrooms and the average house price, as shown in Fig. 3.35, you can simply set the "Aggregation method" to "Average."

At this moment, a rather peculiar situation arises. From Fig. 3.35, it is apparent that houses with the highest prices are those with (6–8] bedrooms, while houses with more bedrooms actually have lower average prices. Why is this? Let's see if we can identify the reason from the histogram. It's clear that there is a correlation between the house price and its area. Let's try to visualize the relationship between the number of bedrooms and the area. By setting "Aggregation columns" to "sqft_living" as shown in Fig. 3.36, we can observe that houses with (6–8] bedrooms have the highest average area, which explains the earlier question about their high prices. Why do houses with more bedrooms tend to be smaller? This calls for a

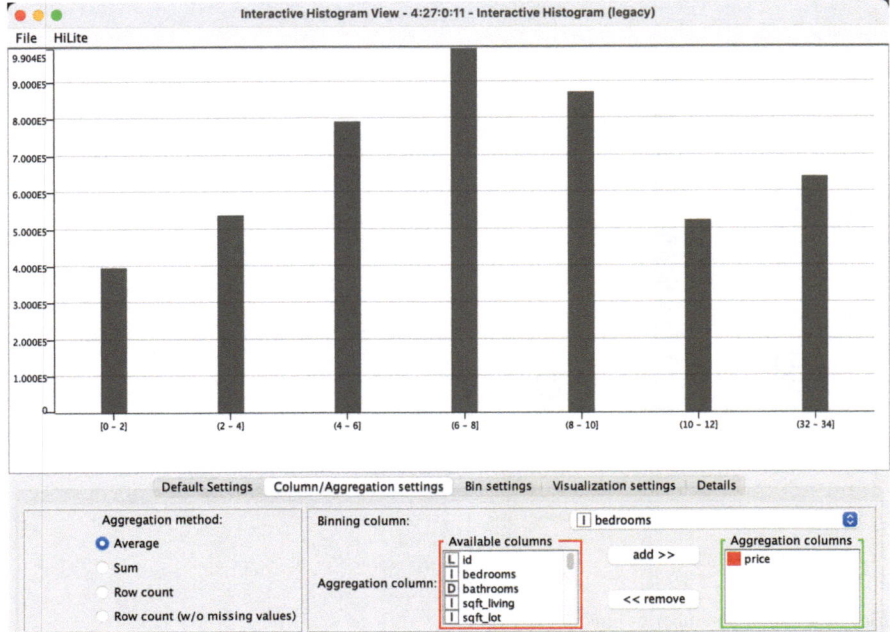

Fig. 3.35 Relationship between house prices and number of bedrooms

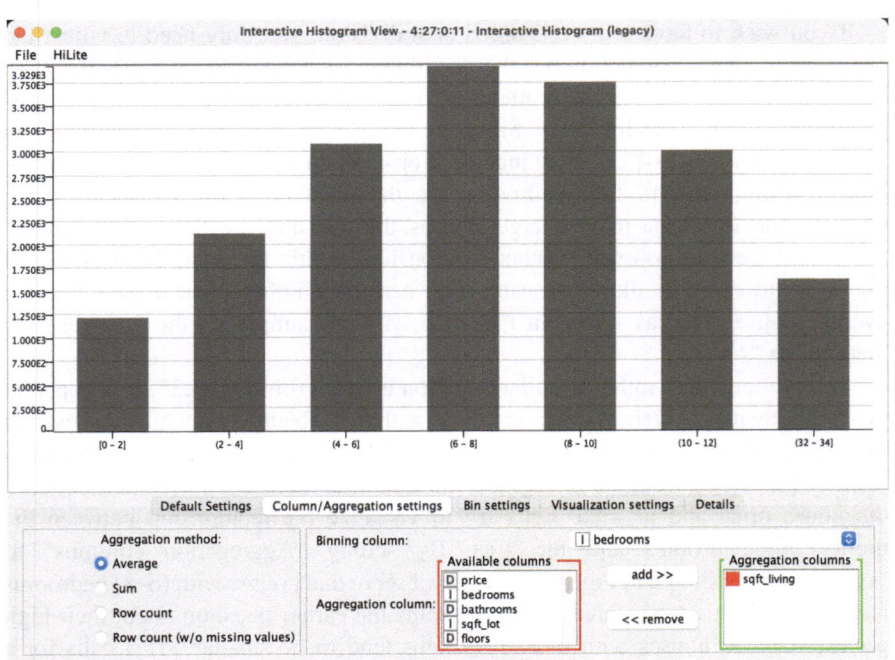

Fig. 3.36 Relationship between area and number of bedrooms

gradual exploration combining data analysis with industry or domain knowledge. Everyone can embark on their own exploration.

3.2.3.4   Correlation

One important goal of data analysis is to understand the relationship between features and targets. Although we have observed the data relationship using histograms, we still lack a quantitative value. We can use correlation coefficients to measure the relationship. Visualizing the correlation coefficients in the form of a matrix plot allows us to easily grasp the relationship between the data. The correlation matrix can be utilized to examine the correlation of all data, including features and target variables.

To achieve this, create a new "Linear Correlation" node, connect it to the "CSV Reader" node, and execute it. Then, click on the "Open view" button of the node to view the results, showcasing the correlation matrix heatmap as depicted in Fig. 3.37.

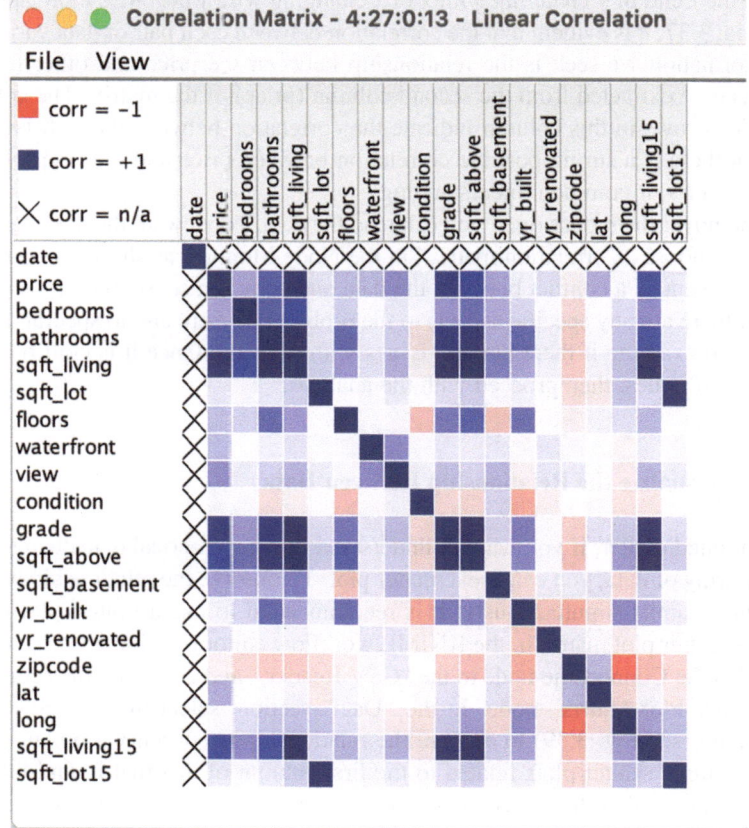

Fig. 3.37 Heatmap of the correlation matrix

▶ 1: Correlation measure ▶ 2: Correlation matrix ▦ 3: Correlation model Flow Variables

Rows: 20 | Columns: 20 Table Statistics

#	Row...	date Number (dou...	price Number (dou...	bedrooms Number (dou...	bathrooms Number (dou...	sqft_living Number (dou...	sqft_lot Number (dou...	floors Number (dou...	waterfront Number (dou...	view Number (dou...	condition Number (dou...	grade Number (dou...	sqft_abo... Number (dou...	sqft_ Numbe
1	date	1	?	?	?	?	?	?	?	?	?	?	?	?
2	price	?	1	0.308	0.525	0.702	0.09	0.257	0.266	0.397	0.036	0.667	0.606	0.324
3	bedr...	?	0.308	1	0.516	0.577	0.032	0.175	-0.007	0.08	0.028	0.357	0.478	0.303
4	bathr...	?	0.525	0.516	1	0.755	0.088	0.501	0.064	0.188	-0.125	0.665	0.685	0.284
5	sqft_...	?	0.702	0.577	0.755	1	0.173	0.354	0.104	0.285	-0.059	0.763	0.877	0.435
6	sqft_...	?	0.09	0.032	0.088	0.173	1	-0.005	0.022	0.075	-0.009	0.114	0.184	0.015
7	floors	?	0.257	0.175	0.501	0.354	-0.005	1	0.024	0.029	-0.264	0.458	0.524	-0.246
8	wate...	?	0.266	-0.007	0.064	0.104	0.022	0.024	1	0.402	0.017	0.083	0.072	0.081
9	view	?	0.397	0.08	0.188	0.285	0.075	0.029	0.402	1	0.046	0.251	0.168	0.277
10	cond...	?	0.036	0.028	-0.125	-0.059	-0.009	-0.264	0.017	0.046	1	-0.145	-0.158	0.174
11	grade	?	0.667	0.357	0.665	0.763	0.114	0.458	0.083	0.251	-0.145	1	0.756	0.168
12	sqft_...	?	0.606	0.478	0.685	0.877	0.184	0.524	0.072	0.168	-0.158	0.756	1	-0.052
13	sqft_...	?	0.324	0.303	0.284	0.435	0.015	-0.246	0.081	0.277	0.174	0.168	-0.052	1

Fig. 3.38 Observing correlation coefficient values

Upon observing the heatmap as shown in Fig. 3.37, the bluer the color, the more positively correlated the data is, while the redder it is, the more negatively correlated it is. In simple terms, for two columns of data, x and y, if they are positively correlated, as x increases, y also increases. Conversely, in the case of negative correlation, as x increases, y decreases. When the absolute value of the correlation is high, the trend of y changing with x or x changing with y becomes more apparent. From Fig. 3.37, it is evident that the correlation between each pair of data varies. The key information we seek is the relationship between the price and other features, which can be extracted from the second column (price) of the matrix. The colors of the various rows in this column indicate the correlation between them. It is noticeable that there is a strong positive correlation between price and area (sqft_living), aligning with our common understanding.

If you hover the mouse over, you can read the data. To view all the data, select the node and choose "Correlation matrix" in the Node Monitor, as shown in Fig. 3.38.

In the event of a conflict between the data and common sense, the first step is to check if there are any specific aspects to the problem. If there are no specific factors, we need to examine if there are any issues with the data. Once it is confirmed that there are no issues, then proceed with the analysis.

3.2.3.5 Visualize the Relationship Between Data

As mentioned earlier, if you want to understand how a numerical data changes with another array of data, you can use a scatter plot. To observe the relationship between different columns simultaneously, it is recommended to use a scatter plot matrix. Create a scatter plot matrix in the KNIME workflow editor by adding a "Scatter Plot Matrix" node. Connect the node to the "CSV Reader," and configure the settings for the "Scatter Plot Matrix" node. In the "Data" section, select the desired data for observation (see Fig. 3.39) to explore the relationships between various numerical data. The three scatter plots related to the first column of this matrix and the price clearly indicate that the price generally increases as the number of bedrooms increases. Similarly, there is a noticeable price increase with an increase in the

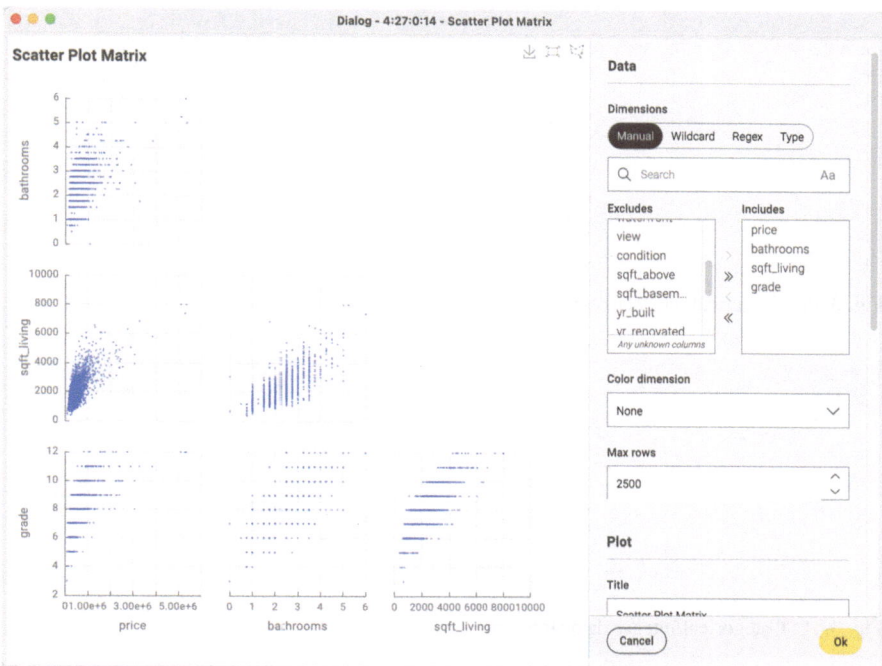

Fig. 3.39 Scatter plot matrix

area (sqft_living). A scatter plot matrix makes it easy to observe more data relationships of this kind.

Furthermore, it is evident that these findings align with the correlation information. However, the scatter plots also reveal more detailed insights, such as house prices not simply increasing with the area in a linear way but, instead, becoming more dispersed within a specific range and beginning to converge in another featuring larger area.

In general, the scatter plot provides us with more information compared to the correlation matrix heatmap:

- Whether there is a trend of quantitative correlation between variables.
- If there is a correlation trend, whether it is linear or nonlinear.
- If there are any points that deviate from the majority, known as outliers, it can be easily seen through a scatter plot. This allows for further analysis of whether these outliers might have a significant impact on the overall modeling analysis.

While scatter plots offer more information, correlation enables us to quantitatively compare the relationship between different data points. Hence, it is recommended to use both together to help with our analysis.

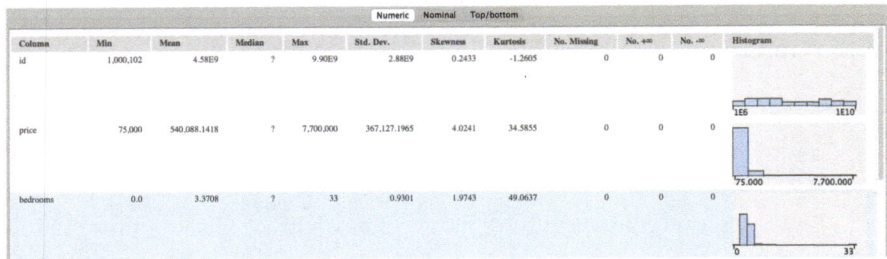

Fig. 3.40 Numerical data statistics

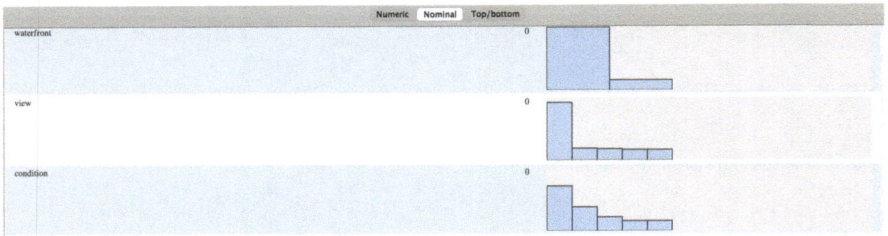

Fig. 3.41 Categorical/integer data statistics

3.2.3.6 Statistical Values

We often need to understand the statistical values of the data, such as the maximum and minimum values, as well as the distribution histogram of the numerical data. These statistical functions can all be implemented through the "Statistics" node.

We can create a "Statistics" node, connect it to the "CSV Reader," and observe the results after executing it. We can see statistical data such as the mean, variance, as well as histograms of various numerical data and bar charts of categorical data.

Under the "numeric" label (see Fig. 3.40), you can see the maximum, minimum, and average values of the numerical data, as well as their histograms.

Under the label of "nominal" (categorical/integer), it primarily consists of histograms of limited range integers or categorical data, as shown in Fig. 3.41. Although the expressiveness of the histogram here is slightly worse than that of a dedicated histogram, it is still sufficient for observing the overall distribution.

The last label "top/bottom" represents the display of the most and least frequent statistics in each column and their respective quantities, as shown in Fig. 3.42.

So far, our workflow can be shown in Fig. 3.43.

Give It a Try

- KNIME also has many other nodes for data visualization, and it is recommended to explore them on your own.

Numeric   Nominal   [Top/bottom]

waterfront	view	condition	grade	sqft_above	sqft_basement	yr_built	yr_renovated	zipcode	sqft_living15
No. missings: 0	No. missings: 0	No. missings: 0	No. missings: 0	No. missings: 0	No. missings: 0	No. missings: 0	No. missings: 0	No. missings: 0	No. missings: 0
Top 20: 0 : 21450 1 : 163	Top 20: 0 : 19489 2 : 963 3 : 510 1 : 332 4 : 319	Top 20: 3 : 14031 4 : 5679 5 : 1701 2 : 172 1 : 30	Top 20: 7 8981 8 6068 9 2615 6 2038 10 : 1134 11 : 399 5 242 12 : 90 4 29 13 : 13 3 : 3 1 : 1	Top 20: 1300 : 212 1010 : 210 1200 : 206 1220 : 192 1140 : 184 1400 : 180 1060 : 178 1180 : 177 1340 : 176 1250 : 174 1320 : 172 1100 : 164 1080 : 161 1040 : 160 1240 : 160 1150 : 159 1330 : 158 1260 : 155 1440 : 155 1120 : 154	Top 20: 0 : 13126 600 : 221 700 : 218 500 : 214 400 : 184 800 : 206 1000 : 149 900 : 144 300 : 142 200 : 108 530 : 107 480 : 106 750 : 105 450 : 103 720 : 102 620 : 91 840 : 85 580 : 85 420 : 81 860 : 80	Top 20: 2014 : 559 2006 : 454 2005 : 450 2004 : 433 2003 : 422 1977 : 417 2007 : 417 1978 : 387 1968 : 381 2008 : 367 1967 : 350 1979 : 343 1959 : 334 1990 : 320 1962 : 312 2001 : 305 1954 : 305 1987 : 294 1989 : 290 1969 : 280	Top 20: 0 : 20699 2014 : 91 2013 : 37 2003 : 36 2005 : 35 2000 : 35 2007 : 35 2004 : 26 1990 : 25 2006 : 24 2002 : 22 1989 : 22 2009 : 22 1991 : 20 1994 : 19 1998 : 19 2001 : 19 1993 : 19 2010 : 18 2008 : 18	Top 20: 98103 : 602 98038 : 590 98115 : 583 98052 : 574 98117 : 553 98042 : 548 98034 : 545 98118 : 508 98023 : 499 98006 : 498 98133 : 494 98059 : 468 98058 : 455 98155 : 446 98074 : 441 98033 : 432 98027 : 412 98125 : 410 98056 : 406 98053 : 405	Top 20: 1540 : 197 1560 : 192 1500 : 181 1460 : 169 1580 : 167 1800 : 166 1610 : 166 1720 : 166 1620 : 165 1510 : 164 1760 : 163 1480 : 160 1410 : 159 1550 : 158 1680 : 157 1670 : 157 1820 : 157 1520 : 155 1660 : 155
Bottom 20:	Bottom 20:	Bottom 20:	Bottom 20:	Bottom 20: 1613 : 1 2587 : 1 2623 : 1 894 : 1 1606 : 1 2244 : 1 2026 : 1 2238 : 1 2517 : 1 2708 : 1 2555 : 1 1405 : 1 4450 : 1 6420 : 1 2531 : 1 1333 : 1	Bottom 20: 176 : 1 225 : 1 1275 : 1 266 : 1 283 : 1 65 : 1 2310 : 1 1770 : 1 2120 : 1 295 : 1 207 : 1 915 : 1 556 : 1 417 : 1 143 : 1 508 : 1	Bottom 20: 1905 : 74 1911 : 73 1937 : 68 1907 : 65 1915 : 64 1931 : 61 1913 : 59 1917 : 56 1914 : 54 1938 : 52 1903 : 46 1904 : 45 1936 : 40 1932 : 38 2015 : 38 1933 : 30	Bottom 20: 1960 : 4 1974 : 3 1945 : 3 1957 : 3 1953 : 3 1955 : 3 1956 : 3 1976 : 3 1971 : 2 1950 : 2 1962 : 2 1940 : 2 1946 : 2 1967 : 2 1954 : 1 1948 : 1	Bottom 20: 98105 : 229 98045 : 221 98002 : 199 98077 : 198 98011 : 195 98019 : 190 98108 : 186 98119 : 184 98005 : 168 98007 : 141 98188 : 136 98032 : 125 98014 : 124 98070 : 118 98109 : 109 98102 : 105	Bottom 20: 3402 : 1 3494 : 1 2156 : 1 3236 : 1 2612 : 1 2323 : 1 2409 : 1 2354 : 1 2616 : 1 1427 : 1 1516 : 1 2456 : 1 2844 : 1 1495 : 1 2594 : 1 2604 : 1

Fig. 3.42 The most and least frequent statistics

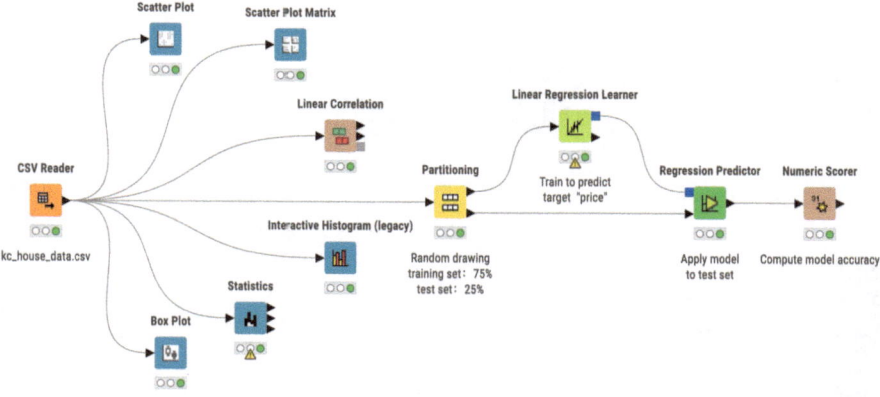

Fig. 3.43 Current workflow

3.2.4   Integrating Nodes

We already know that machine learning projects involve several different steps. In KNIME, each step requires several nodes. In this case, the entire workflow may become messy, and people may get caught up in myriads of intricate details. For example, the current workflow, as shown in Fig. 3.43, already has many nodes. To alleviate this burden, Metanodes or Components can be used. Simply put, a Metanode is a large integrated node composed of several individual nodes, while a Component is a multifunctional module composed of several nodes, and it can have its own configuration window and custom interactive composite view. In general, Component is more suitable for the integration of visual nodes.

For example, we want to collapse the nodes related to data view. We can select the icon [Create component] in the toolbar above as shown in Fig. 3.44. At this point, you will be asked if you want to reset the nodes, which means that all previous node operations will be reset. Click the "OK" button to reset, as shown in Fig. 3.45. Next, fill in the name, and click the "√" button when finished, as shown in Fig. 3.46.

This way, all the nodes related to visualization become a new Component called "Data View," as shown in Fig. 3.47.

If you want to open the "Data View" and check all the nodes inside, you only need to right-click on the Component, as shown in Fig. 3.48. From the popped-up context menu, you may select "Component" > "Open component" sequentially. All contained nodes will be shown in a new window.

All individual nodes can be displayed, as shown in Fig. 3.48.

Now we notice that these visualization nodes are already configured and a path "Workflow Name>Node Name" has been added to the toolbar above. To view the

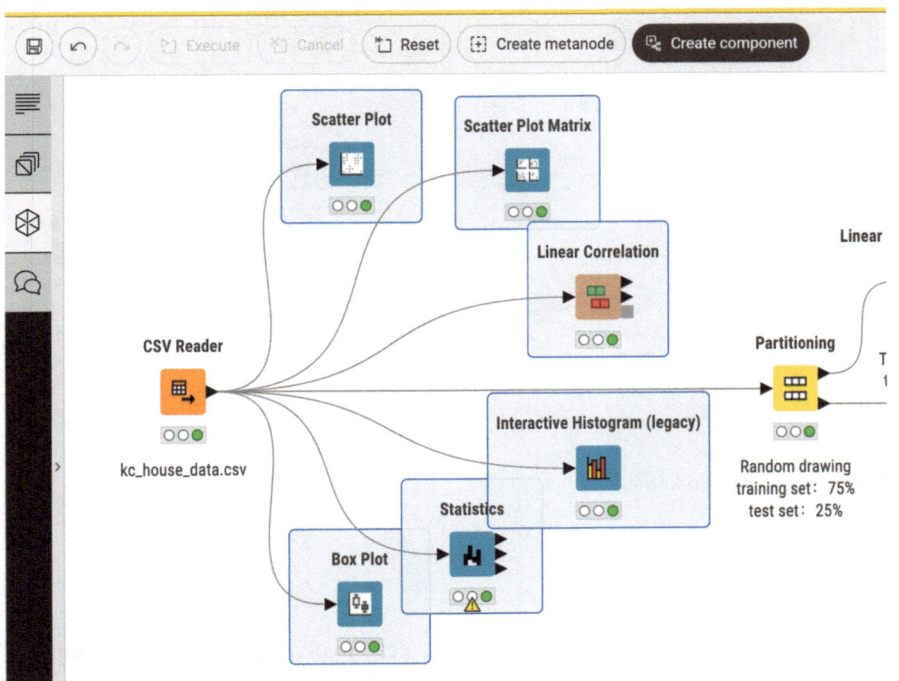

Fig. 3.44 Collapse a set of nodes into a Metanode

Fig. 3.45 Select whether to reset the executed node

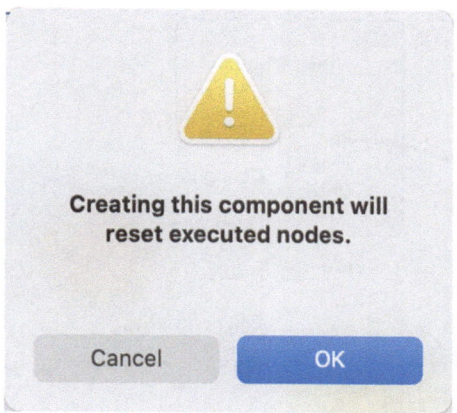

Fig. 3.46 Fill in the name for the Component

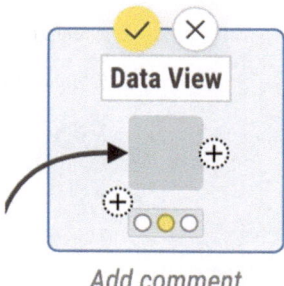

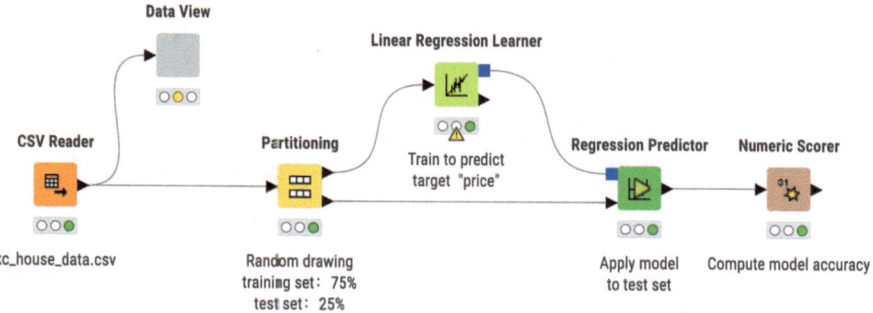

Fig. 3.47 Workflow after Metanode creation

data again, simply run these nodes; if you want to exit the "Data View" node, click on the workflow name with the newly added path in the toolbar (Chap. 3 "Multi-variate Linear Regression" in Fig. 3.49) to exit.

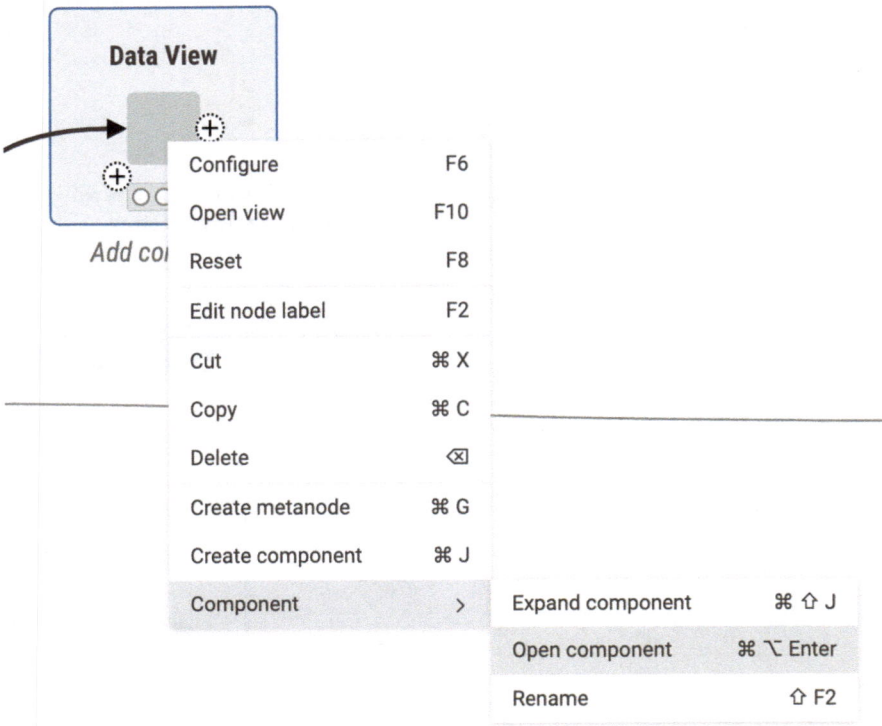

Fig. 3.48 Open component

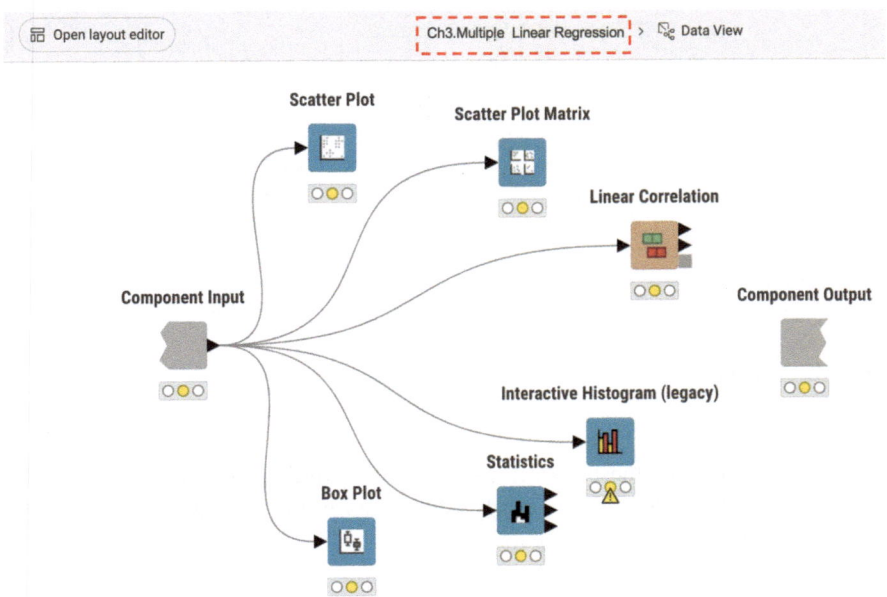

Fig. 3.49 All nodes related to data visualization are collapsed

Give It a Try

- Click on the "Open View" button of the Component. What do you see?
- It's similar to use Metanode to integrate nodes. You may try it out.
- Please compare the ways in which Component and Metanode expand nodes.

3.3   Advanced Multiple Linear Regression

Building on the previous section, we will study how to further optimize the model to make it better.

3.3.1   Optimize the Model

While we have constructed a model, we have not yet carried out any optimization. Our next step will involve optimizing the model based on data exploration.

3.3.1.1   Feature Engineering

Feature engineering is the process of choosing, modifying, and generating important features from raw data to enhance the performance of a machine learning model. It includes grasping the data, pinpointing useful variables, and using different methods like encoding categorical variables, scaling numerical features, dealing with missing values, and crafting new features through feature extraction or selection techniques. The aim is to extract valuable information from the data to assist the model in capturing the underlying patterns more effectively and making precise predictions.

We start learning feature engineering from the most basic aspect, which is feature selection. Feature selection, as the name suggests, is to choose the features we want and remove the ones we don't want. In other words, feature selection is the process of selecting relevant feature subsets to build a model. There are three reasons for using feature selection techniques:

After establishing the baseline workflow, we often aim to engage in feature engineering to extract features from the raw data to the fullest extent for algorithm and model utilization.

There is a saying in the machine learning industry: "Data and features determine the upper limit of machine learning, while models and algorithms merely approximate this limit."

We begin our journey into feature engineering with the fundamental aspect of feature selection. As the name implies, feature selection involves choosing the desired features and discarding the undesired ones. In essence, feature selection is the process of selecting relevant subsets of features to construct a model. There are three reasons for employing feature selection techniques:

- Simplify models to make them easier to understand for researchers or users.
- Reduce training time.
- Improve generalizability and reduce overfitting!

3.3.1.2 Methods for Feature Selection

Three commonly employed methods for feature selection include forward selection, backward elimination, and bidirectional elimination, which combines the principles of both forward selection and backward elimination.

- Forward selection. Forward selection begins with an empty model and adds in variables one by one. In each forward step, you add the one variable that gives the single best improvement to your model. This process will be repeated until adding any new variables no longer improves the evaluation results.
- Backward elimination. The backward elimination process begins by fitting a multiple linear regression model with all the independent variables. Then the model is evaluated to identify the variable with the poorest evaluation result, which will then be removed. This process will be repeated until removing any variables no longer improves the evaluation results.

For example, if we use p-value as the feature selection criterion, we can follow the step to carry out forward selection:

1. Select the most relevant variable and fit it into the model.

   - Add in variables one by one to the empty model.
   - Select the variable with the smallest p-value.

2. Keep the selected variable and fit other remaining ones into the model again to choose the most relevant variable.

   - If the completion condition is not met, repeat step 2.
   - If the completion condition is met, stop the process.

Completion condition: p-value is greater than threshold (0.05), or the number of variables added reaches the limit.

Similarly, if we use p-value as the feature selection criterion, we can follow the step to carry out backward elimination:

1. Set a threshold ($p < 0.05$) for the variables to be retained in the model.
2. Fit all variables in the model.
3. Run the model to find the variable with the highest p-value.

   - If the p-value is greater than the threshold, then delete it and repeat step 3.
   - If the p-value is smaller than the threshold, then stop the process.

In the above example, we used p-value as the feature selection criterion. We can also use loss functions or other indicators as selection criteria.

3.3.1.3 Overfitting

Feature selection serves to mitigate the risk of overfitting, but what exactly is overfitting? In the realm of machine learning, overfitting is a prevalent issue that can significantly impede a model's performance. It arises when a model perfectly or very accurately fits a specific subset of a dataset yet struggles to predict other parts of the dataset. To put it simply, it's akin to a student who becomes so engrossed in dealing with a certain type of problem, but when faced with a slightly altered one, he or she is unable to solve it.

For instance, the data shown in Fig. 3.50 comprises training data points denoted by dots and test data points denoted by crosses.

Points for Discussion

- What are some examples in everyday life that are similar to overfitting?
- What are the dangers of overfitting?

3.3.1.4 Filtering Irrelevant Columns with KNIME

We can insert a "Column Filter" node (which can be found in "Manipulation") between the "Partitioning" node and the "Linear Regression Learner" node as shown in Fig. 3.51.

Set the "Column Filter" as shown in Fig. 3.52 We use the "Column Filter" node to filter out irrelevant data. In this case, assuming that "id" and "date" are not related to house prices, we can filter out these two variables.

Why do we assume that "id" and "date" are unrelated to house prices? For "id," it is something assigned by data collectors and generally has no relation with other data. As for "date," it is a type of variable slightly complicated to be processed, so let's not worry about it for now because we don't plan to delve too deeply into feature processing.

Points for Discussion

- Why do we keep other variables in this case? Can we delete them?

3.3.1.5 Manual Implementation of Forward Selection

Let's start by manually implementing the forward selection algorithm. After reviewing the previous correlation matrix, it's evident that area has the highest

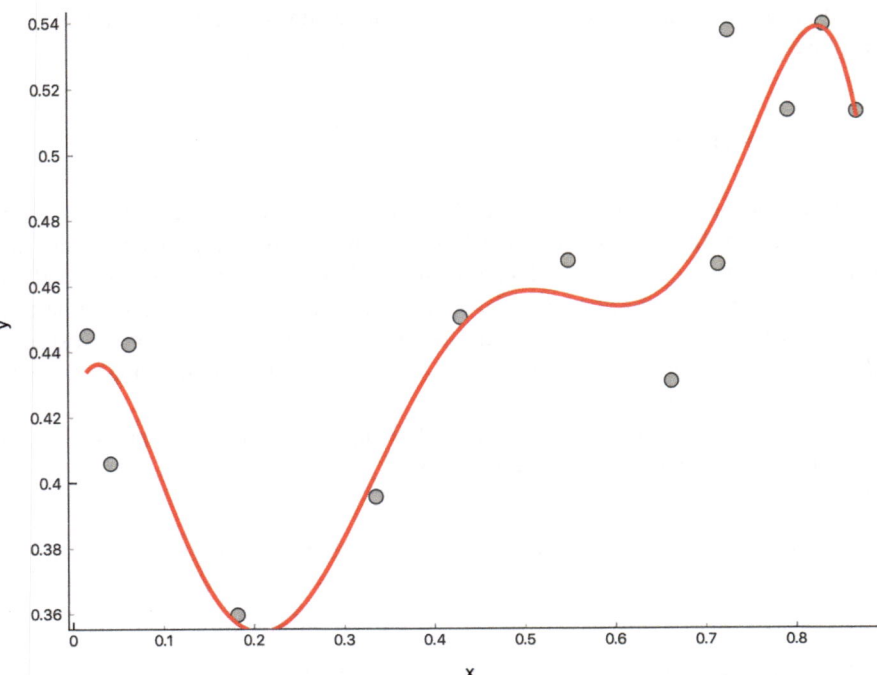

Fig. 3.50 Overfitting

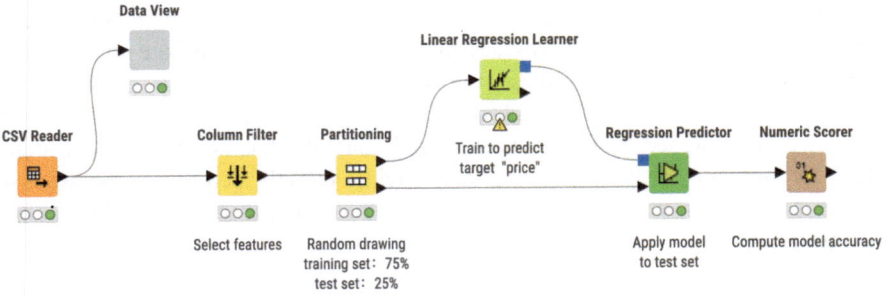

Fig. 3.51 Add a column filter

correlation with price. Therefore, we will solely utilize the area (sqft_living) to predict the price.

Insert a "Column Filter" node in the workflow as shown in Fig. 3.53, double-click on the node, and in the opened interface shown in Fig. 3.54, set it to keep only the columns for price and area (price and sqft_living).

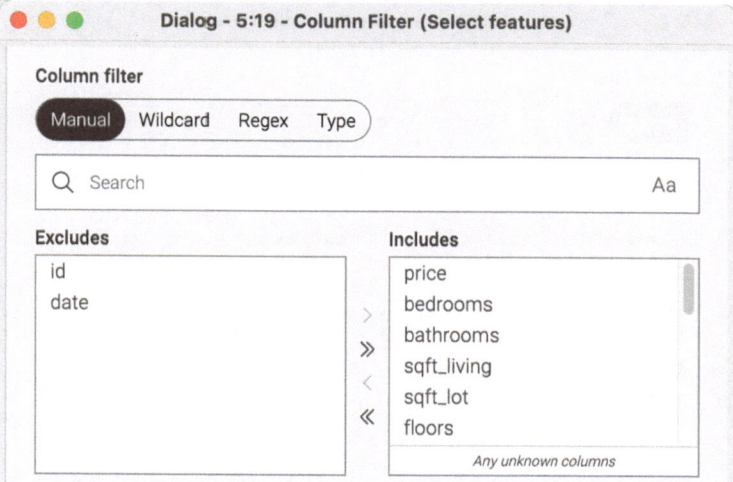

Fig. 3.52 Set the "Column Filter"

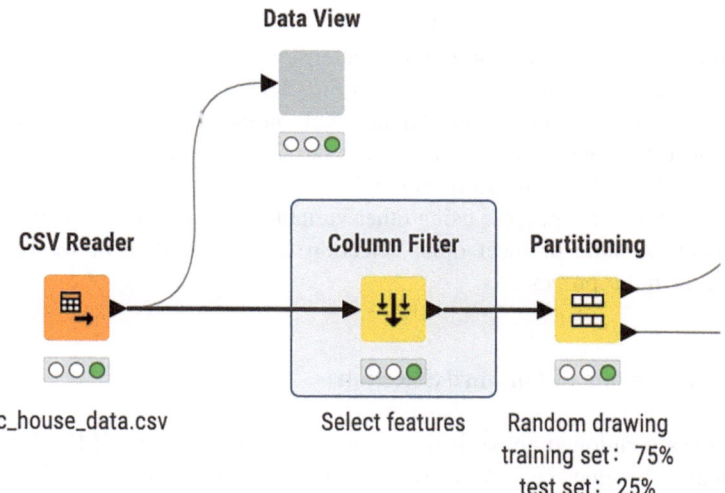

Fig. 3.53 Insert "Column Filter" in the workflow

As illustrated in Fig. 3.53, we can insert a "Column Filter" node into the workflow. Then, double-click on this node and include only the columns for price and area (sqft_living) in the dialog as shown in Fig. 3.54.

Execute the entire workflow and view the "Coefficients and Statistics" data of the "Linear Regression Learner" node in the node monitor. Or you can switch back to the classic KNIME UI to right-click on the "Linear Regression Learner" node and

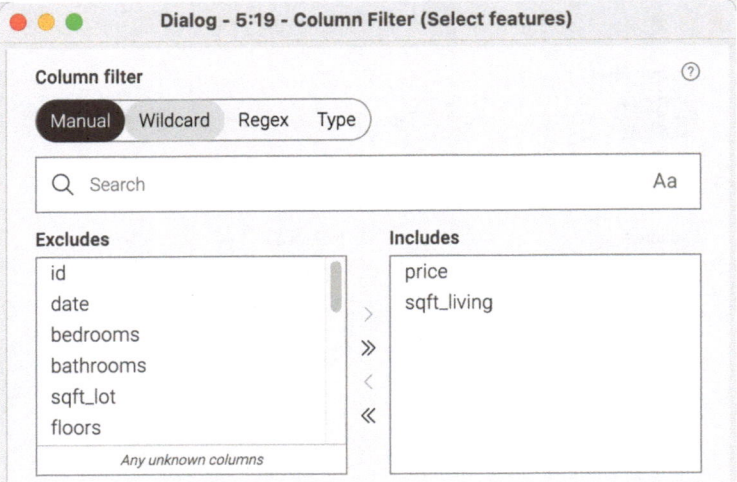

Fig. 3.54 Include price and area (sqft_living) only

select the "View: Linear Regression Scatterplot View" option from the shortcut menu to visualize the results, as shown in Fig. 3.55.

Then we can go to the modern UI and click on the "Linear Regression Learner" node. Select the "Open view" button to see the fitting parameters, and you will find that the *p*-value is 0, as shown in Fig. 3.56.

We can repeat this process using other variables until we select the features with the smallest *p*-value or meet other selection criteria. And then, we keep it and continue a similar process.

3.3.1.6   Metanode of Forward Selection

The process mentioned above is quite complex, but luckily KNIME already has a functionality for that, i.e., "Metanode: Forward Feature Selection." However, this node is not available in the node repository of the modern UI of KNIME 5.1 version. You can switch back to the classic UI to locate and drag the node to the workflow editor, as illustrated in Fig. 3.57. Then, return to the modern UI and rearrange the workflow as shown in Fig. 3.58.

After importing the data, use the "Column Filter" node to get rid of the features that we definitely don't need to consider (such as id, date). Then, use the "Partitioning" node to split the variables into training and test sets. Next, connect the training set data to the "Forward Feature Selection" node. The selected variables are then fed into the "Linear Regression Learner" node, and we are familiar with the rest of the nodes. The only thing to note is that the "Regression Predictor" node will

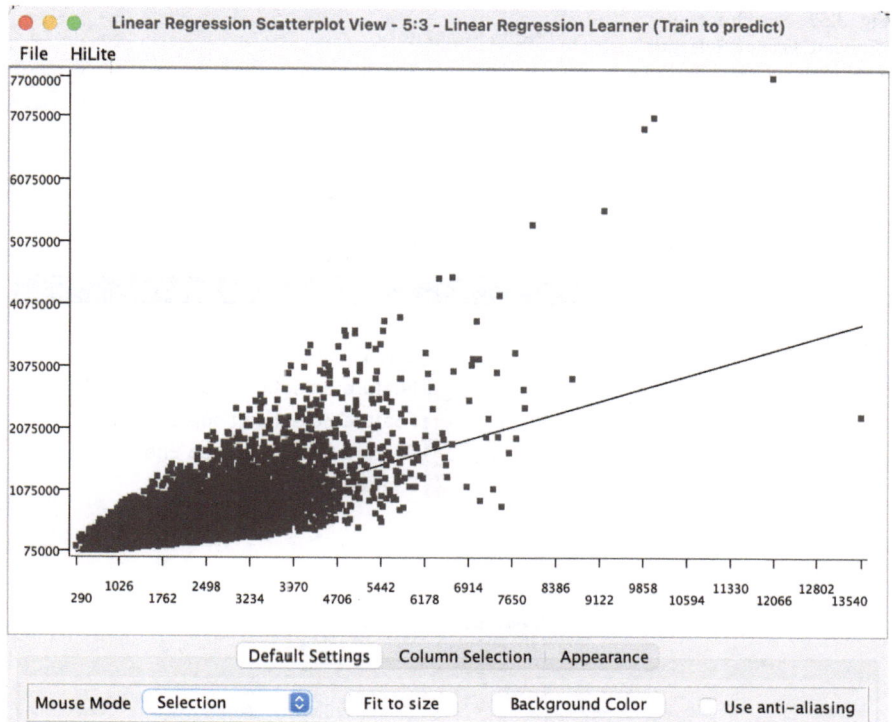

Fig. 3.55 Price model fitted only by area (sqft_living) in the classic KNIME UI

Statistics on Linear Regression

Variable	Coeff.	Std. Err.	t-value	P>|t|
sqft_living	280.0812	2.24	125.0355	0.0
Intercept	-41,303.1839	5,077.3784	-8.1347	4.44E-16

R-Squared: 0.491
Adjusted R-Squared: 0.491

Fig. 3.56 View the fitting result

only use the features already used by the model ("Linear Regression Learner"), so there's no need to filter its input.

Please note that this Metanode must be used with the training data, not all data (i.e., after the "Partitioning" node, not before).

Fig. 3.57 Search for "Forward Feature Selection" node in the classic UI

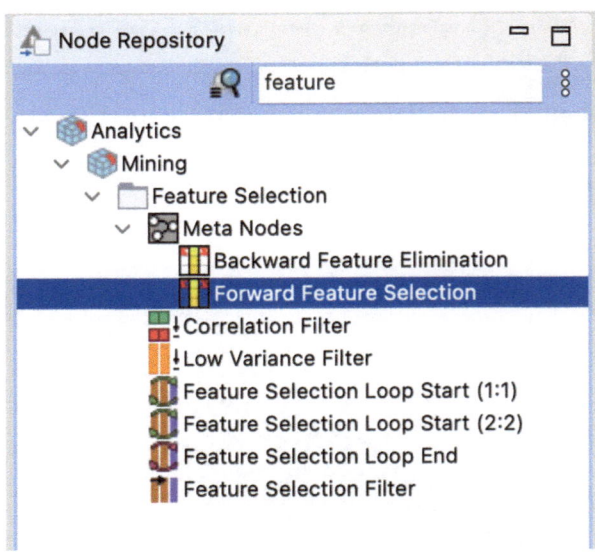

Fig. 3.58 Add Metanode: Forward Feature Selection

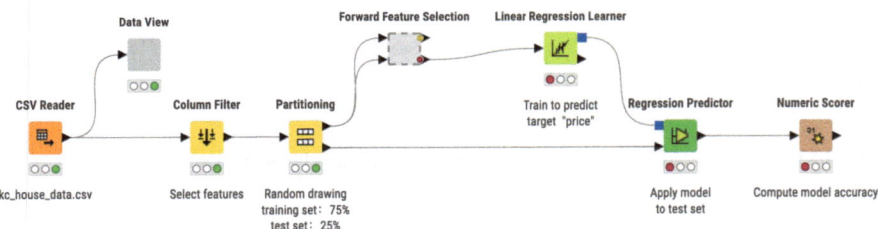

3.3.2   *Forward Selection by KNIME*

Different from the way we use to expand nodes in Component, here, we can double-click to open the "Forward Feature Selection" Metanode, and you'll see the workflow as depicted in Fig. 3.59. Both of the inputs here represent training data. This node will output the detailed information and data table about the selected features. The "Feature Selection Loop Start" indicates the beginning of a selection loop, while the "Feature Selection Loop End" marks its end. The steps in between involve making selection for specific features. The first modification needed for this Metanode is the choice of model. Since we are conducting regression analysis, the model used for feature selection should also be a regression model. Replace the model learner with "Linear Regression Learner" and the predictor with "Regression Predictor." Additionally, replace "Scorer" with "Numeric Scorer."

Combining the methods of forward selection mentioned above, let's look at the approach and algorithm illustrated in Fig. 3.59.

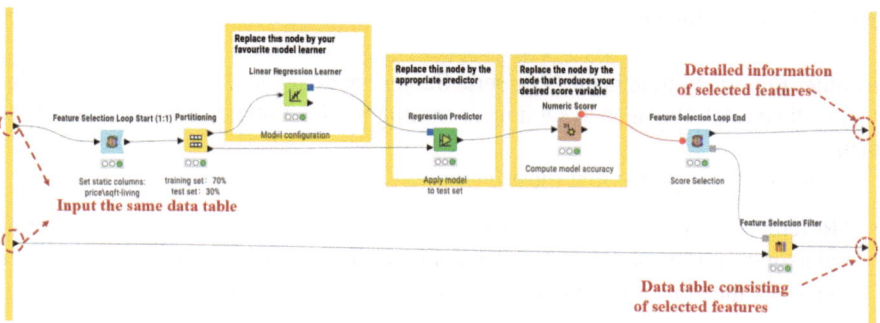

Fig. 3.59 "Forward Feature Selection" Metanode

1. At the start of the loop, select the price (target) and area (most relevant variable).

   - Partition the data into training and testing sets.
   - Input the data into the model learner and predictor.
   - Score the model.

2. Add a new variable to the model.

   - Score the model again.
   - If the number of variables reaches the threshold, stop the process.

The "Feature Selection Filter" at the end of the workflow is used to determine which features to use. Now let's take a closer look at how each step works.

3.3.2.1 Loop Begins

Double-click the "Feature Selection Loop Start" node to open the settings as shown in Fig. 3.59.

Two options need to be set here:

- "Static Columns" refer to columns that do not participate in the forward selection process and remain in the model throughout. Here, we chose the variable which is the most relevant to the target (price), i.e., area (sqft_living).
- "Use threshold for number of features" refers to setting an upper limit for the number of variables in a model or algorithm. Once this threshold is reached, the loop or process stops. For example, select "Yes" (i.e., check this option), and then set "19" as the threshold in the input box on the right. This option can be left unselected or set to another value. You can try it out by yourself.

3.3.2.2   Score the Model

Another important node is the "Numeric Scorer" node, which, as the name suggests, scores the model. We need to set it up according to Fig. 3.60 before we can use these metrics.

There are two options that must be set here:

- "Reference column" and "Predicted column." The reference column is the true value of the target, and the predicted column is the predicted value of the target. In this case, they are actual price and predicted price, respectively.
- "Output scores as flow variables." Once selected, we can use these scores going forward. An important concept in KNIME that we haven't introduced yet is flow variables. Familiarize yourself mainly with its form of expression, i.e., a red circle (see Fig. 3.61). In general, once the nodes are set up, the data flowing in the workflow is deterministic. Selecting price will output the price to the next node. If we don't select data, then date will not be fed to the next node. But how do we decide which variables to use? Or what if we want to select more variables so as to output more data to the next node? We can use flow variables (Fig. 3.62).

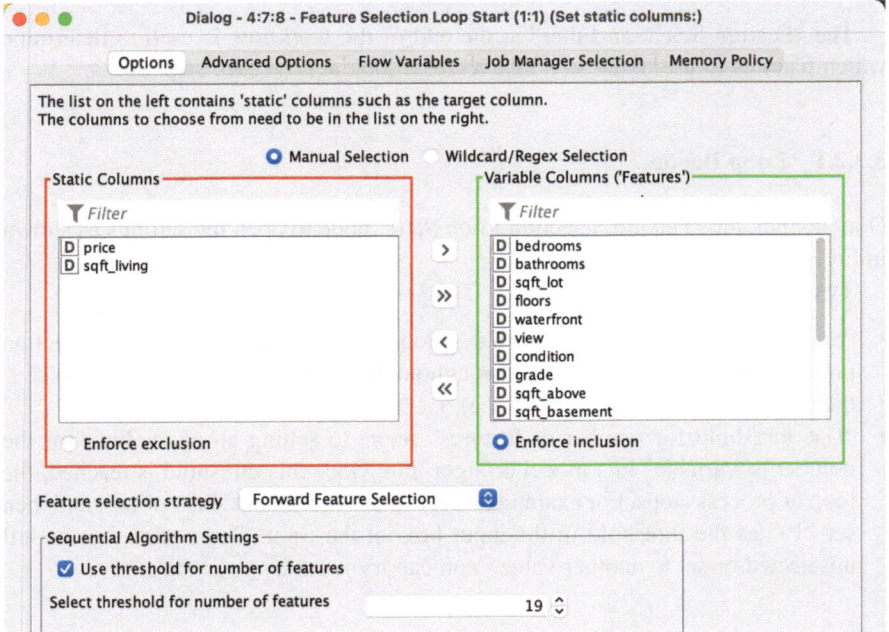

Fig. 3.60 Loop start

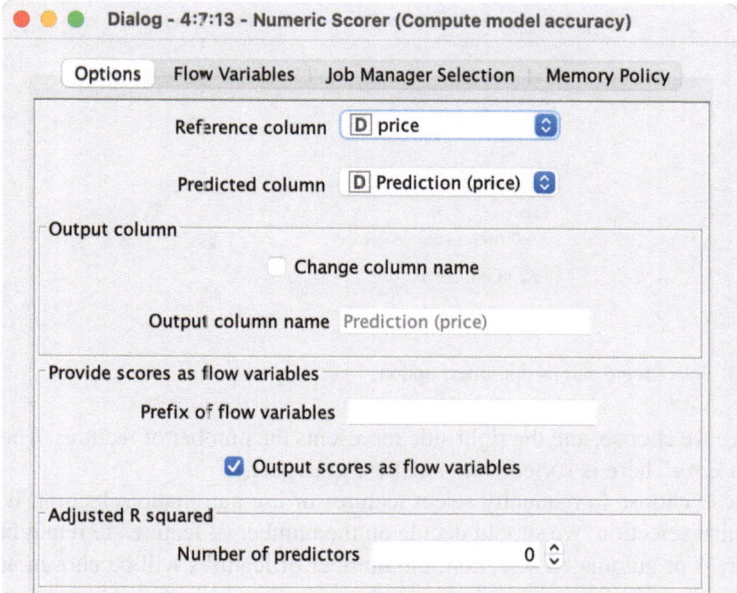

Fig. 3.61 Model scorer

Fig. 3.62 Node with flow variables

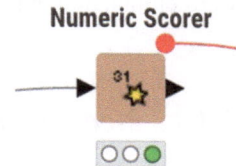

3.3.2.3   Loop Ends

Next, set the "Feature Selection Loop End" node to determine which score to use for evaluation and whether it should be maximized or minimized.

We can choose "mean squared error" (MSE) as shown in Fig. 3.63, and we minimize it by checking "Minimize score." In this way, feature selection will be based on reducing MSE.

3.3.2.4   Feature Selection

We can now execute the entire workflow, then double-click the "Feature Selection Filter" node, and open the settings as shown in Fig. 3.63. The left column in the selection box on the left side of this figure represents "mean squared error," which is

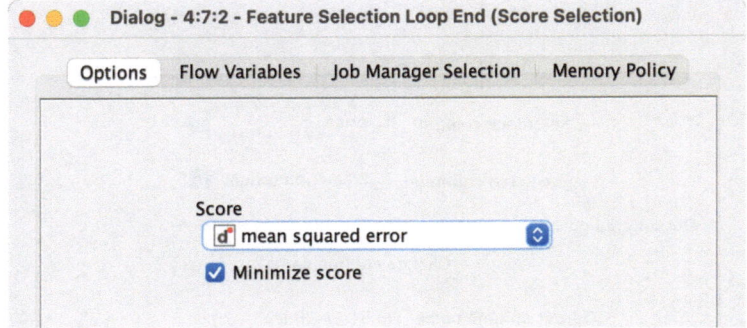

Fig. 3.63 Select score and its optimized option

the score we choose, and the right side represents the number of features. The "mean squared error" here is sorted from smallest to largest.

We can choose to manually select features or use automatic selection. If we opt for manual selection, we should decide on the number of features to retain based on the score. For automated selection, the number of features will be chosen automatically based on the feature threshold. As shown in Fig. 3.61, manual selection is used here, retaining 13 features. You can experiment with changing the relevant options to optimize your own model.

Similar to the forward selection method, we can use KNIME to perform backward elimination, but we won't delve into that here.

3.3.3   Model Explanation

In many data analysis scenarios, we often need to answer questions about the impact and significance of each feature. Let's explore how we can tackle these types of questions. Return to the main workflow editor (not in the Metanode), and click on the "Linear Regression Learner" node. Choose "Coefficients and Statistics" in the node monitor to see the coefficient table of the model. The result is displayed in Fig. 3.64.

The most important thing here is the content in the second column, where the absolute value of "Coeff" (coefficient) indicates the magnitude of the corresponding variable's impact on the result. A positive sign indicates a positive influence, while a negative sign indicates a negative influence. From this graph, we can see that the coefficient for area (sqft_living) is 182.088, which has a significant impact.

The influence of this coefficient can be conveniently seen from the regression formula $y = b + w_1x_1 + \ldots + w_nx_n$. The greater the w value, the larger the impact of the corresponding change in x on y. For example, in $y = b + 100{,}000x_1 + x_2$, for a change of 1 in x_1, y changes by 100,000, while for a change of 1 in x_2, y only changes by 1.

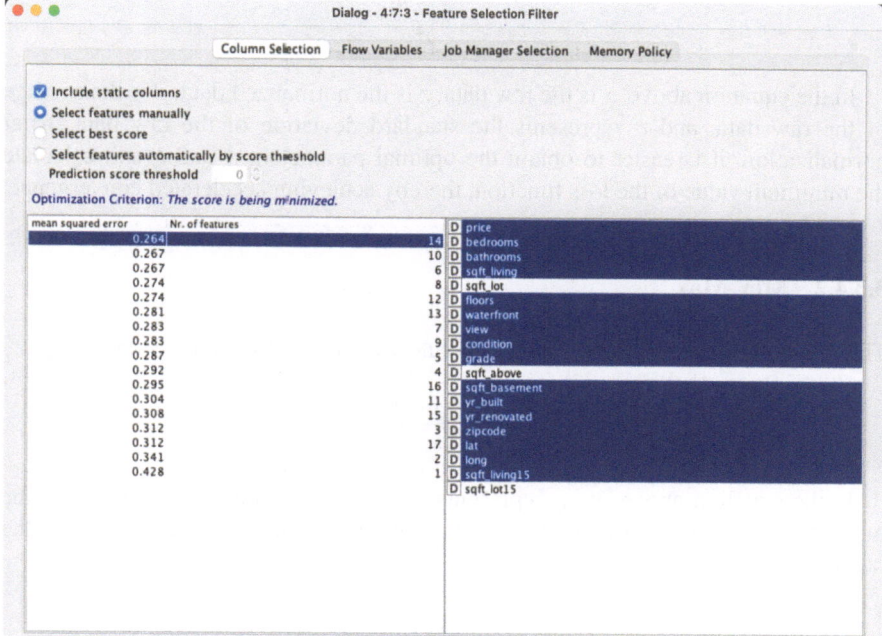

Fig. 3.64 Select features

3.3.4   Feature Normalization

The example above, $y = b + 100{,}000x_1 + x_2$, has a significant problem where the model may become unstable due to its high sensitivity to changes on x_1. Even a slight variation can lead to a substantial impact to x_2. For instance, we want to determine a person's gender based on their height and weight. If a person is 1.7 m tall and weighs 70 kg, a 0.1 m change in height would make a significant difference, while a 0.1 kg change in weight would not have a major impact. Therefore, it's crucial to scale the data to a similar range. In this case, we can convert the height from 1.7 m to 170 cm. This adjustment would ensure that the sensitivity of height and weight to data changes is roughly equivalent.

To address this issue, feature normalization can be employed to process the data. Normalization typically involves mapping the data to a specified range to remove the scale and units of different dimensions.

3.3.4.1   Z-Score

Z-score is a commonly used normalization method that attempts to transform data into a standard normal distribution.

$$z = \frac{x - \mu}{\sigma}$$

In the equation above, x is the raw data, z is the normalized data, μ is the average of the raw data, and σ represents the standard deviation of the raw data. After normalization, it is easier to obtain the optimal parameters w and b and calculate the minimum value of the loss function, thereby achieving accelerated convergence.

3.3.4.2 Min–Max

This method ensures that the data is absolutely distributed within a certain range.

$$x_{norm} = \frac{x - x_{min}}{x_{max} - x_{min}}$$

In the equation above, x_{norm} represents the normalized data, x_{min} represents the minimum value of the raw data, and x_{max} represents the maximum value of the raw data.

3.3.4.3 Choice of Normalization Method

The specific choice of normalization method requires accumulated experience and continuous experimentation. But here is a simple one:

If there's no specific range requirement, opt for the Z-score. For instance, in the context of housing prices, if a particular variable doesn't necessarily need to fall within a specific range, Z-score can be utilized. However, in the case of an image-related issue where the pixel values must be distributed between 0 and 1, the min-max scaler can be more relevant.

3.3.5 Normalization by KNIME

Extra caution is required when implementing normalization. In Fig. 3.65, you may notice that in KNIME, we not only add the "Normalizer" node but also the "Normalizer (Apply)" node. Why's that?

First, let's look at the "Normalizer" node, where we input the training data and obtain the normalized training data for the model to use. Next, we have the "Normalizer (Apply)" node, which has two input ports: the test set and the normalizer. The purpose of the "Normalizer (Apply)" node is to apply the same normalizer to the test data, which is derived from the output of the "Normalizer" node. Why is it made so complex?

#	Row... Variable	Coeff.	Std. Err.	t-value	P>|t|
	String	Number (double)	Number (double)	Number (double)	Number (double)
1	Row1 bedrooms	-0.062	0.005	-11.446	0
2	Row2 sqft_living	0.466	0.008	58.802	0
3	Row3 floors	0.027	0.005	5.095	0
4	Row4 waterfront	0.147	0.005	30.781	0
5	Row5 view	0.106	0.005	20.877	0
6	Row6 grade	0.358	0.008	47.451	0
7	Row7 yr_built	-0.216	0.005	-39.563	0
8	Row8 lat	0.209	0.005	45.997	0
9	Row9 Intercept	0	0.004	0	1

Fig. 3.65 View the model

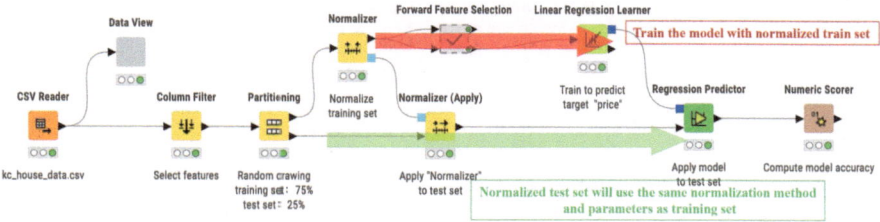

Fig. 3.66 Normalization by KNIME

It's because you don't know what distribution pattern the unknown data follows. You simply don't have that data yet, so how can you normalize them then? Therefore, normalizing unknown data can only rely on the existing data patterns.

The training set consists of the data you already know, so you can normalize them confidently. While you may also know the test set, it's important to act as if you are not in order to closely mimic future unknown data. The dataset for future prediction is genuinely not known to you. As you can never normalize unknown data with information from the unknown data themselves, the parameters from the training set can be used then. That's why we must add the "Normalizer (Apply)" to normalize the test set.

Double-click on the "Normalizer" node and select "Z-score Normalization (Gaussian)" as shown in Fig. 3.66 in the dialog below.

After normalization, execute the entire workflow again and see if the model improves.

3.3.6   Correlation Coefficient

Finally, let's learn something about correlation coefficients. This statistical measure was originally developed by the statistician Karl Pearson. It is utilized to examine the extent of the linear relationship between variables, typically represented by the letter "r." Due to the diverse nature of research subjects, there are various methods to define correlation coefficients, with the Pearson correlation coefficient being the most widely used.

The absolute value of the correlation coefficient approaching 1 indicates a stronger linear relationship between the data. Conversely, a value closer to 0 suggests a weaker linear relationship, as illustrated in Fig. 3.67.

However, by observing the last row of Fig. 3.68, it can be seen that a correlation coefficient of 0 does not imply no relationship between the data but only indicates no linear relationship.

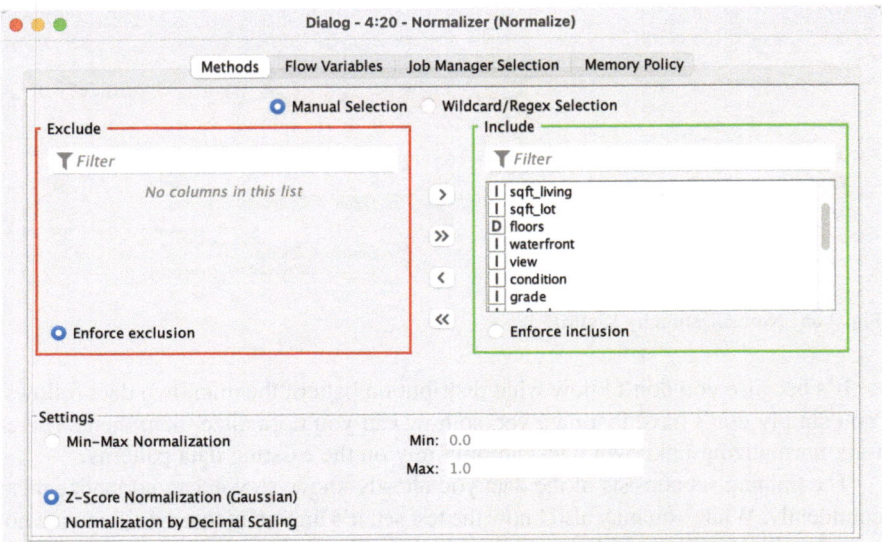

Fig. 3.67 Normalization

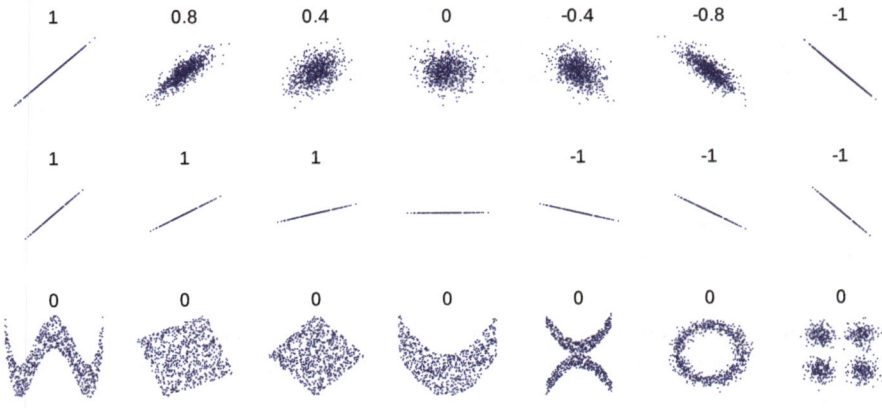

Fig. 3.68 Linear correlation (By DenisBoigelot, original uploader was Imagecreator—Own work, original uploader was Imagecreator, CC0, https://commons.wikimedia.org/w/index.php?curid=1 5165296)

3.4 Practice

1. What do training and test sets refer to? Please provide an example from your learning or work to explain.
2. If a feature has a p-value of 0.1, can it be concluded that this feature has no relationship with the model's label?
3. Viewing from the coefficients of the model, what are the major factors influencing house prices?
4. Try backward elimination method to optimize the model in KNIME.
5. Try min-max normalization method.
6. Why is it necessary to use both the "Normalizer" and "Normalizer(Apply)" nodes together?

Reference

https://www.deeplearning.ai/courses/machine-learning-specialization/

Chapter 4
Logistic Regression

To be, or not to be, that is the question.

—*William Shakespeare*

Abstract This chapter delves into logistic regression, a fundamental machine learning technique for classification problems, using KNIME as the primary analytical tool. It begins with basic concepts, explaining how logistic regression differentiates from linear regression by using the Sigmoid function to predict binary outcomes, such as determining eligibility for a scholarship based on exam scores. The chapter emphasizes the creation of decision boundaries and illustrates the use of dummy variables for categorical data handling.

Substantial focus is given to model evaluation techniques, including ROC (receiver operating characteristic) curves and F1 scores, enhancing the understanding of model performance. Practical implementation is anchored through a step-by-step guide to building a logistic regression model in KNIME, covering aspects from data reading and preprocessing to model training, testing, and evaluation.

A significant part of the chapter is dedicated to a case study: predicting survival on the Titanic using logistic regression. This real-world application provides a comprehensive understanding of the logistic regression process, from handling data types and missing values to interpreting model coefficients. The chapter concludes with discussions and practice exercises, consolidating the concepts learned and their application in logistic regression analysis.

Keywords Logistic regression · Classification · KNIME · Decision boundary · ROC curve · Dummy variable · Model evaluation · F1 score

This chapter covers

- **Classification.**
- **Determining boundaries.**
- **Model evaluation.**
- **ROC.**

- **F1.**
- **Dummy variable.**

While we are already familiar with regression and related solutions using KNIME, let's now delve into another common question—classification—which we can also use KINME to solve.

4.1 Basic Concepts of Logistic Regression (https://www. deeplearning.ai/courses/machine-learning-specialization/)

4.1.1 Classification

Supposing we aim to predict whether a student is eligible for a scholarship (hon) based on their exam scores, we only have two possible outcomes to expect: yes or no. As depicted in Table 4.1, the dataset comprises columns for reading scores (read), writing scores (write), math scores (math), and whether the student can get the scholarship status (hon).

Label 1 represents receiving the honor, while label 0 indicates no honor. If a regression method is employed to address this issue, we can observe the relationship between hon (scholarship status: "Yes/No") and read (reading score), as illustrated in Fig. 4.1.

The diagonal line depicted in Fig. 4.2 represents the regression line. If the regression line exceeds the boundary of 0.5, it means the student can get the honor; otherwise, it is categorized as no honor. Prediction involves identifying the read value that corresponds to this boundary line, setting this value as the boundary, and classifying it as "Yes" if it surpasses the boundary, or else as "No."

Points for Discussion

- What is the meaning of the regression line depicted in Fig. 4.2? If we regard the values of the regression line ranging from 0 to 1 as probabilities of winning the honor, certain concerns will naturally emerge. Firstly, how should we interpret the values predicted by the regression line that exceed 1? Additionally, how should we understand regression line predicting values below 0?
- If there is a point in the top right corner that is significantly distant from the other points (Fig. 4.3), the regression line will adjust itself to accommodate this point. If

Table 4.1 Data table

Read	Write	Math	Hon
30	41	33	0
20	46	44	0
36	43	44	1
34	46	32	0
42	34	34	0
45	35	45	1

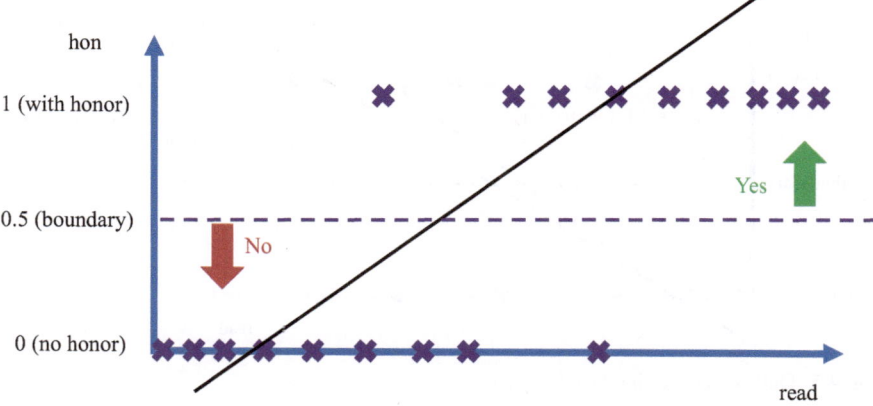

Fig. 4.1 Using regression

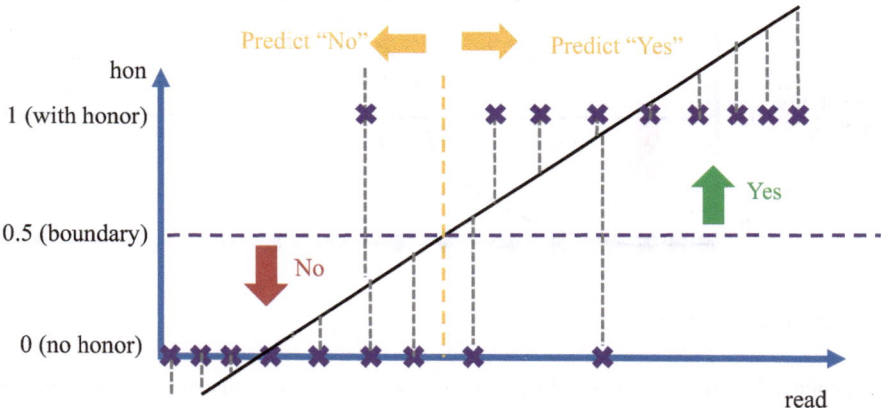

Fig. 4.2 Predict with regression

an outlier point causes instability in a model, what do you think of this model? Are there any other issues that you observe?

To address these issues, it would be ideal to achieve results similar to those depicted in Fig. 4.4. If feasible, predict "Yes" when the computed probability exceeds a certain boundary; otherwise, predict "No."

This type of binary classification problem is a typical example of classification. Logistic regression is one of the preferred methods for binary classification. It calculates a probability value with the range of [0–1] and, based on a boundary, determines whether the final result should be 0 or 1.

For instance, the scholarship prediction example represents a binary classification problem, distinguishing between 0 and 1. Similar scenarios include identifying the

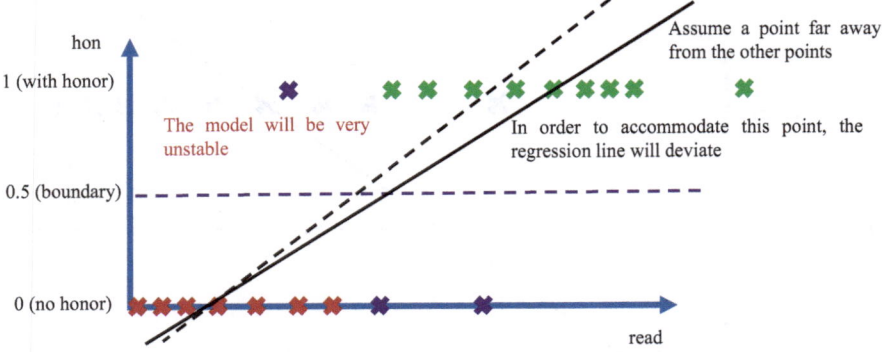

Fig. 4.3 Outliers have a significant impact

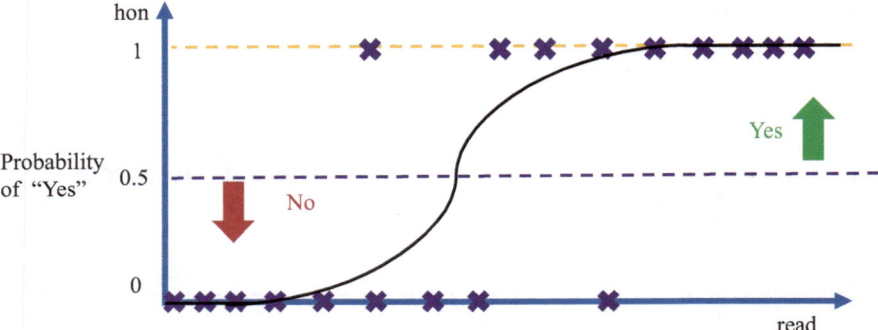

Fig. 4.4 Ideal classification

presence of a disease, determining customer loyalty, and discerning whether an animal in an image is a cat or a dog.

This transforms the problem into how to obtain a curve resembling Fig. 4.4. In this curve, the minimum value is 0, and the maximum value is 1, enabling straightforward prediction of results based on a boundary.

4.1.2   From Linear Regression to Logistic Regression

We use the Sigmoid function to convert the linear regression line into a logistic regression line. The Sigmoid function is

$$\sigma(z) = \frac{1}{1 + e^{-z}}$$

It can be visualized as shown in Fig. 4.5.

Fig. 4.5 Sigmoid function (By Qef (talk)—Created from scratch with gnuplot, Public Domain, https://commons.wikimedia.org/w/index.php?curid=4310325)

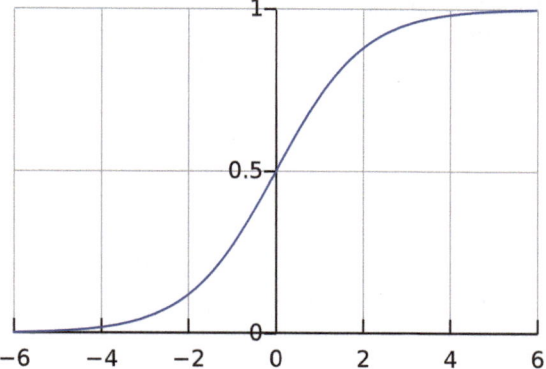

Fig. 4.6 Decision boundary

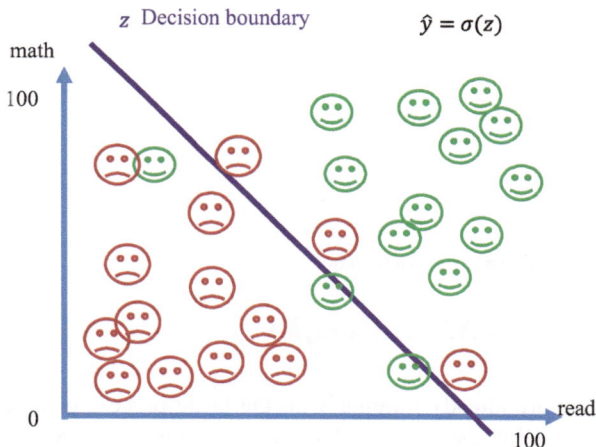

It can transform $z = b + w_1x_1 + \ldots + w_nx_n$ as $\widehat{y} = \sigma(z)$, $\sigma(z) > 0.5$ when $z > 0$; $(z) < 0.5$ when $z < 0$. So here in this case, 0.5 is the boundary.

4.1.3 Decision Boundary

In this example, there are different honor status predictions on either side of the line, which is called the "decision boundary."

In the example in Fig. 4.6, we can visualize the correlation between math grades, reading grades, and the honor status. The right side of the boundary predicts "Yes," while the left side predicts "No." Shifting the boundary to the right would increase the number of correctly predicted scholarship recipients but may also lead to missing out on some actual scholarship recipients. Conversely, shifting the boundary to the left could encompass some actual scholarship recipients, but it also raises the risk of misclassifying those who haven't received scholarship as "Yes."

Fig. 4.7 Workflow overview

Rows: 200 | Columns: 4

Table Statistics

#	Row...	read Number (integer)	write Number (integer)	math Number (integer)	hon Number (integer)
1	Row0	57	52	41	0
2	Row1	68	59	53	0
3	Row2	44	33	54	0
4	Row3	63	44	47	0
5	Row4	47	52	57	0
6	Row5	44	52	51	0
7	Row6	50	59	42	0
8	Row7	34	46	45	0
9	Row8	63	57	54	0
10	Row9	57	55	52	0
11	Row10	60	46	51	0
12	Row11	57	65	51	1
13	Row12	73	60	71	0
14	Row13	54	63	57	1
15	Row14	45	57	50	0

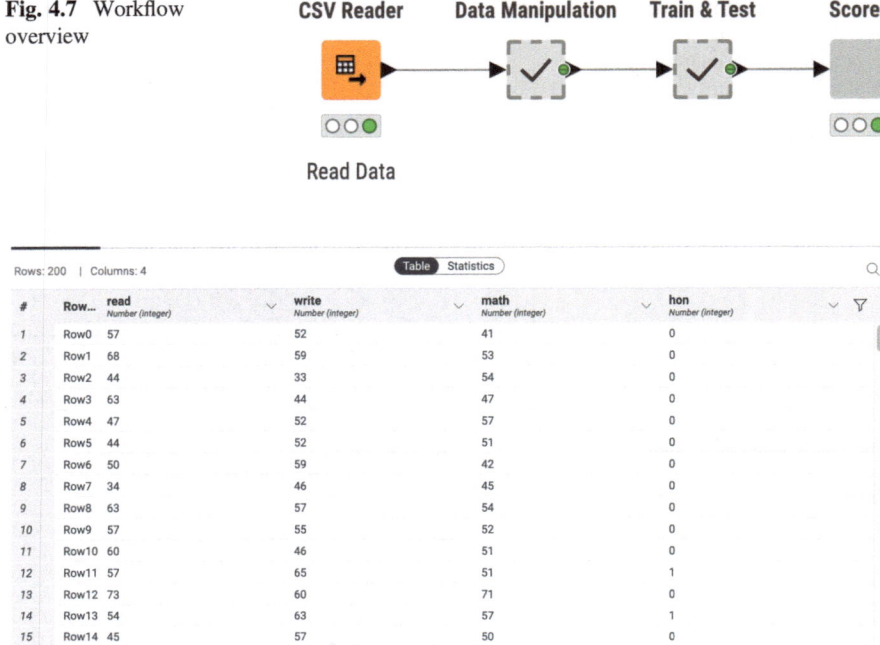

Fig. 4.8 Data overview in hon.csv

4.1.4 KNIME Workflow

We are already familiar with the basic process of machine learning: data reading, data preprocessing, model training and testing, and model evaluation. This process can be easily set up using KNIME, as illustrated in Fig. 4.7.

In Chap. 3, we introduced the use of Metanodes and Components to collapse complex sets of nodes, making the workflow more organized with more apparent functionality. In this workflow, we have collapsed sets of nodes related to data preprocessing, model training and testing, and model evaluation into separate Metanodes and a component. Our task is to build and execute the workflow.

4.1.5 Read Data

Reading data is the same as before; just create a "CSV Reader" node to read the data from the "hon.csv" file. The output results of the node data are shown in Fig. 4.8.

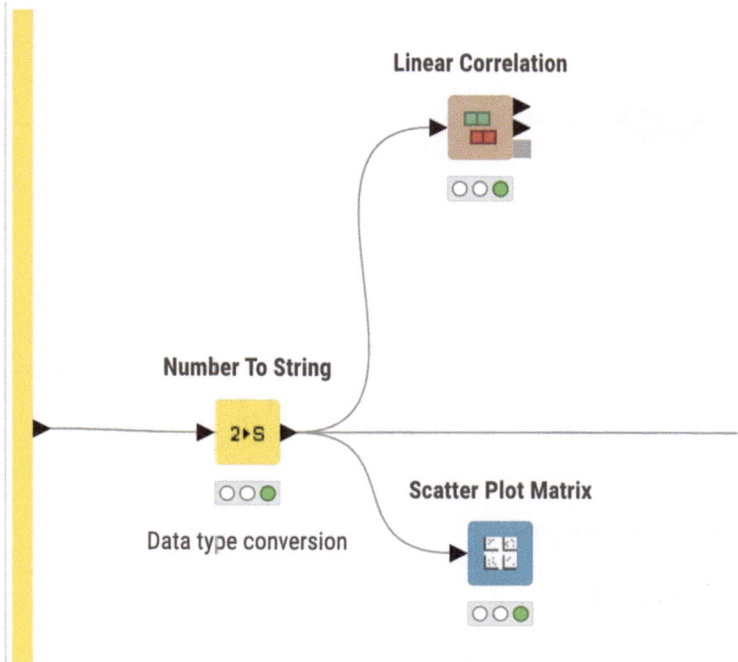

Fig. 4.9 Data processing

4.1.6 Data Processing

In this example, data processing mainly completes two tasks: one is data type conversion, and the other is data visualization. You can establish a data processing workflow as shown in Fig. 4.9 and then collapse the nodes into a Metanode.

4.1.6.1 Data Type Conversion

Based on the data overview in Fig. 4.8, it is evident that the values 0 and 1 in the "hon" column hold no numerical significance. They simply indicate the presence or absence of the honor. Therefore, we need to convert it into categorical data. In KNIME, we achieve this by converting this numerical data (Number) into character data (String). To do this, create a "Number to String" node and choose "hon" for type conversion, as demonstrated in Fig. 4.10.

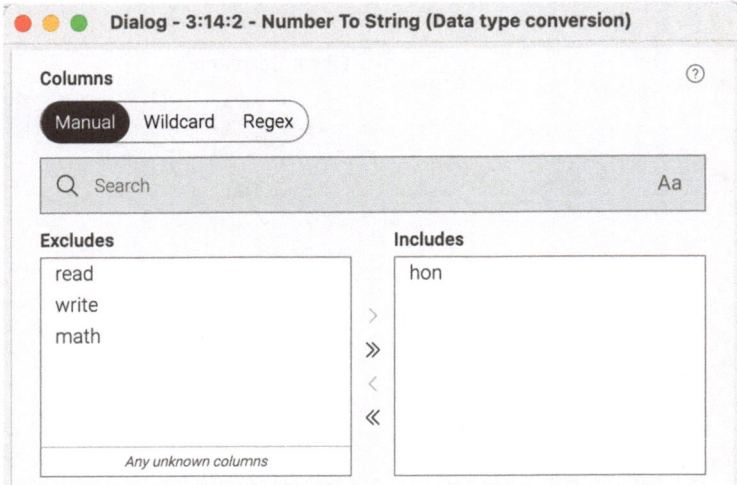

Fig. 4.10 Data type conversion

4.1.6.2 Visualization

We can visualize the relationships between different data by using correlation matrix and scatter plot matrix. Similar to linear regression, we can analyze the data relationship from the graphs, which will not be further elaborated here.

Give It a Try

- Try using different visualization nodes to view the data.

4.1.7 Model Training and Testing

Just like linear regression, the data is initially split into training and test sets. Next, the training set is fed into the "Logistic Regression Learner" node, and the test set along with the trained model is input into the "Logistic Regression Predictor" node (Fig. 4.11). Once this workflow is set up, it can be collapsed into a Metanode for modularization.

4.1.7.1 Set to Train the Model

Click on the settings button of the "Logistic Regression Learner" node to open the dialog and set the model training parameters. Our target is to "hon," so we set the "Target column" to hon. We want to predict honor status, i.e., when "hon" is 1, and

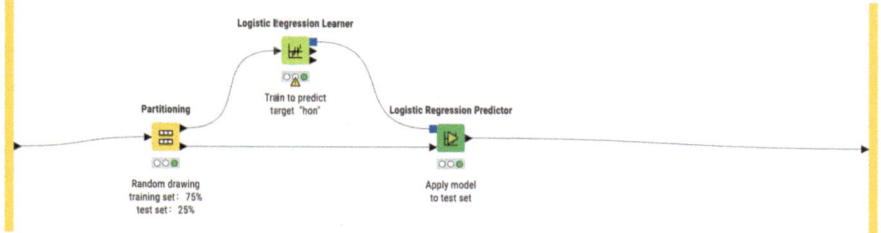

Fig. 4.11 Model training and testing

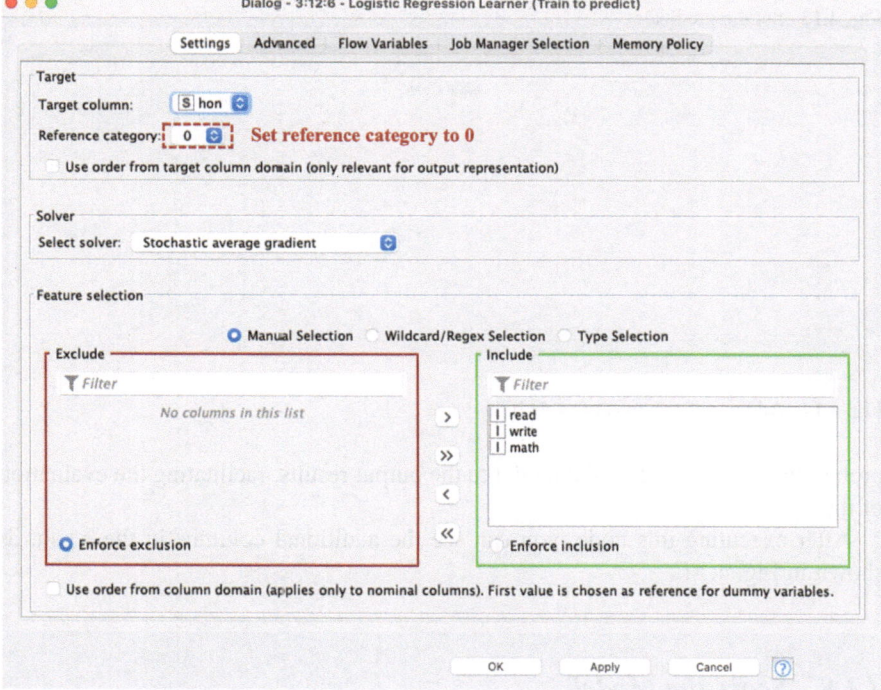

Fig. 4.12 Logistic regression configuration

the "Reference category" should be set as 0. Please note not to fill in 1 here, as shown in Fig. 4.12.

4.1.7.2 Set the Predictor

Next, click on the "Settings" button of the "Logistic Regression Predictor" node to configure the predictor. The only option that needs to be selected here is "Append column with predicted probabilities," as shown in Fig. 4.13. This option will add the

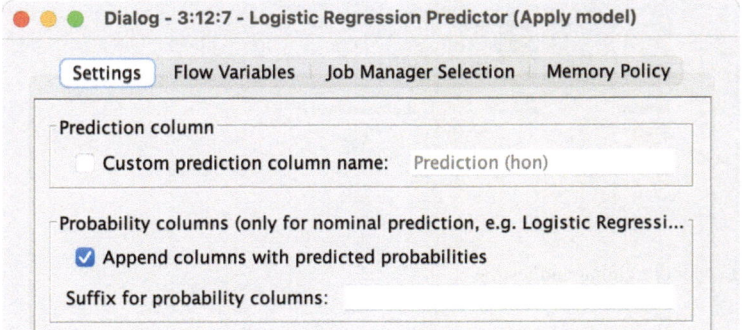

Fig. 4.13 Set the predictor

#	Row...	read Number (Integer)	write Number (Integer)	math Number (Integer)	hon String	P (hon=0) Number (double)	P (hon=1) Number (double)	Prediction (hon) String
1	Row2	44	33	54	0	1	0	0
2	Row4	47	52	57	0	1	0	0
3	Row6	50	59	42	0	1	0	0
4	Row8	63	57	54	0	1	0	0
5	Row11	57	65	51	1	1	0	0
6	Row14	45	57	50	0	1	0	0
7	Row15	42	49	43	0	1	0	0
8	Row17	57	57	60	0	1	0	0
9	Row23	60	52	57	0	1	0	0
10	Row25	34	37	46	0	1	0	0
11	Row30	42	31	57	0	1	0	0
12	Row32	65	67	63	1	1	0	0
13	Row36	68	54	75	0	1	0	0
14	Row37	65	62	68	1	1	0	0
15	Row38	47	31	44	0	1	0	0

Rows: 70 | Columns: 7

Fig. 4.14 Additional columns of probability

probabilities calculated by the model to the output results, facilitating the evaluation of the model later on.

After executing this node, you can see the additional columns in the result, as shown in Fig. 4.14.

4.1.8   Score the Model

Models can be scored using ROC curves or confusion matrices. We can establish the nodes of these scorers as shown in Fig. 4.15. The two nodes can also be collapsed into a Component.

There will be a detailed explanation on these two scorers in the next section. Here, let's first understand how to set the nodes and what the criteria for good are.

4.1.8.1   ROC

We just simply click on the "Settings" button to open the settings of the "ROC curve" node. In the settings as depicted in Fig. 4.16, we can choose the "hon" in the

Fig. 4.15 Score the model

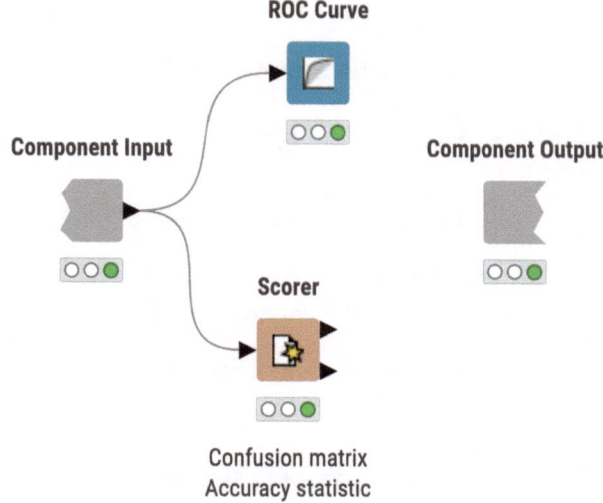

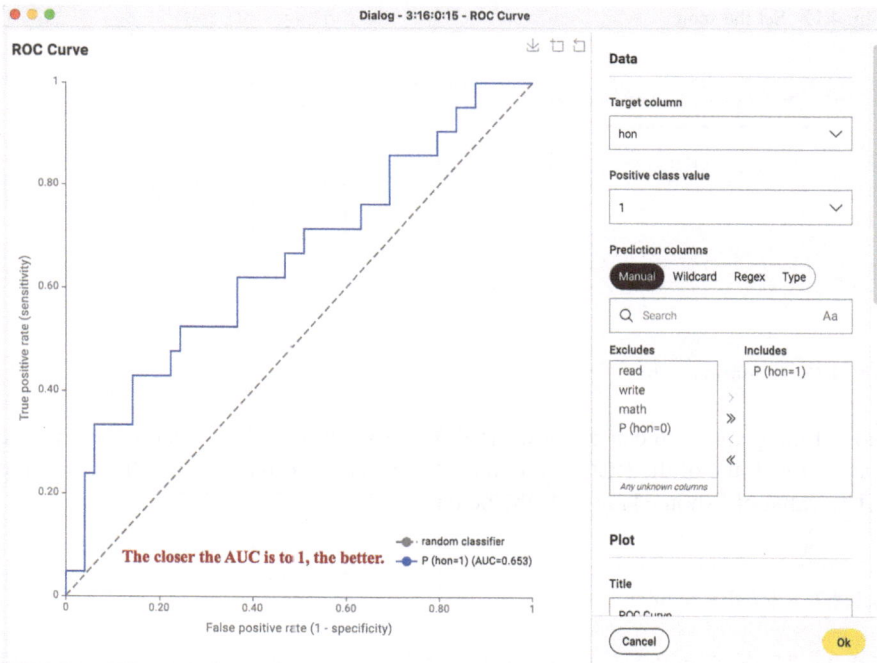

Fig. 4.16 Set the ROC node and preview ROC curve

"Target column" under the "Data" section. Then, select the true value 1 as the "Positive class value," and input the predicted probability, "P(hon=1)," in the "includes" field. Afterward, by clicking the "Save & execute" button on the right

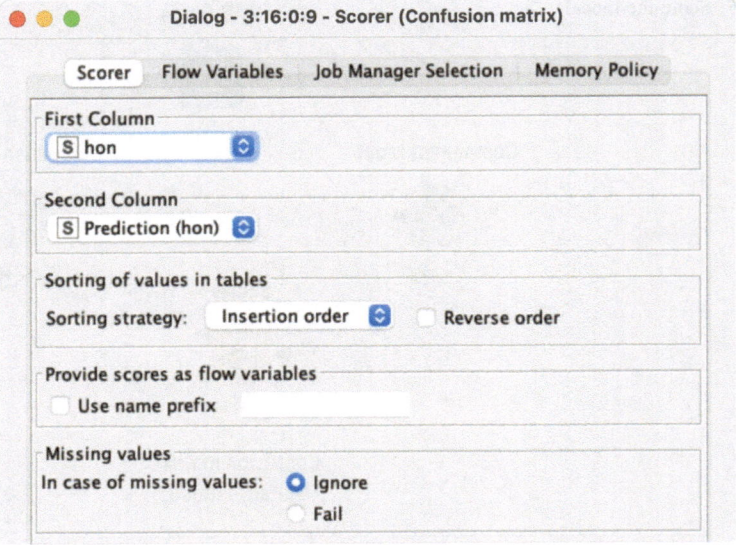

Fig. 4.17 Set the scorer

▶ 1: Confusion matrix   ▶ 2: Accuracy statistics   ☒ Flow Variables

Rows: 2 | Columns: 2

#	Row...	0 *Number (integer)*		1 *Number (integer)*
1	0	47		2
2	1	20		1

Fig. 4.18 Confusion matrix

side of the graph, you can preview the ROC curve. In Fig. 4.17, a larger area above the diagonal line of the ROC curve indicates better performance. And the closer the AUC value of P(hon=1) is to 1, the better.

4.1.8.2   Set the Scorer

Set the "Scorer" node as shown in Fig. 4.17, with "First Column" and "Second Column" set as "hon" and "Prediction(hon)," respectively. After executing it, you can view the results. The "Confusion matrix" is shown in Fig. 4.18, and the "Accuracy statistics" are shown in Fig. 4.19.

The two main indicators we will focus on are

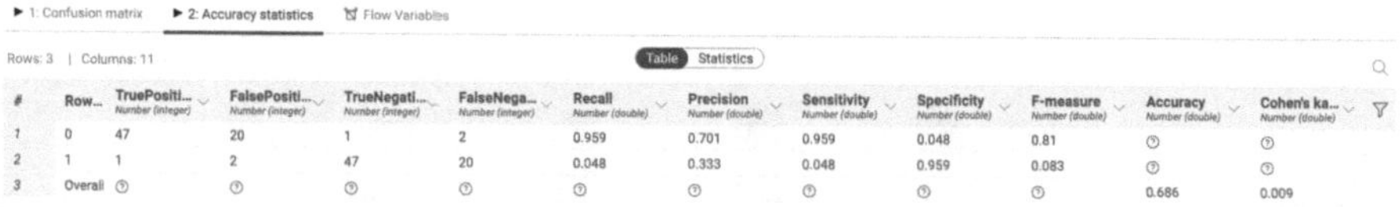

#	Row...	TruePositi... Number (integer)	FalsePositi... Number (integer)	TrueNegati... Number (integer)	FalseNega... Number (integer)	Recall Number (double)	Precision Number (double)	Sensitivity Number (double)	Specificity Number (double)	F-measure Number (double)	Accuracy Number (double)	Cohen's ka... Number (double)
1	0	47	20	1	2	0.959	0.701	0.959	0.048	0.81	?	?
2	1	1	2	47	20	0.048	0.333	0.048	0.959	0.083	?	?
3	Overall	?	?	?	?	?	?	?	?	?	0.686	0.009

Fig. 4.19 Accuracy statistics

- Confusion matrix. The larger the sum of the diagonal from top left to bottom right, the better.
- F1 (F-measure). The closer to 1, the better.

You can find the confusion matrix and other statistics in the "Confusion" and "Accuracy statistics" tables in the node monitor (Figs. 4.18 and 4.19).

Give It a Try

- Which statistics cannot be obtained without selecting the "Append column with predicted probabilities" option as shown in Fig. 4.13? Why?

4.2 Practice of Logistic Regression

In this section, we will work on a classic entry case of Kaggle Competition: Titanic—Machine Learning from Disaster (https://www.kaggle.com/c/titanic). By this case, we can gain a deeper understanding of the steps and methods of machine learning, improve our proficiency in using KNIME, and establish a solid foundation for future learning and work.

4.2.1 Titanic Background

When conducting research in data analysis, it is necessary to have a good understanding of the problem background.

The Titanic (Fig. 4.20) was a British Royal Mail Ship and one of three Olympic-class ocean liners owned by the White Star Line. It held the title of the world's largest ship during its time and was constructed by the Belfast shipyards of Harland & Wolff. It was renowned as the "unsinkable" and "dream" ship. The first-class cabins were designed to offer the utmost comfort and luxury, featuring gymnasiums, swimming pools, reception rooms, upscale restaurants, and luxurious cabins. Additionally, the ship was equipped with a high-powered wireless telegraph machine, providing passengers with telegraph services from Marconi Wireless Telegraph Company and facilitating navigation communication for the Titanic.

On April 10, 1912, the Titanic set off on its first and only journey with passengers, heading for New York. Among the passengers were some of the wealthiest

Fig. 4.20 The Titanic (By Boris Lux—Lux's Type Collection, Ocean liners—Titanic, CC BY-SA 3.0, https://commons.wikimedia.org/w/index.php?curid=3706771)

individuals of the time, as well as many immigrants from England, Ireland, Scandinavia, and other parts of Europe, all seeking new opportunities in the United States. Tragedy struck on the night of April 14th to 15th when the ship collided with an iceberg and sank. Out of the 2224 people on board, 1514 lost their lives, marking it as the deadliest peacetime maritime disaster in modern history. Captain Edward John Smith went down with the ship, and Thomas Andrews, the chief designer of the Titanic, also died in the disaster.

In 1985, US Navy officer Robert Ballard led a team to discover the wreckage of the Titanic, which split into two parts and gradually disintegrated at a depth of 3784 m on the seabed. Thousands of artifacts from the sunken ship have been recovered and exhibited in museums around the world. The Titanic has become one of the most famous ships in history, and her story has been preserved in various forms of popular culture, including books, ballads, movies, exhibitions, and memorabilia. The wreckage of the Titanic is the second largest in the world before World War I, second only to its sister ship Britannic.

However, the Titanic didn't have enough lifeboats, and the crew didn't have proper evacuation training as well. Third class passengers were often kept on the lower decks by the crew and had to find their own way through obstacles, which led to many of them being trapped, and most of the lower decks were filled with seawater. Generally, officers followed the "women and children first" rule during the evacuation, and most male passengers and crew members stayed on the ship.

This question is an entry task for a Kaggle competition, where the goal is to predict whether a passenger will survive based on their information.

Following the general steps of machine learning, we will use KNIME to implement tasks such as data reading, data processing, model training and testing, and model evaluation. The overall workflow can be found in Fig. 4.7.

Points for Discussion

• Why do we need to understand the background of the Titanic?

Rows: 891 | Columns: 12

Table Statistics

#	Row...	Passeng... Number (integ...	Survived Number (integ...	Pclass Number (integ...	Name String	Sex String	Age Number (doub...	SibSp Number (integ...	Parch Number (integ...	Ticket String	Fare Number (doub...	Cabin String	Embarked String
1	Row0	1	0	3	Braund, Mr. O...	male	22	1	0	A/S 21171	7.25	⊘	S
2	Row1	2	1	1	Cumings, Mrs...	female	38	1	0	PC 17599	71.283	C85	C
3	Row2	3	1	3	Heikkinen, Mi...	female	26	0	0	STON/O2. 31...	7.925	⊘	S
4	Row3	4	1	1	Futrelle, Mrs. ...	female	35	1	0	113803	53.1	C123	S
5	Row4	5	0	3	Allen, Mr. Willi...	male	35	0	0	373450	8.05	⊘	S
6	Row5	6	0	3	Moran, Mr. Ja...	male	⊘	0	0	330877	8.458	⊘	Q
7	Row6	7	0	1	McCarthy, Mr. ...	male	54	0	0	17463	51.862	E46	S
8	Row7	8	0	3	Palsson, Mast...	male	2	3	1	349909	21.075	⊘	S
9	Row8	9	1	3	Johnson, Mrs....	female	27	0	2	347742	11.133	⊘	S
10	Row9	10	1	2	Nasser, Mrs. ...	female	14	1	0	237736	30.071	⊘	C
11	Row10	11	1	3	Sandstrom, M...	female	4	1	1	PP 9549	16.7	G6	S
12	Row11	12	1	1	Bonnell, Miss...	female	58	0	0	113783	26.55	C103	S
13	Row12	13	0	3	Saundercock, ...	male	20	0	0	A/5. 2151	8.05	⊘	S
14	Row13	14	0	3	Andersson, Mr...	male	39	1	5	347082	31.275	⊘	S
15	Row14	15	0	3	Vestrom, Miss...	female	14	0	0	350406	7.854	⊘	S
16	Row15	16	1	2	Hewlett, Mrs. ...	female	55	0	0	248706	16	⊘	S
17	Row16	17	0	3	Rice Master. ...	male	2	4	1	382652	29.125	⊘	Q

Fig. 4.21 Overview of the data

Table 4.2 Data description

Variable	Definition	Note
Survived	Survival or not	0: Death; 1: Survival
Pcalss	Cabin class	1, 2, 3 = 1, 2, 3 class
Sex	Gender	
Age	Age	
Sibsp	Number of siblings/spouses aboard	
Parch	Number of parents and children on board	
Ticket	Ticket no.	
Fare	Ticket price	
Cabin	Cabin no.	
Embarked	Embarking port	

4.2.2   Read Data

First, we will read the data in Titanic_train.csv. You'll notice that there are many missing values and the data formatting isn't great (Fig. 4.21). This means that we'll need to put in a lot of effort into preprocessing the features.

The description of the data is shown in Table 4.2.

4.2.3   Data Preprocessing

In this example, data processing is crucial because the data is quite messy and needs a lot of cleaning. This can be handled by the "Data Manipulation" Metanode, as shown in Fig. 4.22. First, we use the "Column Filter" node to get rid of irrelevant columns, and then we use the "Number To String" node to change the data type to string and deal with missing value. Finally, based on the data visualization analysis,

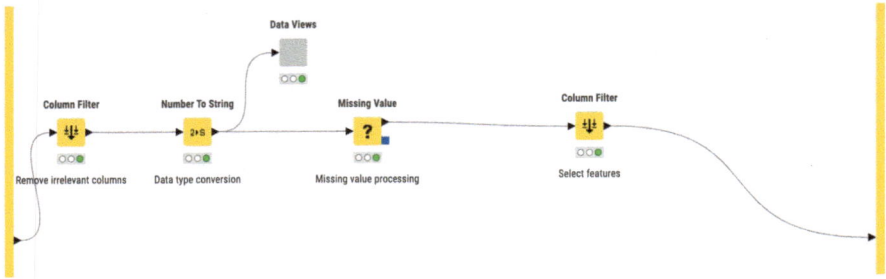

Fig. 4.22 Data preprocessing

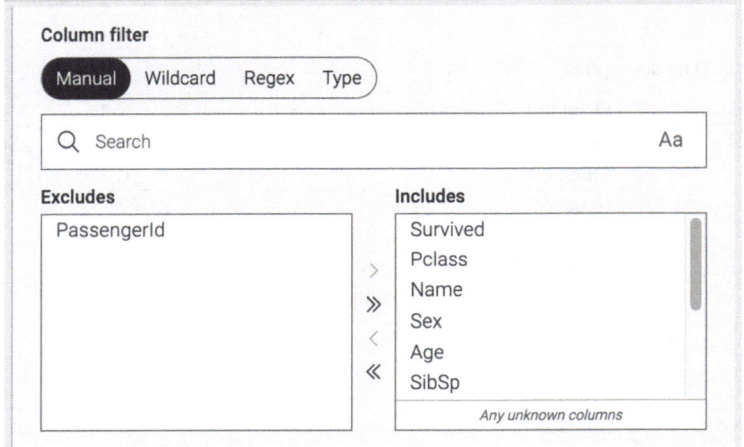

Fig. 4.23 Delete irrelevant columns

we remove irrelevant columns or further preprocess the features. In this case here, we didn't do additional feature preprocessing and just removed the irrelevant columns.

4.2.3.1 Deleting Irrelevant Columns

The first step in data processing is to remove irrelevant columns. Irrelevant columns have no connection to the problem we want to analyze, such as the passenger ID (passengerid), so we simply delete it. Double-click on the "Column Filter" node, and delete the ID in the dialog, as shown in Fig. 4.23.

4.2.3.2 Data Type Conversion

Although some data exists in numerical form, the numerical values don't have any meanings. These data are the categorical data we mentioned before.

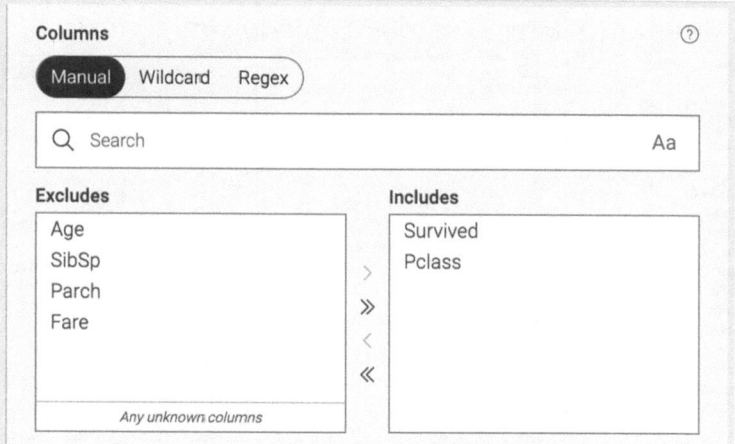

Fig. 4.24 Data type conversion

Here, "Survived" and "Pclass" are categorical data, but they exist in numerical form. We need to inform the model about their actual types. Click the settings button of the "Number To String" node, and select the two options as shown in Fig. 4.24 in the opened dialog to convert the data type to String.

4.2.3.3 Handling Missing Values

Handling missing data is a significant task and can't be done in a snap. Currently, we're only using the most basic method to handle missing data. First and foremost, it's crucial to remember to never discretionally delete a row of data. It means that we shouldn't discard the entire row just because a certain piece of value is missing. Generally, we need to use specific values to fill in these gaps, and commonly used fill-in values include mean, median, and most frequent value. In this example, we're using these basic methods to impute the missing values. Simply double-click on the "Missing Value" node. In the dialog that pops up, choose "Mean" to impute the missing values for numerical data, and select "Most Frequent Value" to impute the missing values for categorical data as shown in Fig. 4.25.

4.2.3.4 Data Visualization

Data visualization is an important part of data preprocessing, which we will elaborate on in the next section.

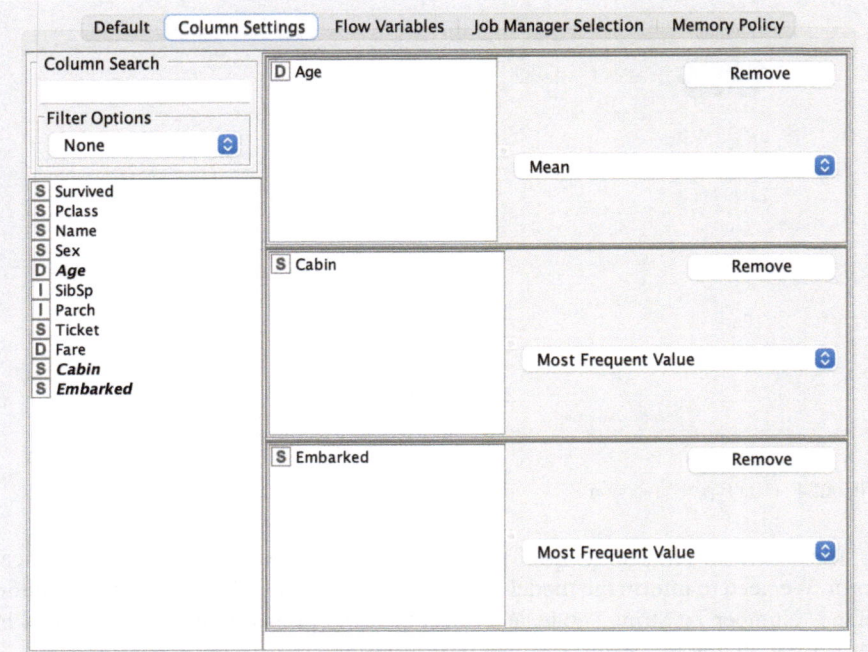

Fig. 4.25 Handling missing values

4.2.4   Data Visualization and Deleting Unrelated Columns

Data visualization can help us gain a deeper understanding of the problem at hand and make it easier to identify certain data relationships. The data visualization in Fig. 4.22 is achieved through the "Data Views" component. As described in Sect. 3. 2.4, when the component is expanded, its composition is shown in Fig. 4.26.

4.2.4.1   Analysis of Survival Percentage

We can utilize the "Pie Chart" node to visualize survival status. As depicted in Fig. 4.27, the "Category dimension" is set as "Survived" and the "Aggregation method" is "Occurrence count." After clicking "Save & execute," a pie chart is generated. Hovering the mouse over each slice of the pie reveals that the number of deaths (0) is significantly higher than the number of survivors (1), and the survival rate is only 38.38%.

To automatically show the count and proportion of each category in a pie chart, you need to follow the setup in Fig. 4.28: select "Both" for "Label value format", which includes both "Absolute" and "Proportion"; then choose "Both" for "Label content," so that both "Category" and "Value" can be displayed in the pie chart.

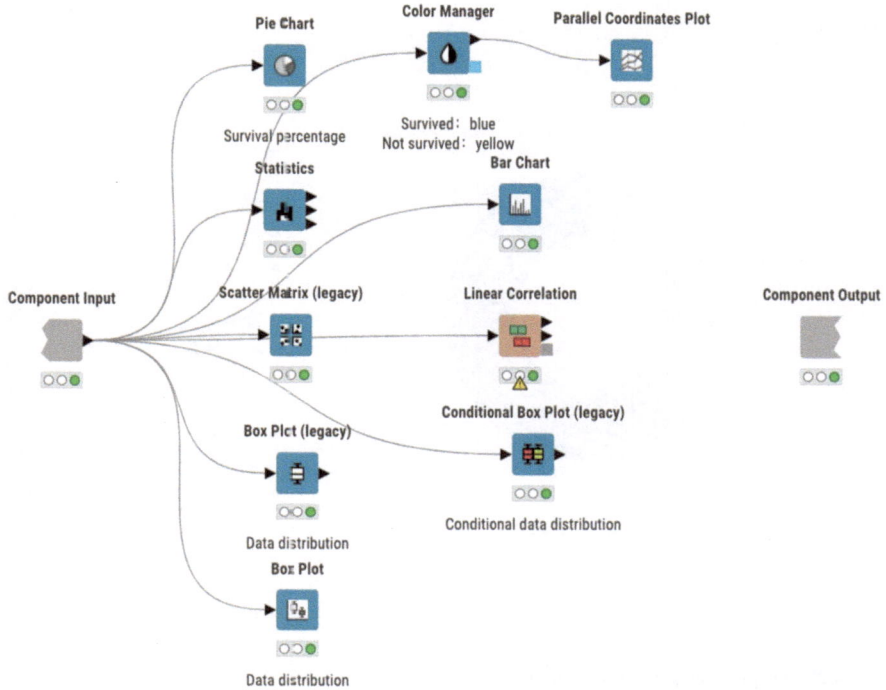

Fig. 4.26 Data visualization

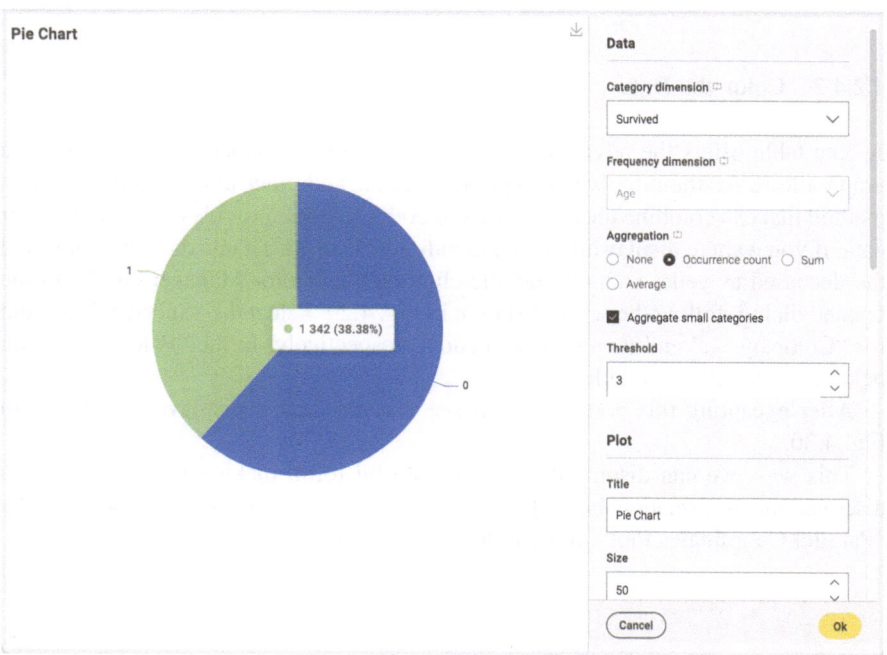

Fig. 4.27 Basic settings for pie chart

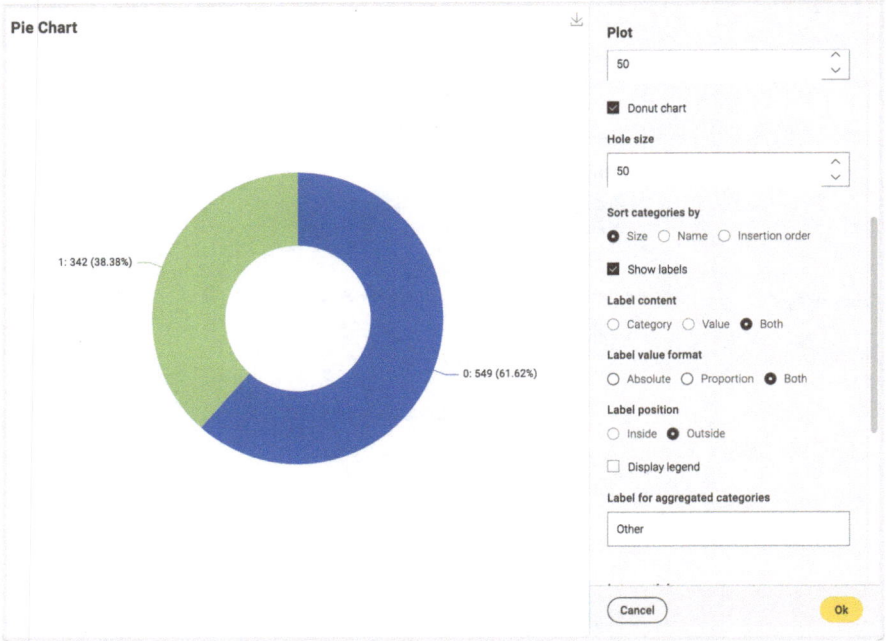

Fig. 4.28 Display settings for pie chart

Lastly, try enabling the "Donut chart" option, and you'll see the original pie chart transformed into a more visually appealing "donut" chart.

4.2.4.2   Color the Data

A data table offers the advantage of displaying precise numerical values, while a graph allows for the intuitive perception of trends and other information. Is there a method that can combine the advantages of both? KNIME provides a feature to color data. If you want to display all surviving individuals in the Titanic dataset as blue and the deceased as yellow, simply double-click on the "Color Manager" node. In the opened dialog, follow the settings shown in Fig. 4.29. Select the "Survived" column for "Color by...," and then choose 0 and 1, respectively. In the "Palettes" section below, select yellow and blue.

After executing this node, you can see that the data is displayed as shown in Fig. 4.30.

This way, we can display the data in colorful form, making it easier for us to discover the information behind the data. (Specific applications are seen in the "Parallel Coordinates Plot" in the following sections.)

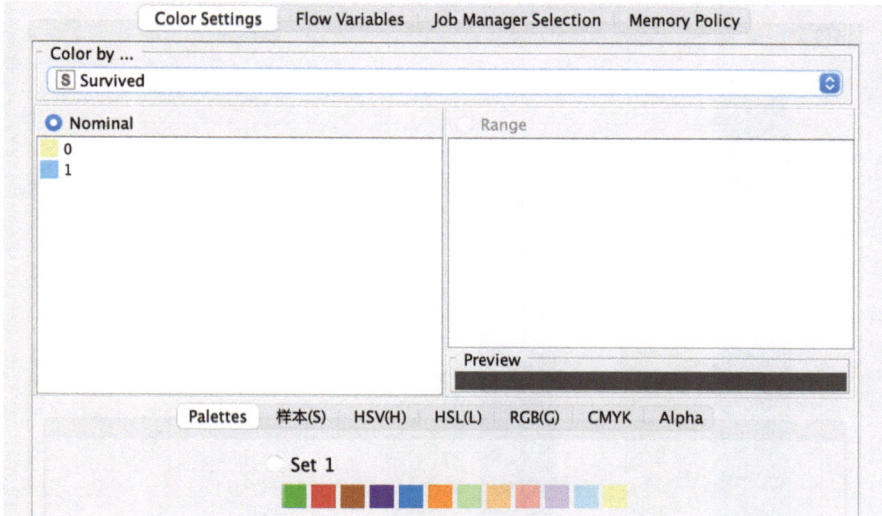

Fig. 4.29 Set the Color Manager node

#	Row...	Survived	Pclass	Name	Sex	Age	SibSp	Parch	Ticket	Fare	Cabin	Embarked
		String	String	String	String	Number (double)	Number (integer)	Number (integer)	String	Number (double)	String	String
1	Row0	0	3	Braund, Mr. Ow...	male	22	1	0	A/5 21171	7.25		S
2	Row1	1	1	Cumings, Mrs. ...	female	38	1	0	PC 17599	71.283	C85	C
3	Row2	1	3	Heikkinen, Miss...	female	26	0	0	STON/O2. 3101...	7.925		S
4	Row3	1	1	Futrelle, Mrs. Ja...	female	35	1	0	113803	53.1	C123	S
5	Row4	0	3	Allen, Mr. Willia...	male	35	0	0	373450	8.05		S
6	Row5	0	3	Moran, Mr. Jam...	male		0	0	330877	8.458		Q
7	Row6	0	1	McCarthy, Mr. T...	male	54	0	0	17463	51.862	E46	S
8	Row7	0	3	Palsson, Master...	male	2	3	1	349909	21.075		S
9	Row8	1	3	Johnson, Mrs. ...	female	27	0	2	347742	11.133		S
10	Row9	1	2	Nasser, Mrs. Ni...	female	14	1	0	237736	30.071		C
11	Row10	1	3	Sandstrom, Mis...	female	4	1	1	PP 9549	16.7	G6	S
12	Row11	1	1	Bonnell, Miss. E...	female	58	0	0	113783	26.55	C103	S
13	Row12	0	3	Saundercock, M...	male	20	0	0	A/5. 2151	8.05		S
14	Row13	0	3	Anderson, Mr. ...	male	39	1	5	347082	31.275		S
15	Row14	0	3	Vestrom, Miss. ...	female	14	0	0	350406	7.854		S
16	Row15	1	2	Hewlett, Mrs. (...	female	55	0	0	248706	16		S
17	Row16	0	3	Rice, Master. Eu...	male	2	4	1	382652	29.125		Q

Rows: 891 | Columns: 11

Fig. 4.30 View the colored result

4.2.4.3 Statistical Data

In the earlier section, we have already discussed the "Statistics" node. This node allows us to access the statistical information of the data, which won't be elaborated on here. You're more than encouraged to explore it independently.

4.2.4.4 Bar Chart

Next, we can utilize the "Bar Chart" node to explore the correlation between the number of siblings/spouses (SibSp) and the number of parents/children (Parch) for

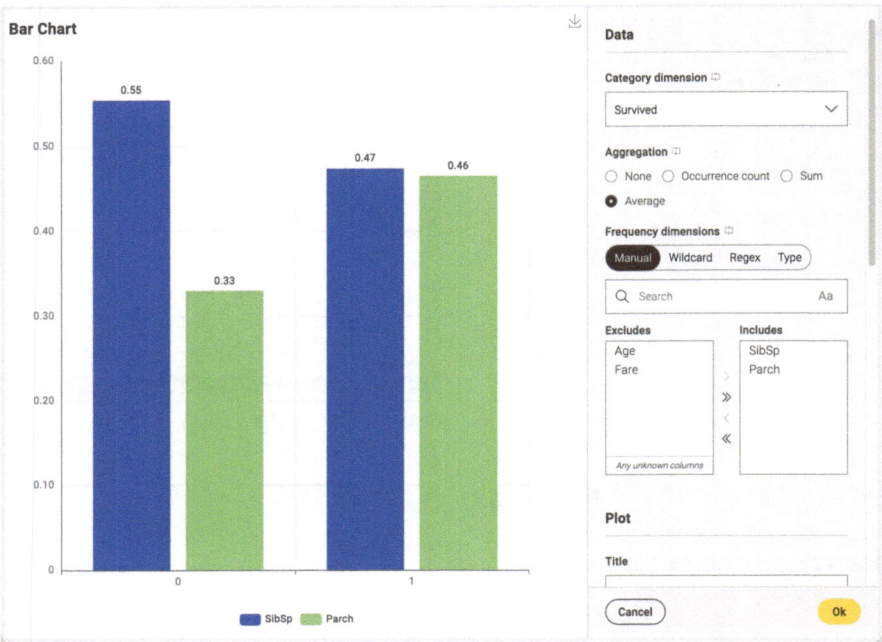

Fig. 4.31 Comparison of bar charts

different survival statuses. As depicted in Fig. 4.31, within the "Bar Chart" node, the "Category dimension" is configured as "Survived," and the "Aggregation method" is set as "Average." In the "Includes" section, SibSp and Parch are chosen. Upon selecting the ☑ Show bar values icon in the "Plot" option below, we can see that among the survivors, the number of siblings/spouses and parents/children is relatively similar, whereas among the deceased, there is a notable disparity between the two.

4.2.4.5   Scatter Plot Matrix

We can utilize the "Scatter Matrix (legacy)" node to visualize the data distribution. Configure a scatterplot matrix as illustrated in Fig. 4.32 to examine the correlation between "survival" and chosen features. However, it becomes apparent that apart from the connection between survival and age, all other features, whether integers or categorical data, are grouped closely together, posing a challenge in assessing the available information. How can this issue be addressed?

Click on the "Appearance" tab, as depicted in Fig. 4.33, and modify the "jitter" to a value that ensures clear visibility of the data details. Following the adjustment, it becomes evident that the count of dead surpasses that of survivors. Specifically, the number of deaths in the third class significantly exceeds the count of survivors. Moreover, the male mortality rate notably surpasses that of females. Additionally,

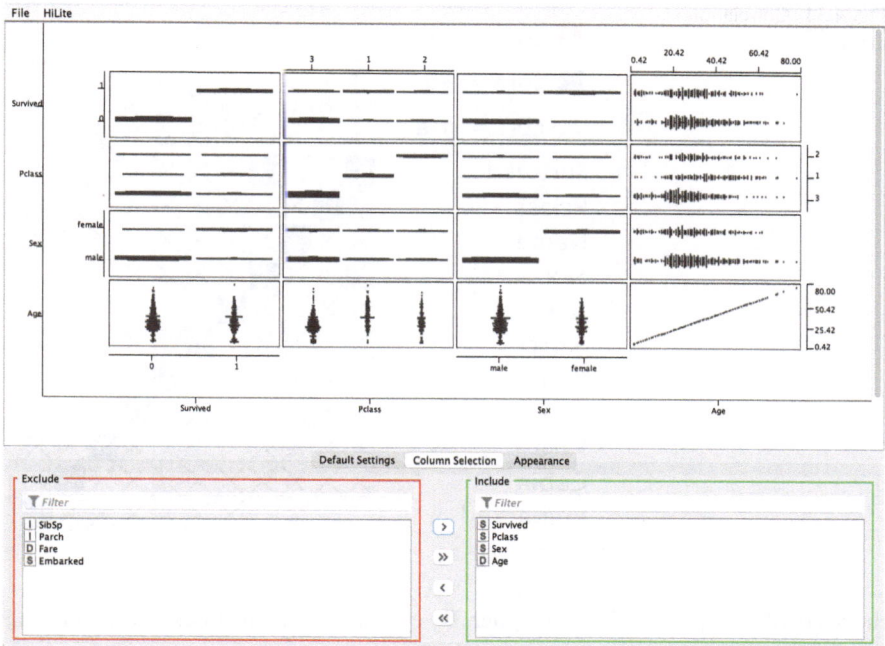

Fig. 4.32 Scatterplot matrix

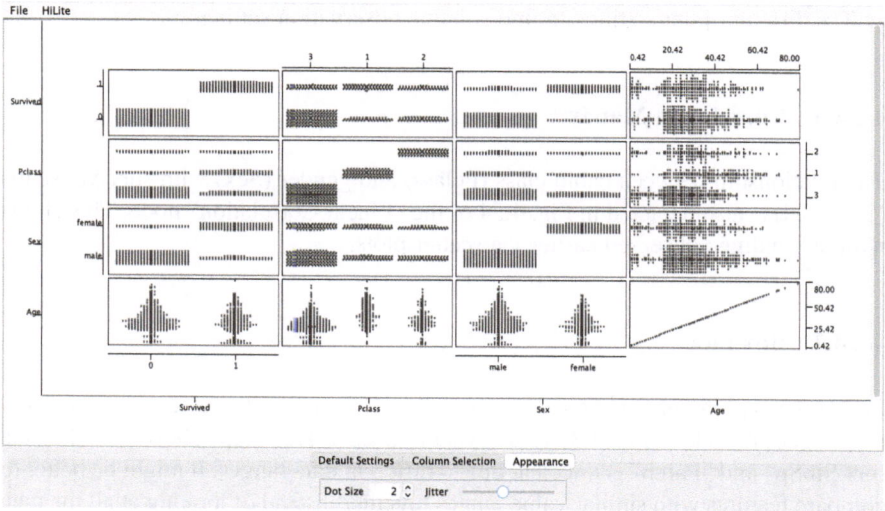

Fig. 4.33 Set the "jitter"

Fig. 4.34 Correlation matrix

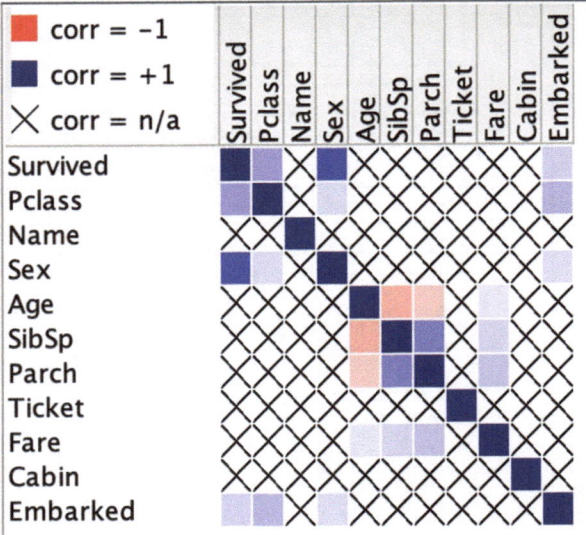

the death of young individuals is higher. By exploring other features, further insights can be revealed.

Give It a Try

• Try different jitter settings to find suitable observation settings.

4.2.4.6 Correlation Matrix

The relationship between cabin class (Pclass) and gender (Sex) with survival status (Survived) is also depicted in Fig. 4.34 of the "Linear Correlation" node. This aligns with the findings observed earlier via scatter plots.

4.2.4.7 Box Plot

Let's now explore what insights we can gather from a boxplot. Check out the "Box plot (legacy)" node in Fig. 4.35 to see what it reveals. This graph shows that the data for "SibSp" and "Parch" is not clear due to different data ranges. It might be better to compare features with similar value ranges together instead of looking at all the data at once. But is this approach too complicated?

Once again, KNIME offers a solution. By selecting the "Appearance" tab, as shown in Fig. 4.36, and checking the "Normalize (with respect to min-max values of the domain)" option, the graph will be displayed in a normalized form, making it easier for us to see the various data (Fig. 4.37).

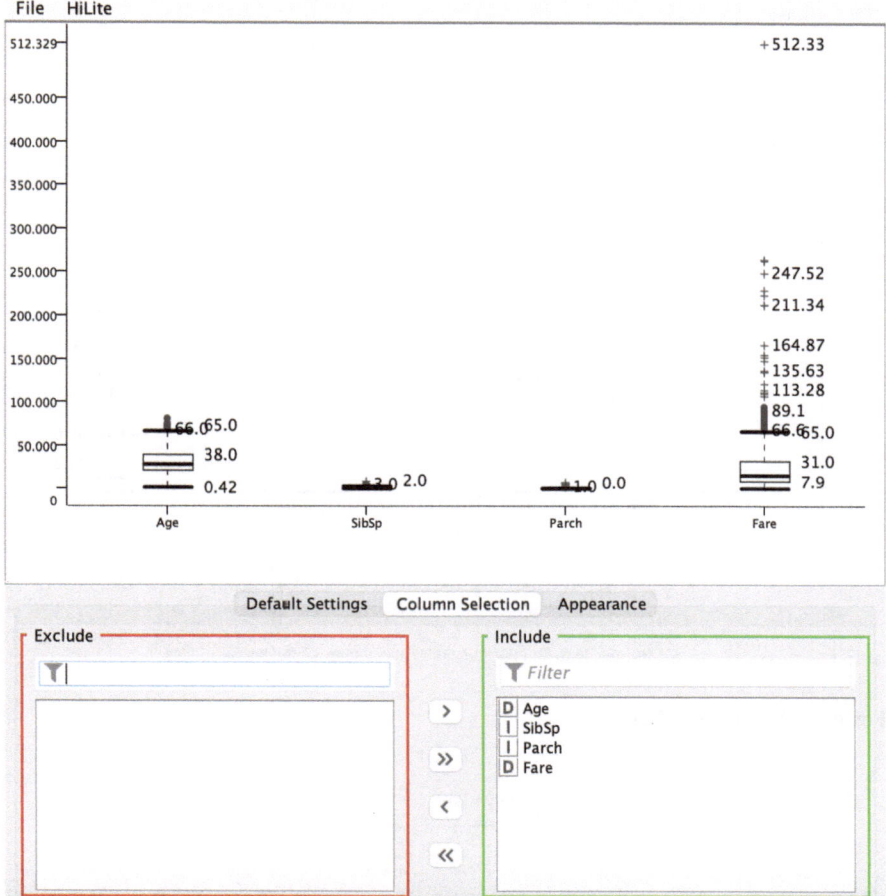

Fig. 4.35 Box plot

Just like a regular box plot, KNIME also offers a node called "Conditional Box Plot(legacy)," which can be set up as shown in Fig. 4.38-(1). Since we want to observe the data under different survival states, we choose "Survived" as the "Nominal Column" and select a feature to view as the "Numeric Column." In the "Box Plot" node of the modern KNIME UI, by setting the "Condition column" to "Survived" and choosing a feature in "Includes," you can achieve the same functionality, as demonstrated in Fig. 4.38-(2).

In Fig. 4.38, the horizontal axis represents death (0) and survival (1), while the vertical axis represents the ticket price (fare). This graph clearly shows that the fare for the survivors is higher.

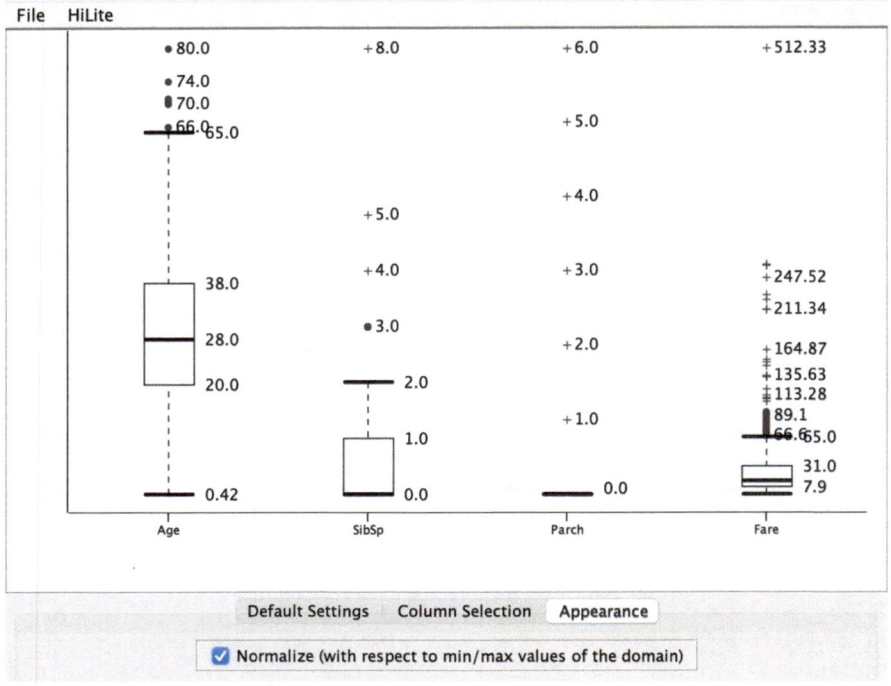

Fig. 4.36 Normalized display

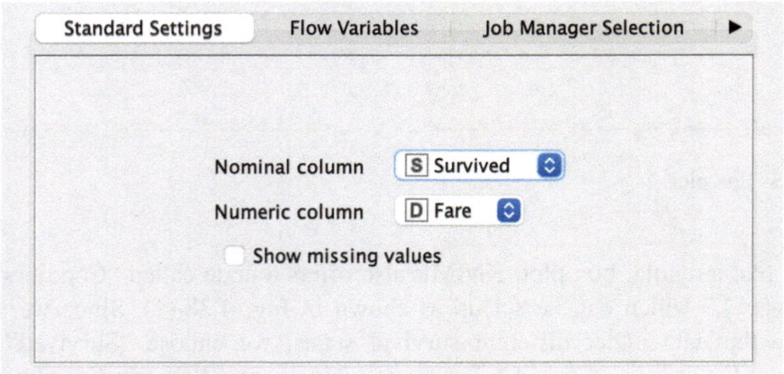

Fig. 4.37 Set the "conditional box plot"

4.2.4.8   Parallel Coordinates Plot

Finally, let's look at a commonly used visualization method for studying high-dimensional data, i.e., parallel coordinates plot. It is achieved through the "Parallel Coordinates Plot" node, which usually needs to be connected with the output of "Color Manager" node. The setting for node "Parallel Coordinates Plot" in Fig. 4.39

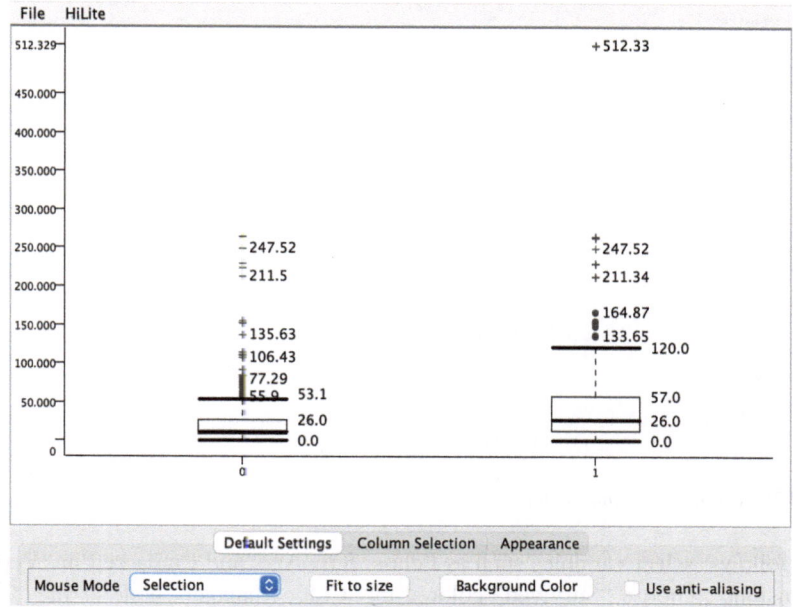

(1) "Conditional Box Plot" node

(2) "Box plot" node in the modern UI

Fig. 4.38 Conditional box plot

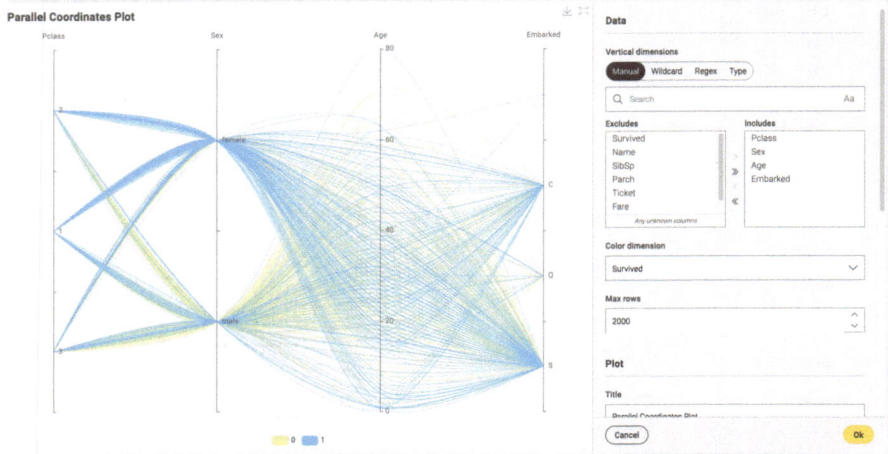

Fig. 4.39   Parallel coordinates plot

is as follows: select variables such as Pclass, Sex, Age, and Embarked in "Vertical dimensions"; choose "Survived" colored by "Color Manager" node as the "Color dimension"; adjust the "Max rows"; and click on "Save & execute" to generate the parallel coordinates plot, where the colors set for "Survived" will be displayed as the colors of the lines in the parallel coordinates plot. To reduce the visual overlap of lines, you can choose to set the lines as curved ○ Straight ● Curved in the "Line shape" option.

In Fig. 4.39, each vertical line (axis) in the parallel coordinates plot represents the variables we have chosen, and each curve represents the data read for each row. The position of the curve on each axis corresponds to the value of that variable for that row. By observing the values, trends, and density of different colored curves on multiple variables, we can intuitively identify data relationships: the number of survivors (blue) is mainly concentrated among females in first and second class cabins, who primarily boarded at ports S and C and are relatively young. By selecting other features, we can discover more information.

Give it a try.

- Please try to interpret the parallel coordinates plot on your own.

4.2.5   Model Training and Testing

This part is implemented within the "Train & Test" Metanode, as shown in Fig. 4.40.

We can firstly divide the data into training and test sets before we train the logistic regression learner. Finally, we may use the test data and the trained model to make predictions.

The "Logistic Regression Learner" node is configured as shown in Fig. 4.41.

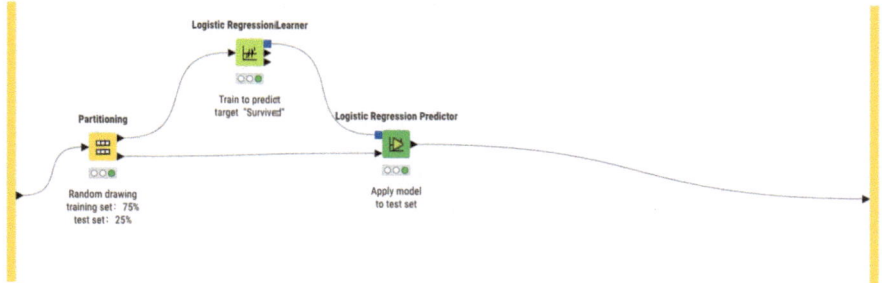

Fig. 4.40 Training and testing of the model

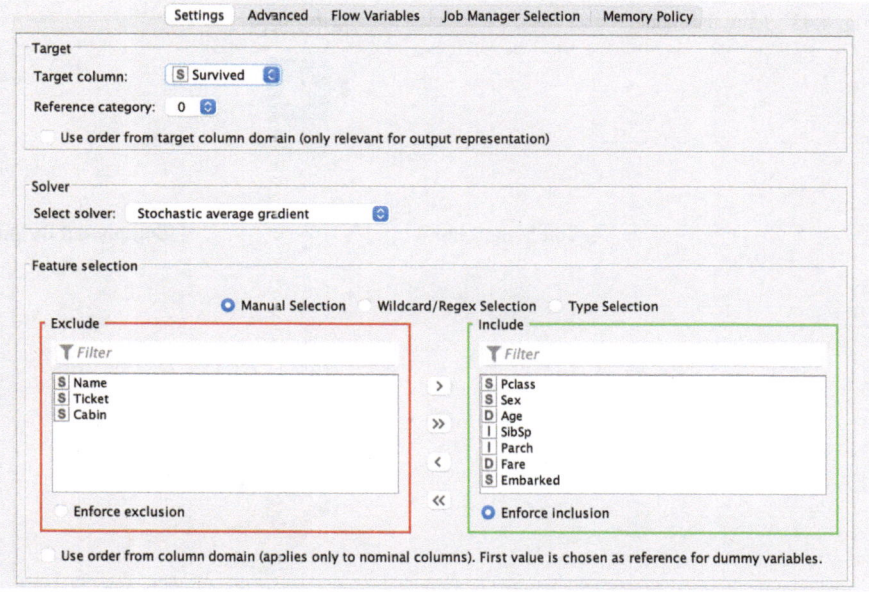

Fig. 4.41 Set the "Logistic Regression Learner" node

The target column is set as "Survived" and the "Reference category" is set as 0. Because the information about "Name," "Ticket," and "Cabin" is complex and not convenient for analysis for beginners, they need to be excluded from the model data for now.

Next, we can proceed with setting the "Logistic Regression Predictor" node, as shown in Fig. 4.42. The main goal is to attach the calculated probabilities to the data table for later use in computing ROC, so we need to select "Append columns with predicted probabilities."

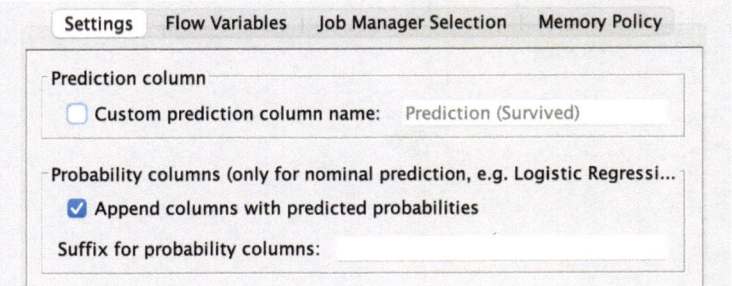

Fig. 4.42 Set the predictor

Fig. 4.43 Score the model

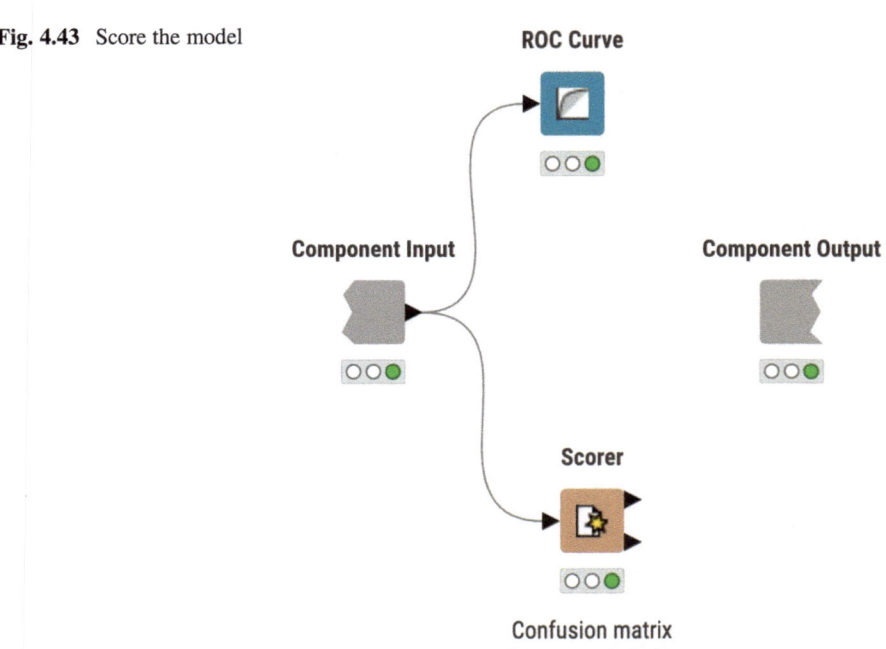

4.2.6 Score the Model

The final step is scoring the model. Here we can use the scorers mentioned earlier, such as ROC curve and confusion matrix, as shown in Fig. 4.43.

4.2.6.1 ROC

Similar to the previous example, we can click on the settings button of "ROC Curve" node, first input the true values, and then add the predicted probabilities, as demonstrated in Fig. 4.44. Once it's done executing, you'll see the ROC curve on the left.

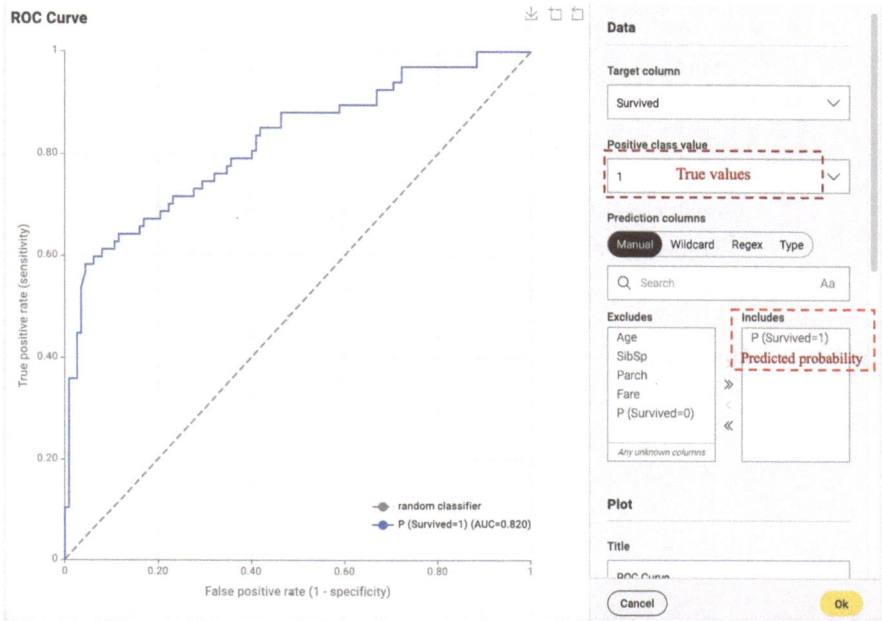

Fig. 4.44 Set the ROC node

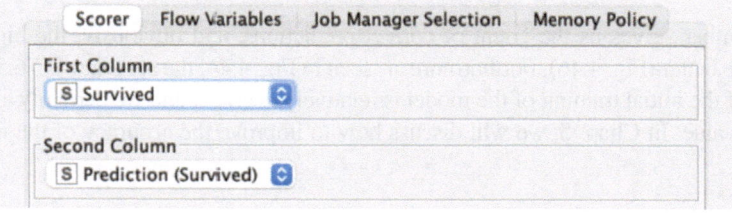

Fig. 4.45 Set other scorers

From the graph, we can tell that the ROC value is 0.8225, which is close to 1, showing that our model has a pretty good predictive performance.

4.2.6.2 Other Scorers

You only need to know the true value and predicted value. Click on the settings button of the "Scorer" node and set it according to Fig. 4.45.

After executing, you can obtain various scorers including the confusion matrix and F1 (F-measure).

Click on the "Open View" button above the node to check out the confusion matrix. As seen in Fig. 4.46, the two columns represent predicted death (0) and survival (1), while the two rows represent actual death (0) and survival (1). The numbers in the top left and bottom right corners are 113 and 78, respectively. These

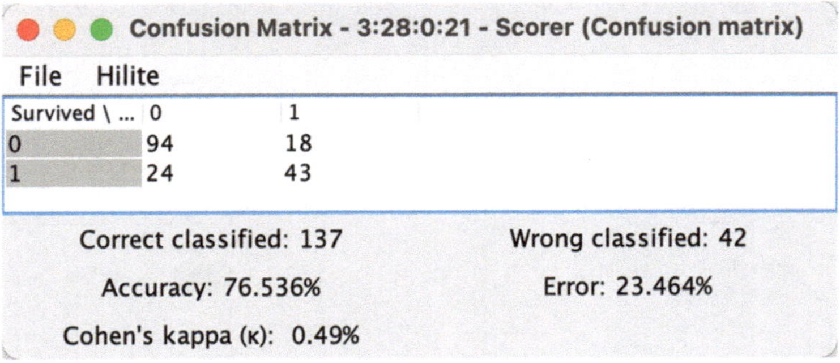

Fig. 4.46 Confusion matrix

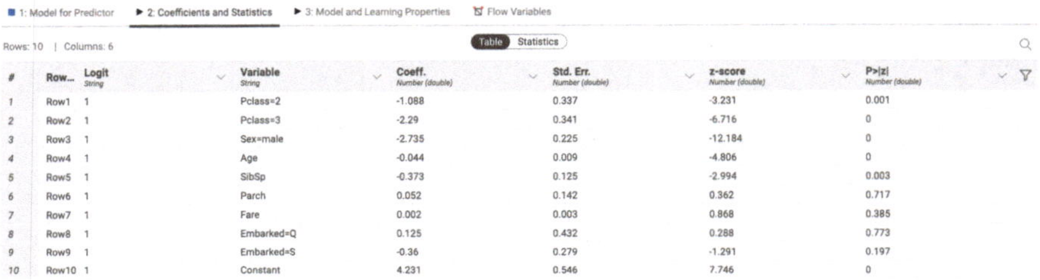

#	Row...	Logit String	Variable String	Coeff. Number (double)	Std. Err. Number (double)	z-score Number (double)	P>|z| Number (double)
1	Row1	1	Pclass=2	-1.088	0.337	-3.231	0.001
2	Row2	1	Pclass=3	-2.29	0.341	-6.716	0
3	Row3	1	Sex=male	-2.735	0.225	-12.184	0
4	Row4	1	Age	-0.044	0.009	-4.806	0
5	Row5	1	SibSp	-0.373	0.125	-2.994	0.003
6	Row6	1	Parch	0.052	0.142	0.362	0.717
7	Row7	1	Fare	0.002	0.003	0.868	0.385
8	Row8	1	Embarked=Q	0.125	0.432	0.288	0.773
9	Row9	1	Embarked=S	-0.36	0.279	-1.291	0.197
10	Row10	1	Constant	4.231	0.546	7.746	0

Fig. 4.47 Coefficients and statistics

two numbers gives us the count of correct predictions, and obviously, the higher the sum, the better (Fig. 4.46). Furthermore, as seen in Fig. 4.46, the accuracy is 76.5%. The result of the initial training of the model is relatively average, and we certainly aim for a higher value. In Chap. 5, we will discuss how to improve the accuracy of the model.

4.2.7 Model Interpretation

The problems in data analysis often require not only the models themselves but also interpretation of the models.

In many application scenarios, a model, no matter how good it is, won't be useful if it can't be explained effectively. A solid model with clear explanation methods will significantly boost its promotion and practical use.

We have already interpreted the linear regression model earlier; now let's look at the logistic regression model.

Click on the "Logistic Regression Learner" node, and select the "Coefficients and statistics" option from "Node Monitor" to view the model parameter results, as shown in Fig. 4.47.

Apart from "Coeff," we can see that there are some strange information appearing in the "Variable" column, such as "Pclass=2" and "Pclass=3." We know these variables represent the cabin class. But where is "Pclass=1"?

4.2.7.1 Dummy Variable

These unusual pieces of information are known as dummy variables. When handling categorical data like occupation and gender, they can't be directly quantified and need a specific method for quantification. This "quantization" is typically done by introducing "dummy variables." We can create variables based on the attribute types of these categorical factors, which only have values of "0" or "1."

For example, in this example, Cabin Class 1, 2, and 3 do not have any numerical significance. If we convert them to dummy variables, it means transforming "Pclass" into "Pclass=2" and "Pclass=3," and only one of these can have a value of 1. The presence of 1 indicates the class of that cabin. What if "Pclass=2" and "Pclass=3" are both 0? This indirectly tells us that "Pclass=1."

So there is one important issue to be aware of in this case. That is, we normally don't include "Pclass=1," "Pclass=2," and "Pclass=3" all together because they satisfy the condition "Pclass=1" + "Pclass=2" + "Pclass=3"=1, and it is required that either "Pclass=1," "Pclass=2," or "Pclass=3" can be equal to 1. This means that knowing two of them will reveal the value of the third one. Therefore, for this problem, KNIME provides dummy variables for "Pclass=2" and "Pclass=3" without including "Pclass=1."

Generally for linear models such as linear regression and logistic regression, if a feature has n possible values, the number of dummy variables corresponding to that feature is $n - 1$.

Knowing this, it is natural to understand other variables like "Sex=male."

4.2.7.2 Meaning of the Coefficients

Let's get to the point: what do the coefficients in logistic regression really mean? Unlike linear regression, the coefficients in logistic regression are not straightforward to interpret because of the transformation it undergoes through the Sigmoid function. In summary,

- A positive sign indicates a positive influence, while a negative sign indicates a negative influence.
- The larger the absolute value of the coefficient, the greater the impact.

Looking at Fig. 4.47, we can see that the coefficient for "Pclass=3" is negative, suggesting a lower probability of survival compared to "Pclass=1." Moving on to the coefficient for "Sex=male," it's a significantly negative value, indicating a higher mortality rate for male passengers compared to female passengers. In conclusion, examining model parameters can provide many valuable insights.

Points for Discussion

- Analyze your model based on the model coefficients.

4.3 Practice

1. Describe the impact of decision boundaries on model accuracy.
2. Describe the scorers that can be used for logistic regression.
3. How do we use KNIME data visualization to gain a deeper understanding of the data?
4. What is a dummy variable?
5. How do we interpret the coefficients of logistic regression?

Reference

https://www.deeplearning.ai/courses/machine-learning-specialization/

Chapter 5
Model Optimization

Life is like climbing a mountain, and finding a way out is a process of learning.

—Xi Murong

Abstract This chapter delves into essential machine learning concepts with a focus on the gradient descent algorithm, a key method for efficient model optimization. The chapter begins by explaining the loss function, illustrating its role in measuring model accuracy. It particularly discusses the mean squared error for linear regression and cross-entropy loss for binary logistic regression. The core concept of gradient descent is then explored, detailing how this algorithm assists in finding the minimum value of the loss function through iterative adjustments based on the learning rate and termination criteria. The chapter proceeds to discuss regularization techniques for preventing model overfitting, maintaining a balance between model complexity and performance. It concludes by evaluating model performance through various methods, including the confusion matrix, F1 score, and ROC curves. These tools are critical in assessing accuracy, precision, recall, and robustness, ensuring the development of reliable and efficient machine learning models.

Keywords Machine learning optimization · Gradient descent · Loss function · Regularization · Model evaluation · Overfitting · Underfitting · Confusion matrix · F1 score · ROC curve

This chapter covers

- **Loss function.**
- **Gradient descent.**
- **Model optimization.**
- **Overfitting and underfitting.**
- **Model scorers.**

In the field of machine learning, having a clear objective to work toward and implementing measures to optimize it are crucial. In this chapter, we'll delve into

© The Author(s), under exclusive license to Springer Nature Singapore Pte Ltd. 2024   133
Y. Geng et al., *Practical Machine Learning Illustrated with KNIME*,
https://doi.org/10.1007/978-981-97-3954-7_5

some key concepts in machine learning, with the most vital one being gradient descent.

5.1 Gradient Descent

During the model training process, there are numerous parameter settings to consider. The way we set these parameters quickly and effectively directly impacts the quality of the model. Having a solid grasp of the gradient descent algorithm can greatly assist us in efficiently debugging and improving models (https://www.deeplearning.ai/courses/machine-learning-specialization/).

5.1.1 Loss Function

Loss function, also known as cost function, was introduced earlier to measure the inconsistency between the predicted values of a model and the true value. Linear regression uses MSE as the loss function. Binary logistic regression uses "cross-entropy loss" instead.

$$L = -[y\log\hat{y} + (1 - y)\log(1 - \hat{y})]$$

Here, L is the loss, y is the true value, and $\hat{y}$ is the predicted value. For binary classification, the target can only be 0 or 1. What does this loss function mean? Why can it be used as a loss function?

We can consider it in different cases. When the true value is 0, the cross-entropy becomes

$$L = -\log(1 - \hat{y})$$

As introduced in Chap. 4, the predicted value $\hat{y}$ is a probability ranging [0,1], which makes the cross-entropy look like Fig. 5.1. To minimize this loss function, we need to have the predicted value as close to 0 as possible.

Similarly, when the true value is 1, the cross-entropy becomes

$$L = -\log\hat{y}$$

Thus, the cross-entropy will look like Fig. 5.2. To minimize this loss function, it is desirable for the predicted value to be as close to 1 as possible.

How can we minimize the cross-entropy loss function? Well, we employ the gradient descent algorithm to calculate the minimum value of the loss function.

The choice of the loss function depends on the specific problem at hand. In simpler terms, for regression, we can use MSE, while for classification, cross-

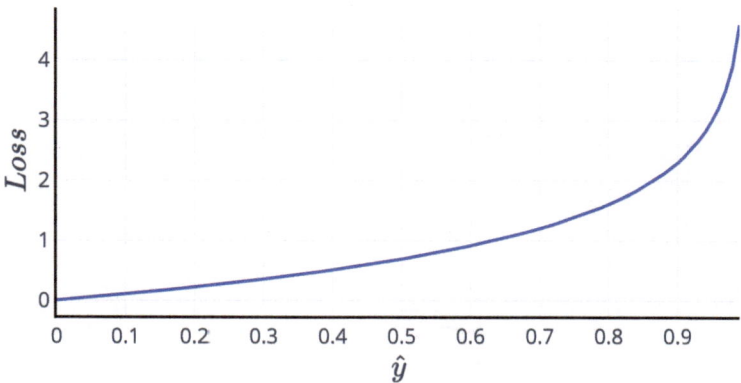

Fig. 5.1 Cross-entropy when the true value y is 0

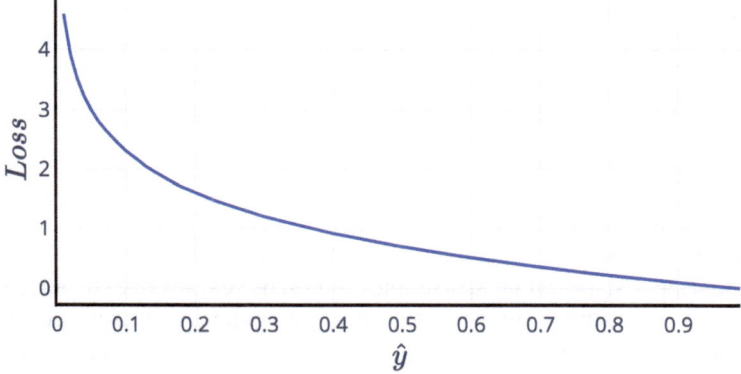

Fig. 5.2 Cross-entropy when the true value y is 1

entropy is a suitable option. Selecting the wrong loss function can hinder model optimization and may even lead to errors.

Points for Discussion

- Do you know any examples of failure in the history caused by choosing the wrong "loss function"?

5.1.1.1   Gradient

Gradients can be easily understood using contour lines. When we reach a specific point, the line that is perpendicular to the contour lines is known as the gradient line. The gradient itself is a scalar value. Take a look at Fig. 5.3, where you can see concentric contour lines surrounding the highest peak on the topographic map of Mount Everest. Starting from the peak and moving downward, the line we draw

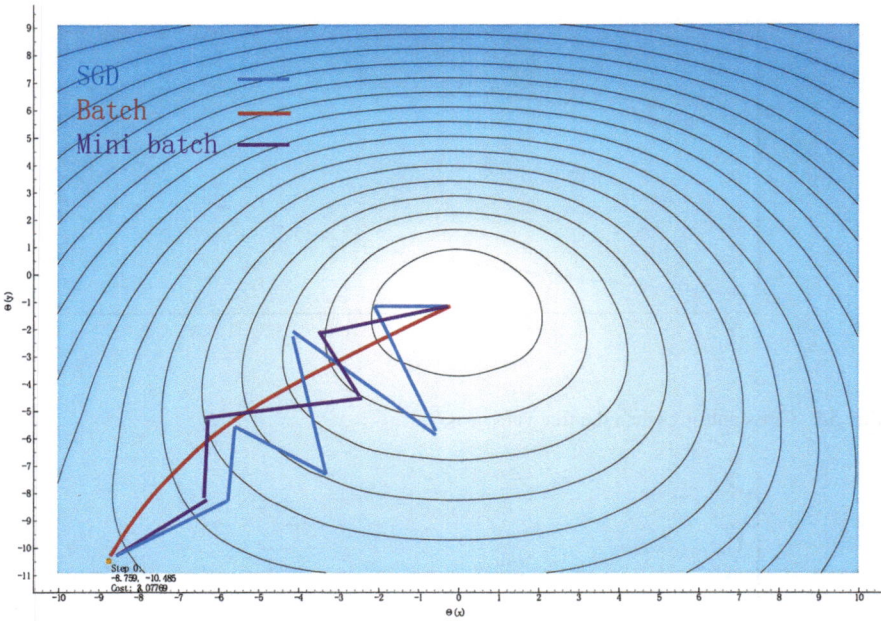

Fig. 5.3 Gradient lines of Mount Everest

perpendicular to the contour line represents the gradient line. Dense contour lines represent steeper slope, while sparse lines represent gentle slope. If our goal is to descend, we will naturally move faster along the gradient line. If we consider this map as our loss function, our objective is to determine the gradient and then follow the gradient line to minimize the loss function.

5.1.1.2   Gradient Descent

How can we identify the lowest point? The gradient descent algorithm is like taking a walk downhill along the gradient—just as shown in Fig. 5.4, where we can effortlessly follow the gradient line from the mountain's peak at 6200 m altitude to a position at 5500 m altitude.

How do we know which direction to go? We can use this equation:

The next position = current position − learning rate × gradient.

Here, the learning rate is a positive number, a hyperparameter often set by our experience. To find the minimum value, this equation is repeatedly computed, constantly updating and iterating to the next position.

Figure 5.5 illustrates the diagram of the gradient descent algorithm. This algorithm is employed to optimize a model parameter, with the gradient being negative on the left and positive on the right. When the initial position is on the right, the updated value is obtained by subtracting a positive number from the current value,

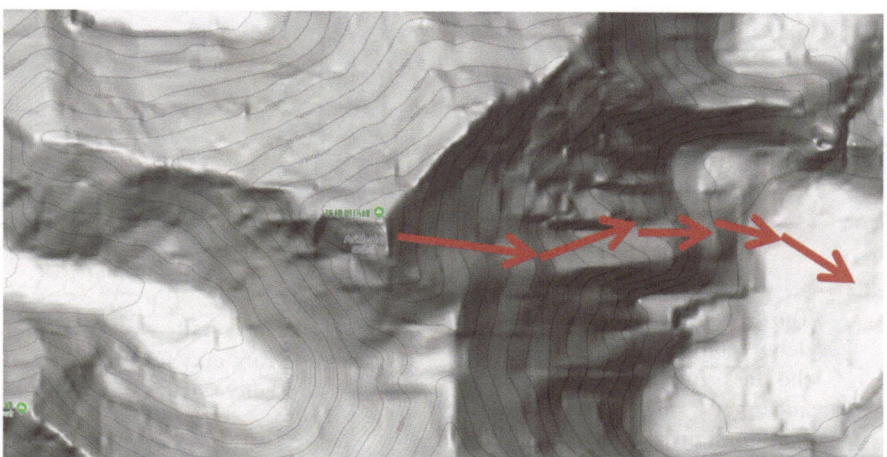

Fig. 5.4 Gradient descent

Fig. 5.5 Illustration of the gradient descent algorithm

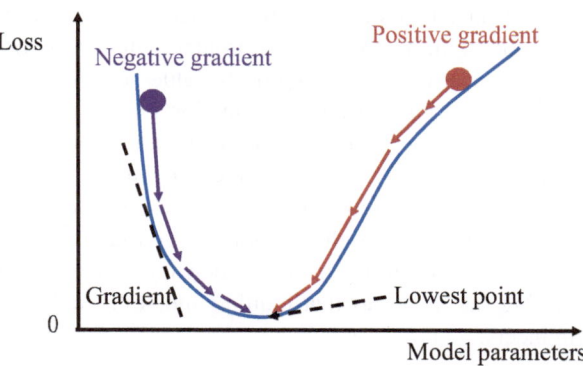

causing the new value to shift toward the left of the curve. Consequently, the next position will be lower than the current position. Similarly, when the initial position is on the left, the updated value is obtained by subtracting a negative number, resulting in a shift toward the right. Again, the next position will be lower than the current position. In ideal conditions, each subsequent position will continuously decrease. By iteratively calculating the next position, we gradually approach the lowest point. Each iteration in this process is referred to as an epoch.

5.1.1.3   Learning Rate

There is an issue of learning rate when using gradient descent algorithm, as either too large or too small is not good. You can observe the illustration of the gradient descent algorithm (see Fig. 5.5) and think about why.

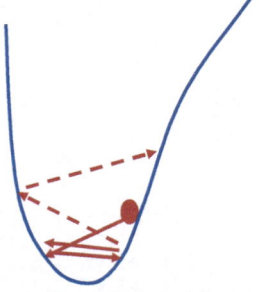 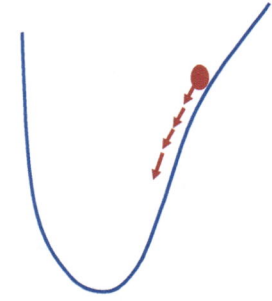

Too large learning rate can deter convergence and cause the loss function to fluctuate and get stuck in a local minimum or even to diverge

A too small learning rate may lead to slow convergence

Fig. 5.6 A too large or too small learning rate

Let's take a look at Fig. 5.6 to better understand learning rate. If the learning rate is set too high, it can lead to a problem where the optimization steps taken are too big. This can cause the parameter values to fluctuate around the minimum point, resulting in a jitter of the loss function gradient and preventing it from decreasing effectively. In fact, with an excessively large learning rate, it is even possible for the gradient to jitter in an upward direction. On the other hand, if the learning rate is set too small, each optimization step becomes overly cautious. While the absolute value of the loss function gradient continues to decrease, the progress made with each step is too small, resulting in low efficiency. Therefore, determining the appropriate learning rate is an art that requires the skills and experience of machine learning engineers.

5.1.1.4 Termination Criterion

How can we determine when to stop the iteration of the gradient descent algorithm? We have two criteria as follows, and satisfying either one is sufficient to terminate the algorithm:

- When the loss function is lower than the threshold (ϵ).
- When the maximum number of epochs has been reached.

5.1.1.5 Is the Gradient Descent Algorithm Functioning Properly?

We've covered learning rate and termination criteria. So, how do we actually apply this knowledge? Well, one of the most practical applications is to assess the performance of the gradient descent algorithm. We can achieve this by plotting the

Fig. 5.7 Normal learning rate

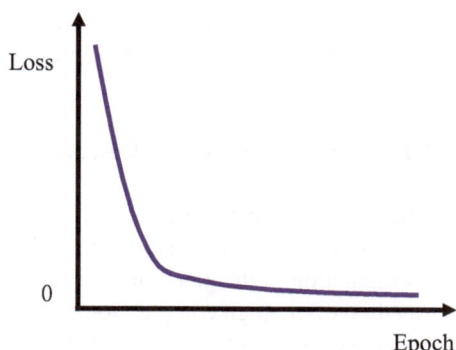

Fig. 5.8 Malfunctioning learning rate

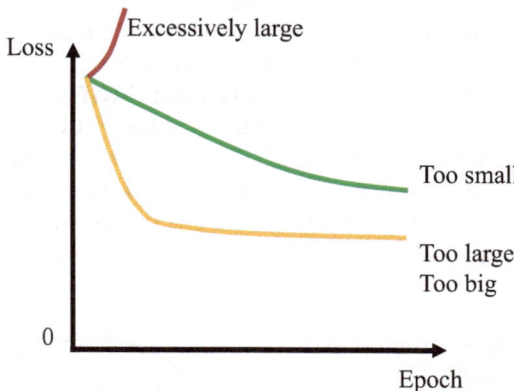

changes in the loss function for each epoch. This visual representation allows us to easily track the algorithm's progress and ensure it is functioning as intended.

In a normal scenario, the loss should decrease continuously with the number of epochs and eventually stabilize around a certain value, as shown in Fig. 5.7.

Why does the loss decrease rapidly at first and then gradually slow down until it reaches a stable point? If we look at Fig. 5.5, we may notice that a larger gradient corresponds to a steeper slope of the hill. Consequently, when the gradient is larger, the step size for each iteration is also larger, resulting in more significant progress in gradient descent. However, as the gradient continues to decrease, the slope becomes flatter, causing the gradient to become smaller. And step size within each epoch decreases accordingly, leading to a more stable descent, as depicted in Fig. 5.7.

How does an unstable loss function behave with respect to epochs? As shown in Fig. 5.8 (we can also look at Fig. 5.6), when the learning rate is set too small, the loss consistently decreases but fails to stabilize. This aligns with the issue of slow convergence, as it continuously struggles to approach the vicinity of the minimum point. Conversely, if the learning rate is set too large, the loss rapidly plummets to a relatively large value and then ceases to change, indicating oscillations in the gradient of the loss function. If the loss demonstrates an upward trend instead of a downward one, it indicates that the learning rate is set excessively large, resulting in each gradient update step moving in the opposite direction of the desired descent.

5.1.1.6   Types of Gradient Descent Algorithms

1. **Batch Gradient Descent.**
   In this approach, known as batch gradient descent, all samples are involved in the computation of the gradient, and all parameters are updated simultaneously. This method offers high computational efficiency but demands a significant amount of memory, as illustrated in Fig. 5.9.
2. **Stochastic Gradient Descent (SGD).**
   In this method, samples participate in the calculation of the gradient one by one, which requires significant computational resources, and the calculations are not very stable, as shown in Fig. 5.10.
3. **Mini-Batch Gradient Descent (MGD).**
   This method combines the above two algorithms, dividing the samples into several batches, participating in gradient calculation one by one, thereby achieving a combination of stability and speed, as shown in Fig. 5.11.
   When comparing these algorithms, we can observe distinct differences in their paths, as depicted in Fig. 5.12. The concentric circles represent contour lines, and

Fig. 5.9 Batch gradient descent. (*This figure has been designed using images from* Flaticon.com)

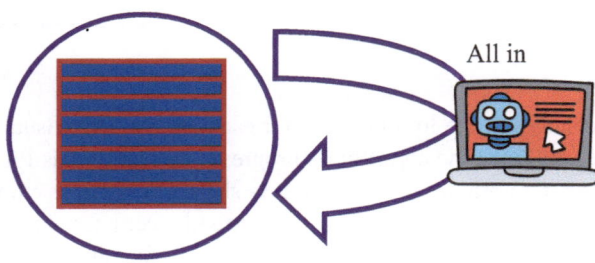

Fig. 5.10 Stochastic gradient descent. (*This figure has been designed using images from* Flaticon. com)

Fig. 5.11 Mini-batch gradient descent. (*This figure has been designed using images from* Flaticon. com)

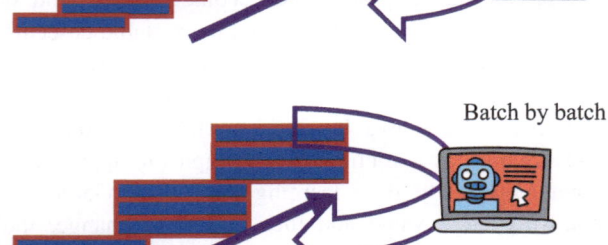

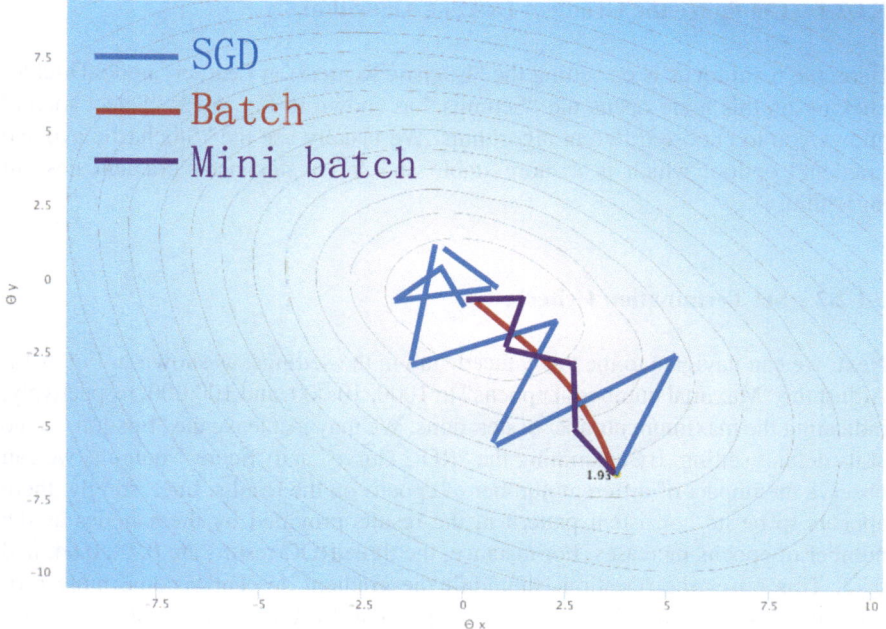

Fig. 5.12 Algorithm comparison

the ultimate objective is to locate the central point surrounded by these contour lines. In batch gradient descent, all parameters are simultaneously involved in the computation, resulting in a consistent direction of the gradient and a smoother curve. Notably, the curve's progression is perpendicular to the contour lines. On the other hand, the SGD algorithm calculates the gradient using each individual sample, which introduces randomness and prevents a guaranteed decrease in the gradient with each computation. Consequently, the resulting curves exhibit erratic behavior. Mini-batch gradient descent combines elements of both methods, making a compromise. Although some jitter still occurs, there is a more pronounced tendency to move in a specific direction. From the graph, it is apparent that the performance of mini-batch gradient descent lies between that of batch gradient descent and SGD. It strikes a balance between the smoothness of batch gradient descent and the adaptability of SGD.

5.1.2 Optimize the Model with KNIME

We'll continue with the example of the Titanic and optimize the model using the algorithms discussed above.

5.1.2.1 Configure the Gradient Descent Algorithm

Here, the main focus is on setting the "Logistic Regression Learner" node. Double-clicking on this node opens the "Settings," as shown in Fig. 5.13. "Select solver" allows you to choose different algorithms. We usually use the "Stochastic average gradient" option, which is a more stable and faster stochastic gradient descent algorithm.

5.1.2.2 Set Termination Criteria

Next, we can navigate to the "Advanced" tab in the settings, as shown in Fig. 5.14. Adjust the "Maximal number of epochs" to 1000, 10,000, and 100,000, respectively, indicating the maximum number of iterations. We may just leave the "Epsilon" value at its default setting. By examining the "ROC Curve" and "Scorer" outputs, we can observe the impact of different number of epochs on the results. Interestingly, there appears to be no consistent pattern in the results provided by these nodes as the number of epochs increases. For instance, the three ROC results are 0.78, 0.64, and 0.73. This raises the question: shouldn't the gradient descent become more pronounced with more training epochs? Is it possible for the gradient not to decrease after several epochs?

Do you notice any hints from Fig. 5.8? Could it be because of the learning rate?

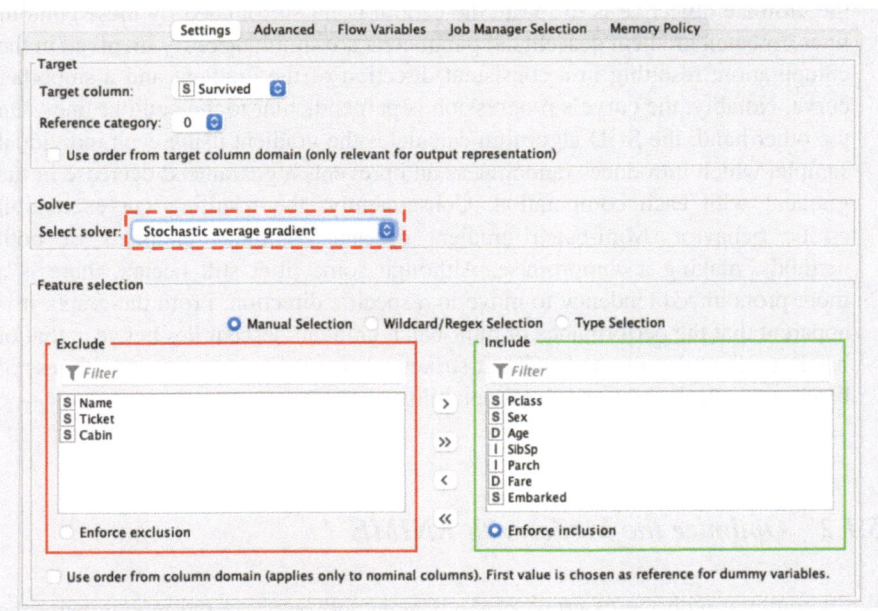

Fig. 5.13 Choosing optimization algorithm

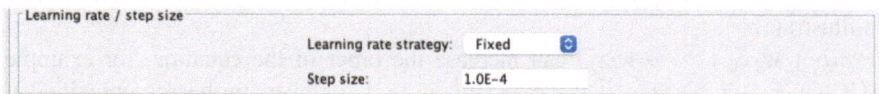

Fig. 5.14 Setting details for the algorithm

Fig. 5.15 Reduce the learning rate

5.1.2.3   Set the Learning Rate

Stay on the "Advanced" tab and set the "Learning rate strategy" to "Fixed" and the "Step size" to 0.1. You can try different learning rates to see what impact it has.

Since we observed that the gradient did not consistently decrease in the previous step, we hypothesized that the learning rate might have been set too high. To address this, let's proceed to Fig. 5.15 and experiment with reducing the learning rate to 0.01, 0.001, and 0.0001.

Upon observation, it becomes apparent that, with the exception of the smallest learning rate of 0.0001, the other learning rates do not lead to an increase in the ROC value. As a result, it can be tentatively concluded that setting the learning rate to 0.0001 yields more favorable results, with ROC values stabilizing at 0.84, 0.85, and 0.85.

Other parameters can be kept default. If you are interested in their meanings, you're encouraged to check them out on your own.

5.2   Regularization

Regularization is an important method to prevent model overfitting. In this part, we will introduce why regularization is useful and how to apply regularization.

5.2.1   Accuracy and Robustness

We have already discussed the assessment of accuracy, but what is robustness? Robustness, in the context of computer systems, refers to the system's capacity to maintain normal operation even when confronted with errors and exceptions during execution.

5.2.2   Complex Models

We already know that linear regression can be understood as $\hat{y} = b + w_1x_1 + \cdots + w_nx_n$, and logistic regression can be understood as $\hat{y} = \sigma(z) = \sigma(b + w_1x_1 + \cdots + w_nx_n)$. Both of these models can be further complexified to improve predictive ability. We'll take logistic regression as an example to illustrate.

$\sigma(b + w_1x_1 + \cdots + w_nx_n)$ can increase the order of the equation, for example, $\sigma(b + w_1x_1 + \cdots + w_nx_n + w_{11}x_1^2 + \cdots + w_{nn}x_n^2)$, or even further complexified to $\sigma(b + w_1x_1 + \cdots + w_nx_n + w_1x_1^2 + \cdots + w_nx_n^2 + w_{12}x_1x_2 + \cdots + w_{1n}x_1x_n + \cdots w_{23}x_2x_3 + \cdots + w_{2n}x_2x_n)$. You might have been intimidated by the lengthy string of formulas. Don't worry, as all you need to know is that this equation is highly intricate. However, can this equation be further intensified, and does greater complexity equate to better outcomes? You may already doubt: can we truly believe in something we fail to understand?

Research has shown that simple models tend to underfit, while complex models tend to overfit. In order to improve the quality of the model, we often tend to build a complex model, but it may actually decrease the model's quality.

5.2.3   Underfitting and Overfitting

As mentioned earlier, overfitting occurs when a model perfectly or very accurately fits one portion of the dataset but fails to predict other parts effectively. On the contrary, underfitting refers to a situation where the model exhibits a low degree of fitting, resulting in a poor match between the data and the fitting curve.

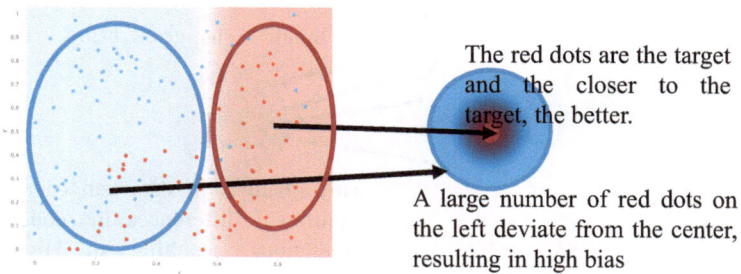

The red dots are the target and the closer to the target, the better.

A large number of red dots on the left deviate from the center, resulting in high bias

Fig. 5.16 Underfitting

5.2.3.1 Underfitting

As shown in Fig. 5.16, the left side represents the data graph of the task, while the right side illustrates the mapping of its results. Let's suppose our objective is to distinguish the red and blue dots in the left part, accurately identifying the red dots as the target. The ideal model would map the red dots in the left to the bullseye on the right. In the left part, the white line that separates the red and blue areas is referred to as the decision boundary. It is evident that all points on the right side of the decision boundary are mapped to the bullseye, while those on the left side are far from it. However, there are numerous misclassifications on both sides of the decision boundary. Specifically, for the red points of interest, a significant number of red points on the left side deviate from the center, resulting in a substantial bias in the model. In this case, the predicted results of the training data exhibit a considerable bias, indicating underfitting, also known as "high bias."

It can also be observed that the points in the left part can vary in a fairly large range without changing their mapping positions on the right. This indicates that the model is not sensitive to data changes and is relatively robust.

5.2.3.2 Overfitting

The decision boundary becomes very complex, as seen in Fig. 5.17, just like in the previous example of underfitting. On the right side, the points inside the green circle on the left may correspond to a large range in the mapping region on the right, leading to significant differences in the model's predictions due to small changes in the data, indicating that the model is not robust enough. This indicates that the data has high variance. Therefore, overfitting is also called "high variance."

5.2.3.3 Underfitting and Overfitting on Loss Function

From the loss function depicted in Fig. 5.18, it is evident that underfitting occurs when the model is excessively simplistic or inadequately trained, leading to

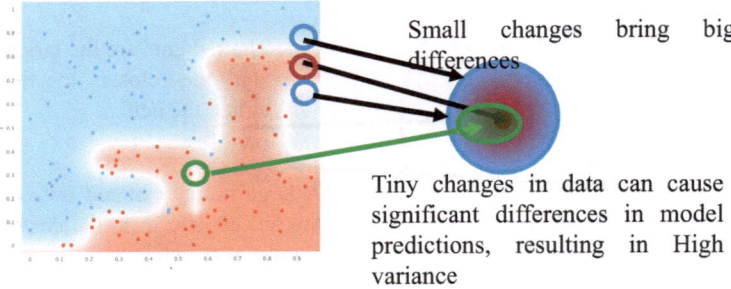

Small changes bring big differences

Tiny changes in data can cause significant differences in model predictions, resulting in High variance

Fig. 5.17 Overfitting

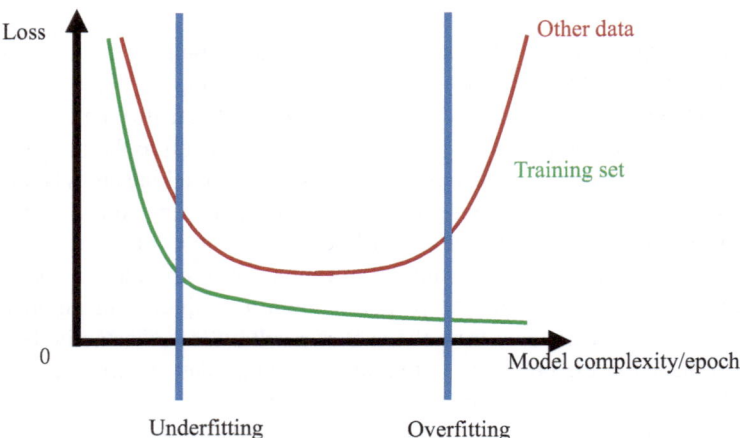

Fig. 5.18 Overfitting and underfitting on the loss function

insufficient decrease in the loss function of the training data. On the other hand, overfitting transpires when the model is overly complex or excessively trained, resulting in a small loss function for the training data but an increased loss function for other data.

5.2.3.4 Underfitting, Overfitting, and Model Complexity

As the model parameters increase, the complexity of the model gradually increases, transitioning from underfitting to fitting and eventually to overfitting. When the model is simple, the loss function for all classes tends to be relatively large. As the model becomes more complex, the loss decreases, as depicted in Fig. 5.19. However, once the model becomes excessively complex, although the loss of the training data may continue to decrease, the loss of other data will start to increase.

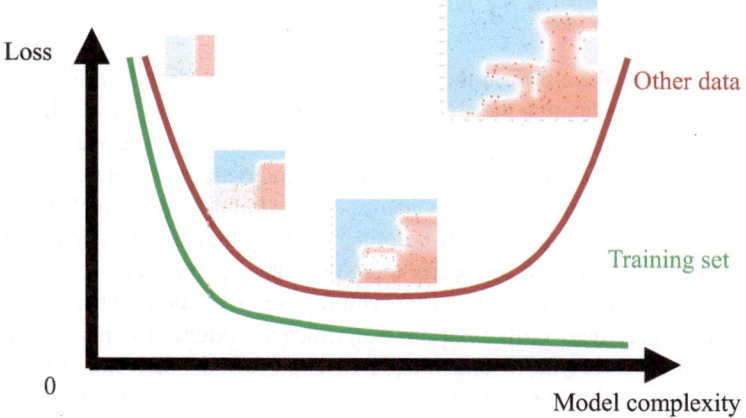

Fig. 5.19 Underfitting, overfitting, and model complexity

5.2.4 *Regularization to Prevent Overfitting*

Increasing the complexity of the model enhances its capability, which is our goal. However, it also brings about the side effect of overfitting. If one chooses to abandon complex models to prevent overfitting, it's just like throwing out the baby with the bathwater. So, what should we do in such a scenario? The solution lies in employing regularization techniques to prevent overfitting while still utilizing complex models.

Now let's take another look at this scary formula $\sigma(b + w_1 x_1 + \cdots + w_n x_n + w_1 x_1^2 + \cdots + w_n x_n^2 + w_{12} x_1 x_2 + \cdots + w_{1n} x_1 x_n + \cdots + w_{23} x_2 x_3 + \cdots + w_{2n} x_2 x_n)$. It's hard to believe in this formula because if any part of it is incorrect, it could lead to significant deviation. What if each part of this formula has little impact? In other words, can we reduce the weights of each part in this intimidating formula? Just like domesticating wolves into dogs, can we tame this wild formula with any method?

Yes, we can use regularization.

The loss function of logistic regression $L = -[y \log \widehat{y} + (1 - y) \log(1 - \widehat{y})]$ can be modified as

$$L = -[y \log \widehat{y} + (1 - y) \log(1 - \widehat{y})] + \lambda \sum w^2$$

Here, λ represents the regularization parameter. Since the objective of model optimization is to minimize the loss function, a large λ value will lead to a significant reduction in the weights (w) to minimize the loss function as much as possible. Consequently, the impact of each feature, including both low-order and high-order features, is diminished due to the reduced coefficients. This can be seen as a way to restrain the unconstrained power of order and transform the model from a wolf into a well-behaved dog.

5.2.5   Regularization by KNIME

In Fig. 5.20, let's take a look at the settings for "Regularization," which is currently set to "Uniform," indicating the absence of any applied regularization. The other two options available are Gaussian and Laplace regularization. While we don't need to delve into the specific workings of these two regularization techniques, the fundamental concept remains the same as mentioned earlier. In the current model, since we haven't incorporated any higher-order parameters, we can confidently opt for the "Uniform" and do not need to specify a regularization parameter.

However, the decision of whether to employ regularization and the selection of regularization parameters often require a significant amount of experience and trial and error. Therefore, you're encouraged to explore the other two regularization methods to see if they yield better results. Through my experiments, I find that utilizing both regularization techniques can enhance the performance of the model slightly.

5.3   Score the Model

5.3.1   Confusion Matrix

First familiarize yourself with these acronyms:

- TP = True Positive.
- FP = False Positive.
- FN = False Negative.
- TN = True Negative.

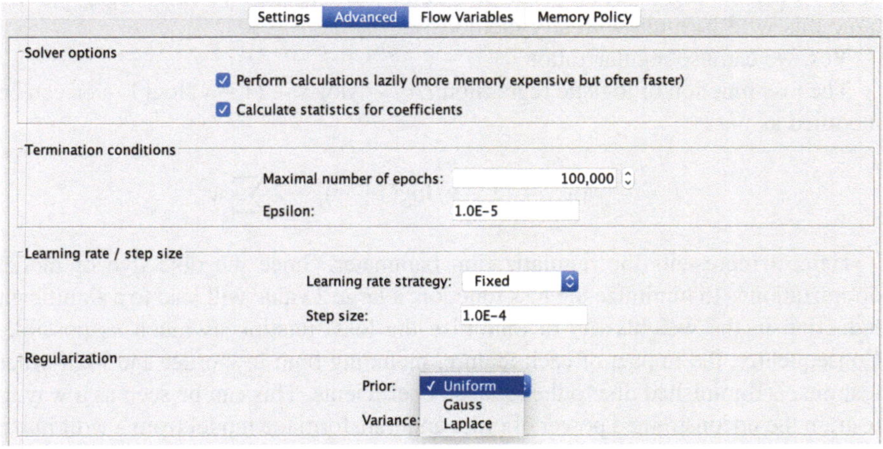

Fig. 5.20 Regularization parameter settings

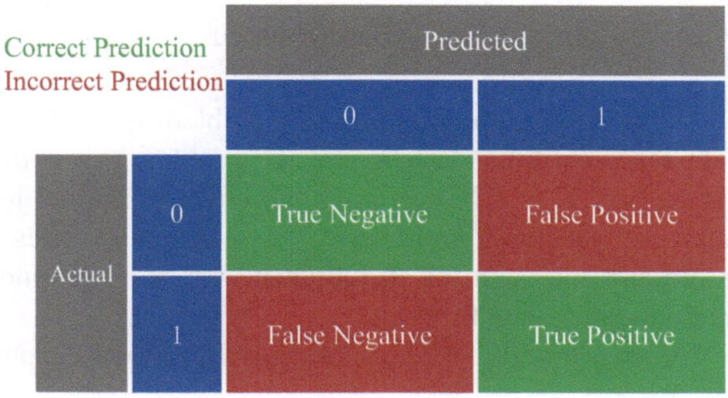

Fig. 5.21 Confusion matrix

The confusion matrix, shown in Fig. 5.21, provides a visual representation of predicted values (in columns) and actual values (in rows). In this matrix, a predicted value of 1 indicates a positive outcome, while a predicted value of 0 represents a negative outcome. When the prediction matches the actual value, it is considered as "True"; otherwise, it is labeled as "False." For example, TP (True Positive) signifies that the prediction is true and aligns with the actual positive value. On the other hand, FP (False Positive) occurs when the prediction is true, but the actual value is negative.

Let's consider the example of product detection, where we define product issues as positive:

- TP (True Positive) = the problem is detected, and it is indeed a real issue that needs to be addressed.
- FP (False Positive) = the problem is detected, but in reality, there was no problem, causing a false alarm.
- FN (False Negative) = the problem goes undetected, posing a latent risk despite its actual existence.
- TN (True Negative) = no problem is detected, and there was actually no problem, resulting in a satisfactory outcome.

In a better-performing model, TP and TN should be larger, while the occurrences of FP and FN should be minimized. If we consider the confusion matrix as our objective, our focus should be on maximizing TP and TN.

Next, let's familiarize ourselves with several commonly used terms closely related to the confusion matrix.

5.3.1.1 Accuracy

Accuracy is the proportion of correctly classified data compared to the total data, and the formula is

$$\text{Accuracy} = \frac{TP + TN}{ALL}$$

TP represents the detection of a problem when there is indeed a problem, while TN represents the detection of no problem when there is indeed no problem. Therefore, TP + TN represents the count of correct judgments, where the model correctly identifies both positive and negative cases. On the other hand, ALL represents the count of all judgments made by the model, including both correct and incorrect classifications.

As shown in Fig. 5.22, accuracy is the ratio of the purple area to the yellow area.

5.3.1.2   Precision

The formula for calculating precision is as follows:

$$\text{Precision} = \frac{TP}{TP + FP}$$

The precision metric aims to address the following question: what proportion of the products flagged as problematic are actually problematic?

As shown in Fig. 5.23, precision is the ratio of the purple area to the yellow area.

5.3.1.3   Recall

Recall can be calculated with the following formula:

Fig. 5.22 Accuracy

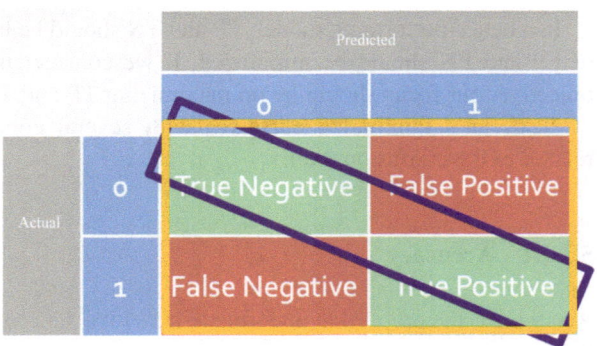

Fig. 5.23 Precision

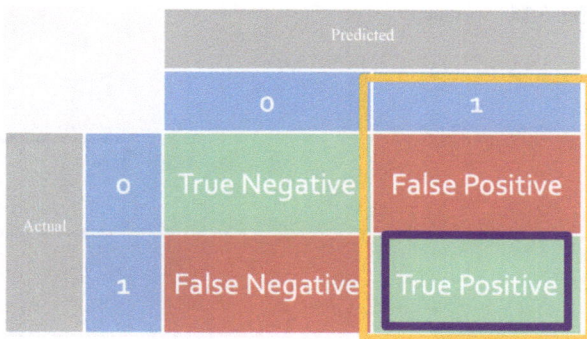

Fig. 5.24 Recall

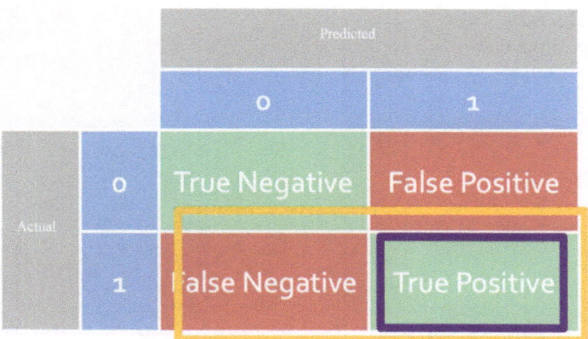

$$\text{Recall} = \frac{\text{TP}}{\text{TP} + \text{FN}}$$

Recall refers to the TP rate, which represents the proportion of problematic products that need to be recalled for repair.

As shown in Fig. 5.24, recall is the ratio of the purple area to the yellow area.

5.3.1.4 Specificity

The calculation formula for specificity is

$$\text{Specificity} = \frac{\text{TN}}{\text{TN} + \text{FP}}$$

Specificity is the TN rate, which refers to the proportion of products without real problems that are also considered to have no problems.

As shown in Fig. 5.25, specificity is the ratio of the purple area to the yellow area.

Fig. 5.25 Specificity

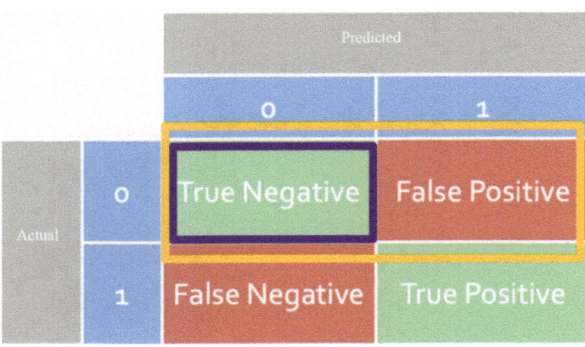

Fig. 5.26 False
positive rate

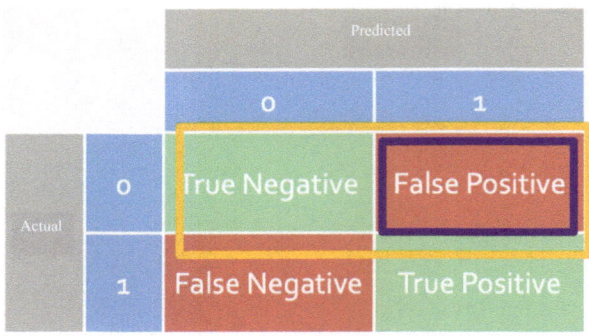

5.3.1.5   FP Rate

The calculation formula for FP rate is

$$FPR = \frac{FP}{TN + FP}$$

FPR is the proportion of products without real problems that are considered to
have problems.

As shown in Fig. 5.26, FPR is the ratio of the purple area to the yellow area.

5.3.2   F1 Score

We often aim to improve both precision and recall simultaneously, but this can be
challenging and usually involves a trade-off. To achieve this, we can adjust the
position of the decision boundary. In Fig. 5.27, which illustrates a simple example of
the Titanic case, the purple line represents the decision boundary. On the right side of
the line, the model predicts death, while on the left side, it predicts survival.

Fig. 5.27 Adjusting the decision boundary

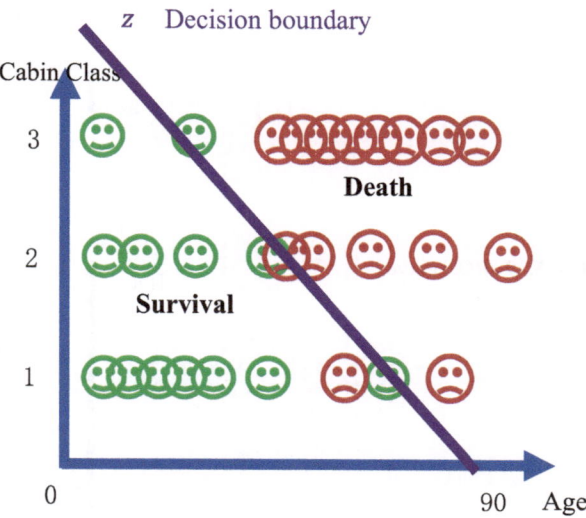

Moving the decision boundary to the left would result in an increase in both predicted and actual survival examples on the left side of the boundary, thereby improving accuracy. However, this would also lead to a large number of actual survival on the right side of the boundary being incorrectly predicted as death, reducing recall. Similarly, moving the decision boundary to the right would result in the left boundary including all the actual surviving people, increasing recall. However, this would also lead to a large number of actual deaths on the left side of the boundary being incorrectly predicted as survival, reducing accuracy.

Here, accuracy represents the proportion of those predicted to survive who actually survived.

Recall represents the proportion of actual survivors who were correctly identified.

How should we evaluate the quality of the model?

Here, we need to introduce the F1 score, which is the harmonic mean of precision and recall:

$$\frac{1}{F1} = \frac{1}{2}\left(\frac{1}{Precision} + \frac{1}{Recall}\right)$$

When training and comparing models, F1 score can be used as a standard to determine the quality of the models.

You can refer to Fig. 5.28 to understand the F1 score. Recall and precision can be imagined as two resistors, and then these two resistors are connected in parallel, and the F1 score is twice the resistance value of this parallel connection.

Fig. 5.28 F1 score

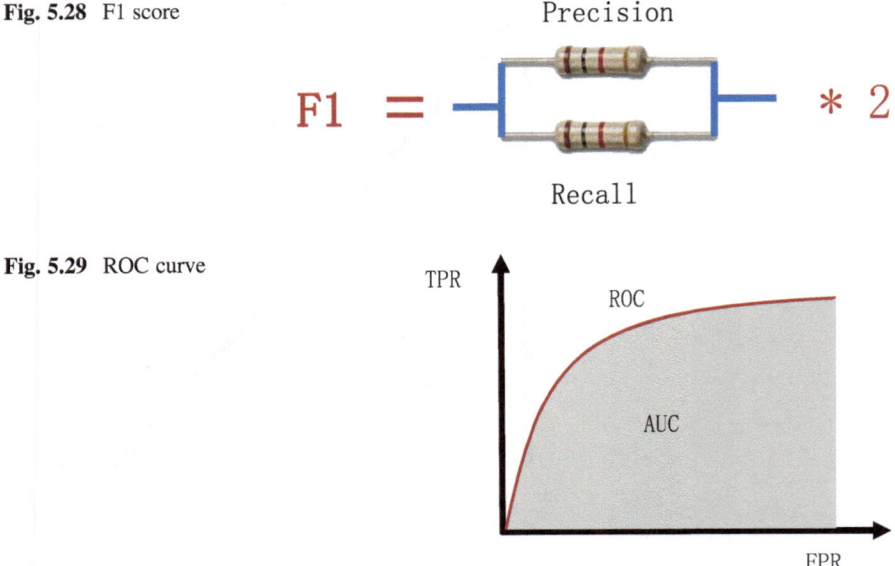

Fig. 5.29 ROC curve

5.3.3   ROC Curve and AUC

There is another evaluation criterion called the ROC curve, which stands for receiver operating characteristic curve. It is a graphical representation that shows how well a classification model performs at different classification thresholds. The performance comparison can be done by calculating the area under the curve (AUC) (Fig. 5.29).

The ROC curve was initially developed by electronic engineers and radar engineers during the World War II to detect enemy vehicles such as airplanes and ships on the battlefield. They needed a way to distinguish between real enemy aircraft and noise points on the radar. Later on, the ROC curve was adopted in psychology for perceptual detection of signals. Over the decades, ROC analysis has been widely used in various fields including medicine, radio, biology, and criminal psychology. More recently, it has made significant advancements in the field of machine learning.

Suppose the predicted data of the model has a probability density distribution (the sum of the AUC is 1) as shown in Fig. 5.30.

To determine whether a problem exists or not, it is necessary to establish a threshold as depicted in Fig. 5.31. The left side of the threshold is considered to be non-problematic, while the right side indicates the presence of a problem. The figure illustrates the range of TN, FN, TP, and FP. In the previous example of product detection, it may be more appropriate for us to refer to TP as recall (or sensitivity in the case of radar).

Now we can try to move the threshold line to the left. It will result in an increase in the area of TP (sens) and a decrease in the area of TN (spec). If moved to the right, the area of TP (sens) would decrease while the area of TN (spec) would increase. In

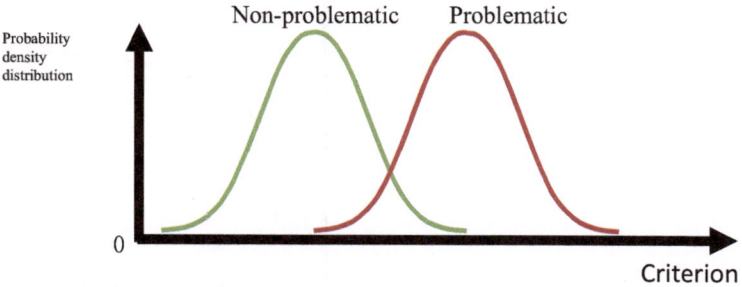

Fig. 5.30 Probability density distribution of predicted data

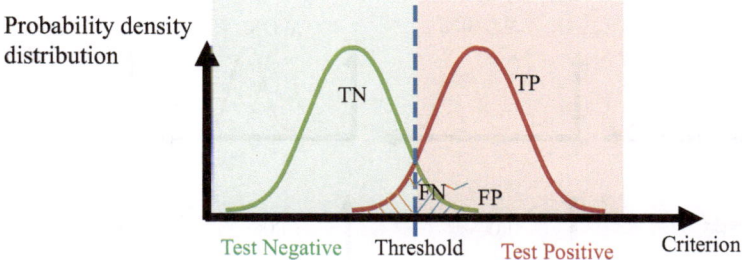

Fig. 5.31 Threshold and testing

Fig. 5.32 Relationship between sens and spec

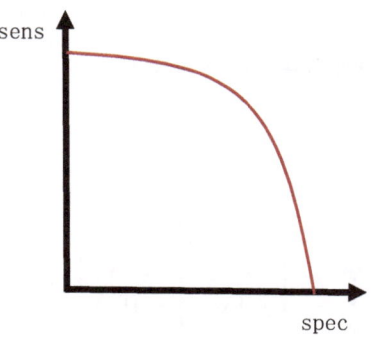

other words, when either of sens or spec increases, the other one will decrease, as shown in Fig. 5.32.

As shown in Fig. 5.33, Fig. 5.32 can easily be converted into a different representation as ROC curve.

As depicted in Fig. 5.35, the closer the area under the ROC curve is to 1, the better the model performs. AUC greater than 0.9 indicates a well-separated classification, where TP and TN are effectively distinguished. On the other hand, if it is impossible to distinguish between the two classes, the AUC value will be 0.5, indicating that the model can only rely on random guessing (Fig. 5.34).

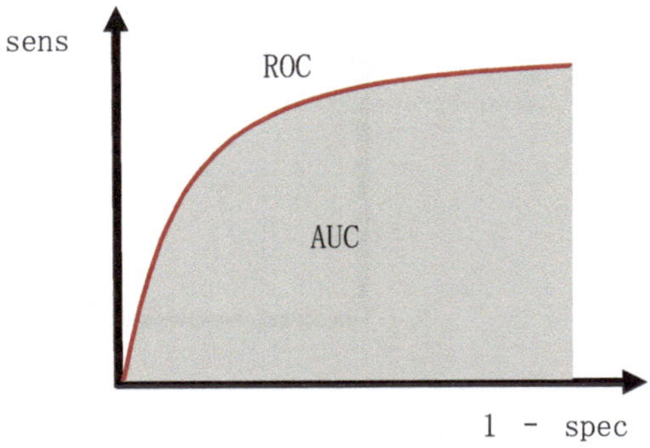

Fig. 5.33 ROC

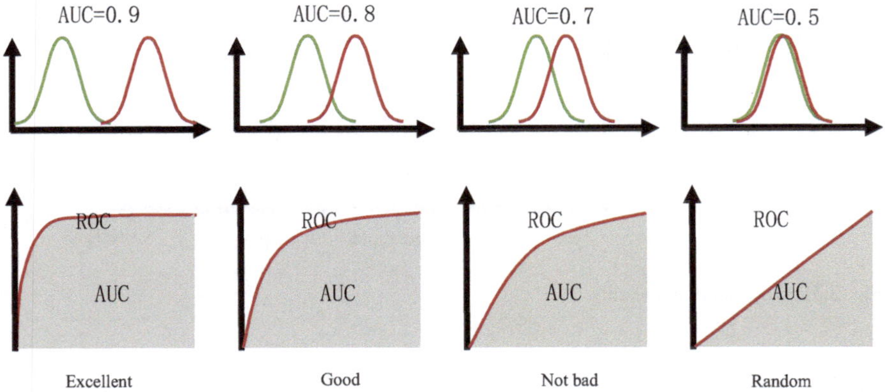

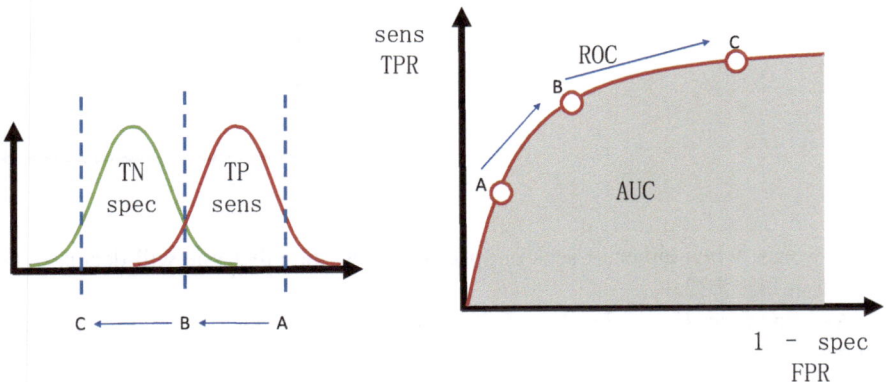

Fig. 5.34 Quality of ROC

Fig. 5.35 Blood test

For instance, in Fig. 5.35, the concentration of blood protein in the human body follows a continuous variable that conforms to a normal distribution. The distribution of blood protein concentration in patients is represented by the red curve, with an average value of A (g/dL), while the distribution in healthy individuals is represented by the blue curve, with an average value of C (g/dL). During medical checkups, the concentration of a specific protein in a blood sample is measured, and if it reaches or exceeds a certain value (threshold), it is diagnosed as a sign of disease. Researchers can adjust the threshold (by moving the vertical line B in the figure to the left or right) to get different FPR or TPR, thereby obtaining varying levels of prediction accuracy.

5.4 Practice

1. Describe the role of learning rate.
2. Which is more likely to occur, underfitting or overfitting? Why?

Reference

https://www.deeplearning.ai/courses/machine-learning-specialization/

Chapter 6
Support Vector Machine

I call it (SVM) the widest street approach.

—*Patrick Henry Winston*

Abstract This chapter provides a comprehensive overview of support vector machines (SVM), a critical algorithm in classification and regression analysis. It begins with a basic introduction to SVM, including its concept, application in binary classification, and the significance of support vectors. The chapter then delves into the comparison of SVM with logistic regression, focusing on their differing sensitivities to data through their respective loss functions. A substantial portion of the chapter is dedicated to explaining the role of the kernel in SVM, particularly the radial basis function (RBF) kernel, and its capability to transform data for effective classification. The chapter also discusses the tuning of hyperparameters in SVM, emphasizing the importance of the C parameter and the influence of the RBF kernel's gamma parameter. Practical applications of SVM are explored through a case study on breast cancer diagnosis and the Titanic Kaggle dataset, highlighting data preprocessing techniques like one-hot encoding, normalization, and model scoring methods.

Keywords Support vector machine · Kernel · Hyperparameter tuning · RBF kernel · One-hot encoding

This chapter covers

- **Concept of support vector machine.**
- **Kernel.**
- **Model tuning.**
- **One-hot encoding.**

In the opening of "The Three-Body Problem: Death's End" by author Liu Cixin, there is a description of a story where an ordinary woman divides three-dimensional space in a higher-dimensional space. She struggles to find the right word, saying, "Those (closed) places... I see them... open." This concept bears some resemblance to support vector machine (SVM). In this chapter, building upon our previous knowledge and skills, we will delve into the SVM and learn to apply SVM with KNIME.

6.1   Basic Concepts of SVM

6.1.1   *What Is SVM?*

Support vector machine (SVM) is an algorithm widely used for classification and regression analysis. It is particularly useful when dealing with labeled training samples that belong to one of two categories. SVM creates a model that can assign new samples to one of these categories, making it a non-probabilistic binary linear classifier. Another key advantage of SVM is its ability to perform nonlinear classification effectively. It achieves this through the use of the "kernel trick," which implicitly maps the input data to a higher-dimensional feature space.

For SVM, data points are viewed as n-dimensional vectors, and we want to know if it is possible to separate these points with a $n - 1$-dimensional hyperplane. This is what is called a linear classifier. In three-dimensional space, $n = 3$, $n - 1 = 2$, and the so-called "hyperplane" is simply the two-dimensional plane that we are familiar with. Similarly in two-dimensional space, the hyperplane will actually be a one-dimensional line.

When classifying, there are often multiple ways to classify the data, as shown in Fig. 6.1a, we may have multiple different lines that can classify different data. SVM will find an optimal hyperplane, as shown in Fig. 6.1b, to classify the data and maximize the distance between the two classes. Or to put it more simply, SVM helps us find a street as a boundary line, and the wider the street, the better.

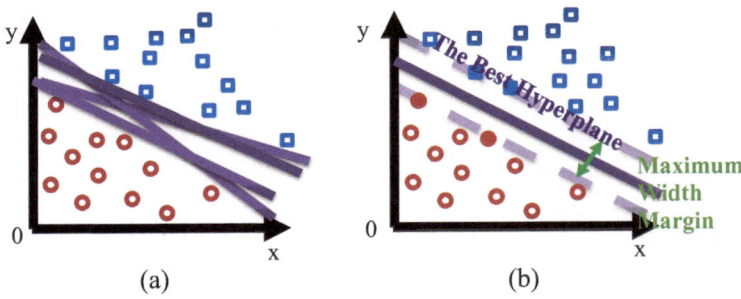

Fig. 6.1 SVM classifier. (**a**) Several lines can classify different data. (**b**) Find a street as the boundary; the wider the street, the better

6.1.2 What Are Support Vectors?

Support vectors are data points that are closer to the hyperplane and influence the position and orientation of the hyperplane. But how did the point become a vector? It is because each point can be seen as a vector pointing from the origin to that point (Fig. 6.2). SVM algorithms use these support vectors as support points to make the margin as wide as possible.

6.1.3 Comparison Between Logistic Regression and SVM

We can understand the difference in data sensitivity between logistic regression and SVM by examining their loss functions. Logistic regression uses the logistic loss function, while SVM uses the hinge loss function, as shown in Fig. 6.3.

Fig. 6.2 What are support vectors

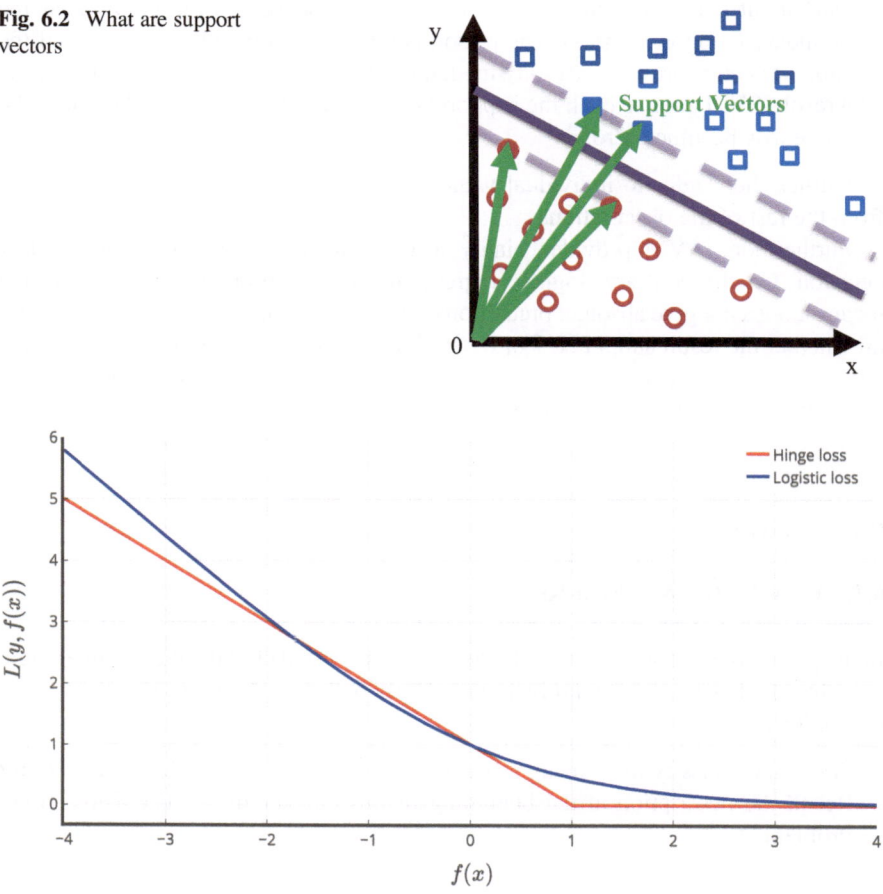

Fig. 6.3 Different loss functions

However, it's important to note that the analysis provided here does not take into account the effects of regularization. Further analysis on regularization will be discussed later.

By observing the two loss functions, we may notice that

- The hinge loss function does not increase as rapidly as the logistic loss function. This means that logistic regression penalizes outliers more severely, resulting in a lower tolerance for outliers compared to SVM.
- Regardless of the choice of loss function, even when the classification is correct, values near the decision boundary are still penalized. This implies that both logistic regression and SVM require better classification to ensure that values are as far away from the boundary as possible.
- Even if a value is confidently classified and is far from the boundary, the logistic loss function does not become zero. This further emphasizes the requirement of logistic regression for all points to move away from the boundary.
- On the other hand, if a value is well classified and far from the boundary, the hinge loss function immediately becomes zero. This indicates that SVM doesn't care about exact position of far points; it only focuses on the points near the boundary (support vectors). The reason is that, as mentioned earlier, even if the support vectors are correctly classified, the hinge loss function is still not zero. As a result, SVM aims to push the support vectors away from the boundary until the hinge loss becomes zero.

Outliers here refer to individual values in the sample that significantly deviate from the rest of the observations.

Furthermore, SVM provides binary results, either 1 or 0, indicating a clear decision. On the contrary, logistic regression provides probability values, which means it does not give absolute predictions. Instead, it requires us to make a decision on whether the result should be 1 or 0 based on the probability. In this way, logistic regression does not make the decision for you, leaving room for interpretation, while SVM directly provides the decision.

6.1.4   Kernel

6.1.4.1   Why Do We Use a Kernel?

In the previous example, we can achieve a relatively good classification with a single straight line, but what if the problem to be classified is as shown in Fig. 6.4? Points for Discussion

- We cannot classify them using a straight line. What should we do in such cases? Recall the description at the beginning of this chapter from "The Three-Body Problem."

Fig. 6.4 How can we do
nonlinear classification

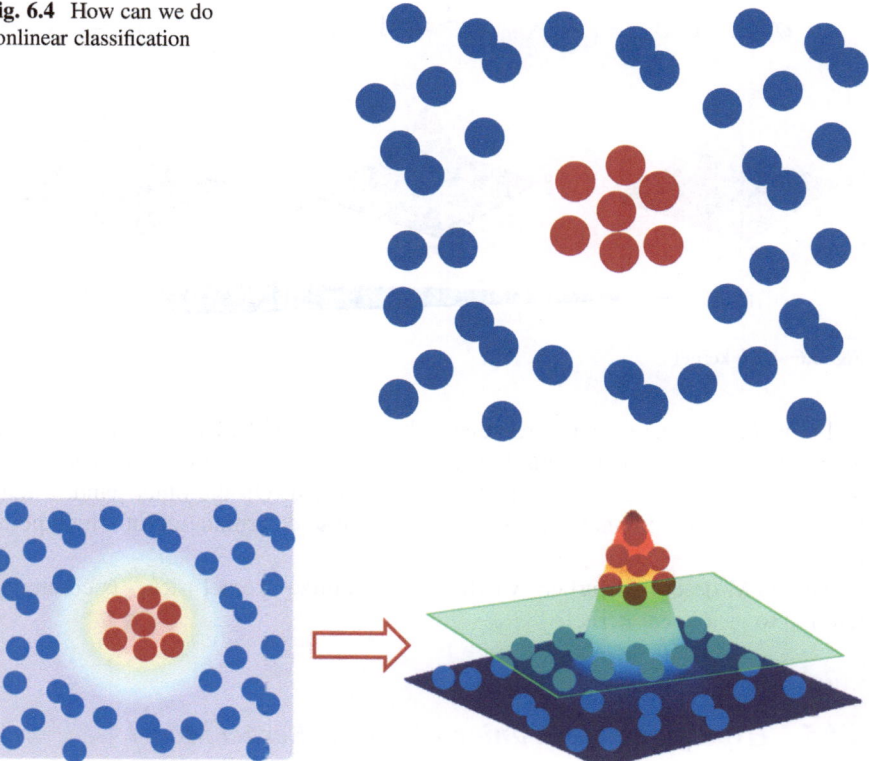

Fig. 6.5 Mapping a low-dimensional plane to a high-dimensional space

As shown in Fig. 6.5, if the middle red points can be lifted, can a plane be used as a boundary to classify them? The answer is clearly yes.

But how to lift these points? This is where "kernel" is needed. The most commonly used kernel is the radial basis function (RBF) kernel, as shown in Fig. 6.6:

6.1.4.2 What Is a Kernel?

In simple terms, using a kernel allows us to transform data into a higher-dimensional space where the boundaries between different categories are more clearly defined. Essentially, it enables us to linearly separate data in a n-dimensional space that cannot be linearly separated in the $n - 1$-dimensional space as shown in Fig. 6.6.

By applying a kernel function, such as the RBF, we can transform the original two-dimensional data into a three-dimensional space. This transformation lifts the relevant portions of the data, allowing us to use a plane (or more precisely, a hyperplane) to linearly classify the data.

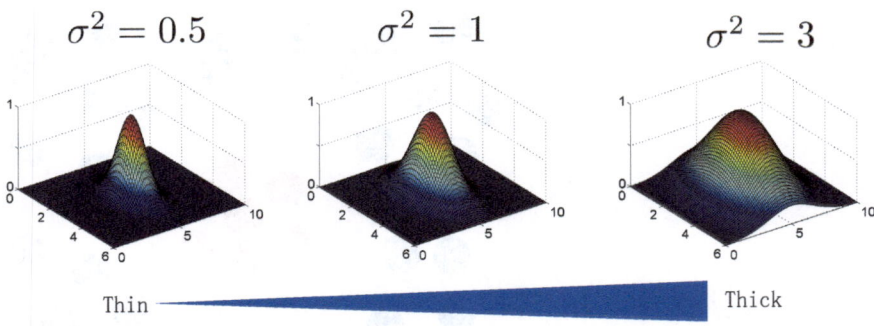

$$\sigma^2 = 0.5 \qquad \sigma^2 = 1 \qquad \sigma^2 = 3$$

Thin ————————————————————————— Thick

Fig. 6.6 RBF kernel

In Fig. 6.6, the parameter σ^2 represents the variance of the RBF. A larger variance makes the function "fatter." Intuitively, a narrower RBF lifts fewer points but allows for more precise control over which points are lifted. On the other hand, a fatter function lifts more points but provides a less precise determination of which points are lifted.

The SVM described earlier, which does not utilize "kernel trick," is commonly referred to as a linear kernel.

6.1.5 Hyperparameter Tuning of Linear Kernel Model

SVM offers a range of hyperparameters that require manual tuning for optimal model performance. To effectively debug and improve the model, it is essential to have a thorough understanding of SVM principles (https://www.deeplearning.ai/courses/machine-learning-specialization/).

Let's begin by exploring the simplest form of SVM, which is linear classification.

In a two-dimensional plane, a straight line can effectively separate data into two classes. For instance, a linear classifier can be represented as

$$f(x) = wx + b$$

As depicted in Fig. 6.7, the decision boundary $f(x) = 0$ effectively separates the data on either side—by normalizing the data and assigning it to respective classes of $f(x) \geq 1$ and $f(x) \leq -1$. The distance between the two parallel lines $f(x) = 1$ and $f(x) = -1$ represents the margin.

$$\text{Margin} = \frac{2}{\|w\|}$$

The primary objective of SVM is to maximize the Margin $= \frac{2}{\|w\|}$, which means we need to minimize $\|w\|^2$. However, it is also crucial to ensure accurate

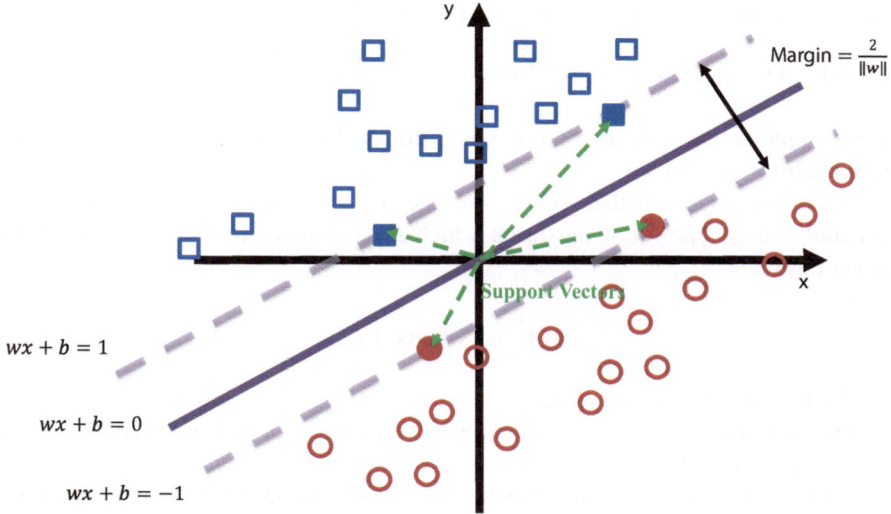

Fig. 6.7 Margin

classification without any errors. This is achieved by minimizing the hinge loss. Combining these two requirements, the regularized loss function for SVM can be expressed as

$$\|w\|^2 + C(\text{Hinge Loss})$$

In the regularized loss function, C is a regularization parameter with $C \geq 0$. The objective of SVM is to minimize this loss function. A wider margin is desirable as it allows for better separation of data points while also bringing more errors. To address this contradiction, we typically weigh the trade-offs by setting the value of C.

- If the C is small, the loss function is less influenced by hinge loss, which means it imposes less penalty on errors or has higher tolerance for mistakes. This makes the model less strict, allowing for wider margin. If $C = 0$, then hinge loss has no impact at all, resulting in no penalty for errors.
- If the C value is large, the loss function will be heavily influenced by hinge loss, which means a higher penalty for errors or less tolerance for mistakes, resulting in a stricter model and narrower margin. If C equals $+\infty$, the loss function becomes infinity as long as hinge loss is not 0, indicating zero tolerance for errors.

C represents how strict the model is trained.

6.1.6   Hyperparameter Tuning for Nonlinear Kernel Models

6.1.6.1   The Strictness of the Model

What would be the manifestation of model strictness in nonlinear cases, specifically when using kernel trick?

It is important to note that regardless of the presence of a kernel, the loss function generally adopts the following form, which aims to make the margin wider with fewer errors:

$$\|w\|^2 + C(\text{Hinge Loss})$$

We'll use RBF as an example.

The effect of C is the same as that of a linear model. You may see a similar result as shown in Fig. 6.8.

A large value of C can result in overly strict training, causing the boundary line to become curved and twisted. This can make the model prone to overfitting, where it becomes too specific to the training data and performs poorly on new, unseen data. On the other hand, a small value of C can lead to issues such as excessive classification errors in the training data. To ensure the model performs optimally, it is important to find a suitable value for C.

6.1.6.2   Level of Hesitance in Data

Now let's look at how "hesitant" the data points are in a SVM when using the kernel trick, where points far away from the decision boundary may hesitate to participate in the classification.

When we look at the graph of the radial basis function (RBF), we can observe that a larger variance leads to a broader spread of the function, giving it a "fatter" shape.

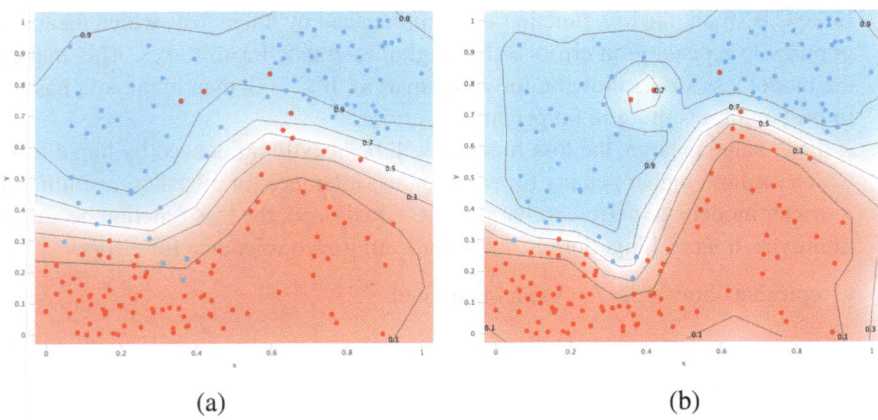

(a)   (b)

Fig. 6.8 Impact of nonlinear kernel C. (**a**) Smaller C, wider margin. (**b**) Larger C, narrower margin

Conversely, a smaller variance makes the function appear "thinner" with a narrower range of influence from its peak. If we consider the base of the mountain as the decision boundary, it suggests that a larger variance allows points that are farther away from the boundary to exert a greater impact. This range of impact is represented by a parameter known as γ (gamma). In the case of RBF, γ is inversely proportional to the variance. By adjusting the value of γ, we can control the extent to which individual data points influence the model's decision boundary.

- If γ is small, the peaks of the mountain are flatter, and even points that are far from the decision boundary have an impact. They actively expand their influence area. In this case, it is necessary to consider the influence of both nearby and distant points because they all want to be farther away from the boundary. The correct classification of nearby points is not the only primary factor to consider.
- If γ is large, the peaks of the mountain are sharper, and only points close to the decision boundary have an impact. Distant points do not have much influence and are more hesitant to participate in the classification. In this situation, it is important to prioritize the correct classification of points near the boundary and not focus too much on the influence of distant points, which can result in a more curved boundary.

As a result, the outcome may resemble what is shown in Fig. 6.9. From the graph, it can be observed that when γ is small, each point is less "hesitant" and actively expands its range. Their influence areas are larger, and in case of disagreement, the side with more points prevails. When γ is large, each point is more hesitant and reluctant to expand its range. In case of disagreement, the side that is closer to the points holds more weight in determining the final classification.

A large γ causes the boundary line to be curved, making it prone to overfitting. However, if γ is too small, it can also result in too many classification errors in training data. Therefore, we need to find a suitable γ to ensure that the model works in an ideal state. γ can be understood as the "hesitance" of the data.

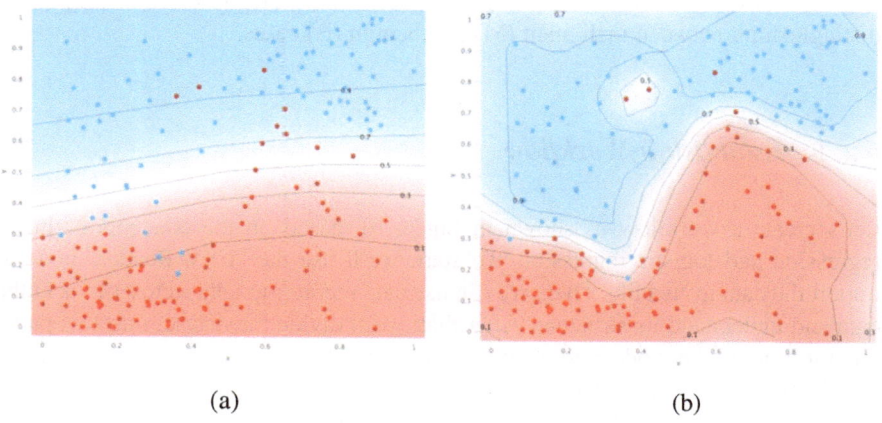

| (a) | (b) |

Fig. 6.9 Influence of γ. (**a**) Smaller γ. (**b**) Larger γ

6.1.7   C and γ

When it comes to performance, smaller values for both variables tend to make the boundary appear "straighter," while larger values make it appear "more curved." However, the reasons behind this behavior are different. Let's say each data point represents a person, and the model is a set of rules. In this scenario, the parameter C represents the strictness of these rules, while γ represents the level of hesitance exhibited by each person. The decision boundary serves as the criterion for determining whether a mistake has been made based on these rules.

Tuning hyperparameter is a technique that tests algorithm understanding and practical experience, requiring a lot of practice. This basic principle applies to various industries.

Points for Discussion

- If we personify C and γ, can you think of any real-life examples that resemble their characteristics?

6.2   First Practice on SVM

Take the breast cancer diagnosis data from Kaggle (https://www.kaggle.com/datasets/uciml/breast-cancer-wisconsin-data), and test your skills using KNIME.

6.2.1   Case Explanation

This dataset contains digitized images of cell nuclei, with features such as radius, texture, and other information. Its main objective is to diagnose tumor properties, distinguishing between malignant (M) and benign (B) cases.

6.2.2   Establish a Workflow

Let's create a workflow as shown in Fig. 6.10, based on the concepts of linear regression and logistic regression. To start, we'll use the "CSV Reader" node to import the data from a file. Then, we can use data monitor to take a closer look at the data and understand its structure. After that, we'll divide the dataset into two parts: one for training the model and the other for testing its performance. Once the model is trained, we'll test and score the model.

Now let's briefly go through each step of the workflow.

6.2.3   View the Data

6.2.3.1   Scatterplot

First, take a brief look at the relationship between the data through a scatter plot matrix. Click on the "Scatter Matrix" node, select the "Configure" button, and observe the relationship between each feature as shown in Fig. 6.11. You may select more other features and see what you can discover.

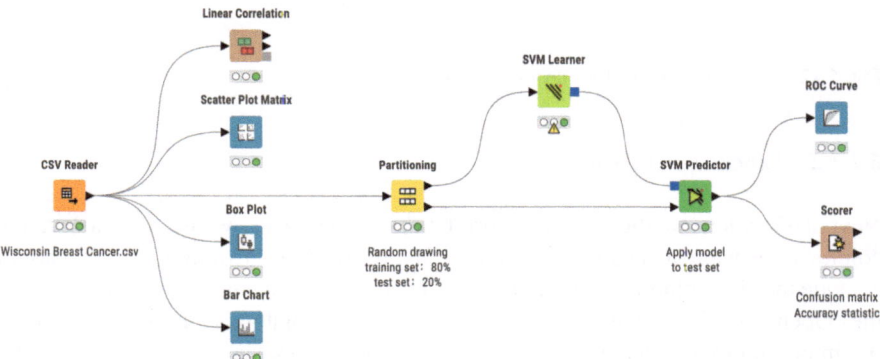

Fig. 6.10 KNIME workflow

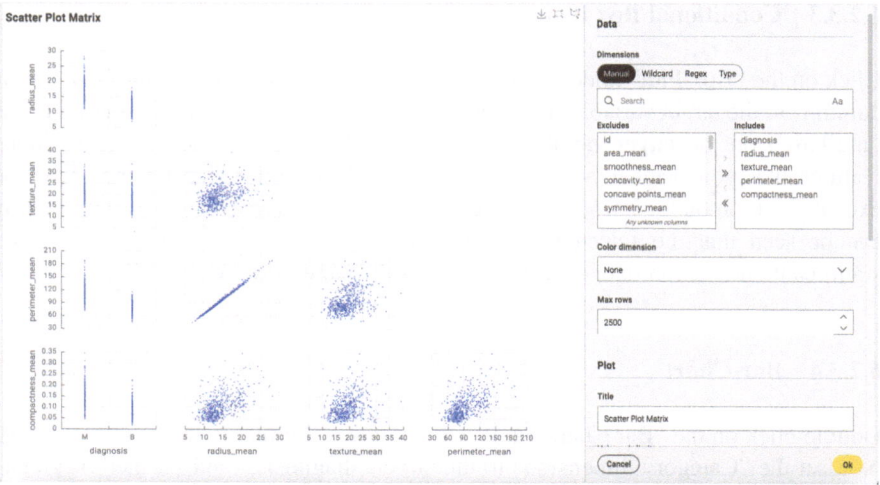

Fig. 6.11 Scatterplot matrix

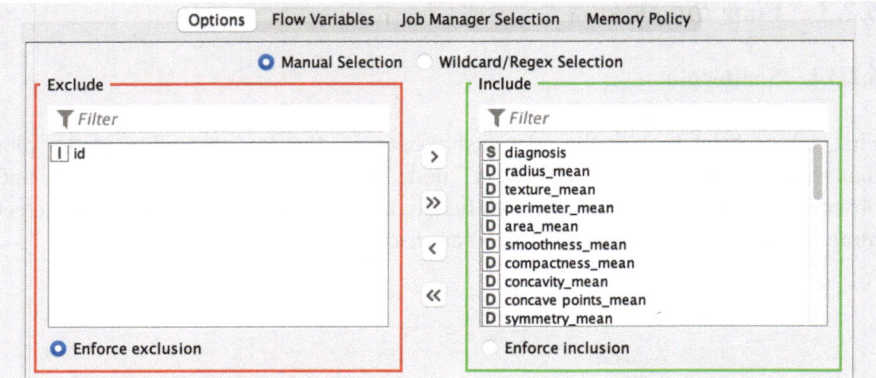

Fig. 6.12 Selecting columns for linear correlation analysis

6.2.3.2   Linear Correlation

Next, let's click on the "Linear Correlation" node. And we need to select the columns we want to analyze, as shown in Fig. 6.12 in "Configure." We should exclude the "id" column from the analysis. Then we can execute the node and choose the "Open View" to see the correlation matrix in a new dialog, as shown in Fig. 6.13. It's important to note that since the target variable "diagnosis" is represented by the letters "B" and "M," KNIME may not be able to understand its relationship with the other variables, so there's no correlation values being displayed.

6.2.3.3   Conditional Box Plot

Click on the "Box Plot" node, select the "Configure" button, and set the "Condition column" to the target variable "diagnosis" as shown in Fig. 6.14. This will divide the data based on the target variable and create a plot. You can select one or more features that you want to view in the "Includes" section. Click on the "Save & execute" button on the left side of the configuration interface to view the box plot. It can be seen that the selected features "radius_mean" and "texture_mean" have significant variance in data distribution under different tumor conditions.

6.2.3.4   Bar Chart

Double-click on the "Bar Chart" node to configure it. Similar to the conditional box plot, set the "Category dimension" to the target "diagnosis," and set the "Aggrega-tion" to the "Average." As shown in Fig. 6.15, we should select "radius_mean" as the object to be displayed in the chart. By doing so, we can observe that the average values of this particular feature vary significantly under different tumor conditions.

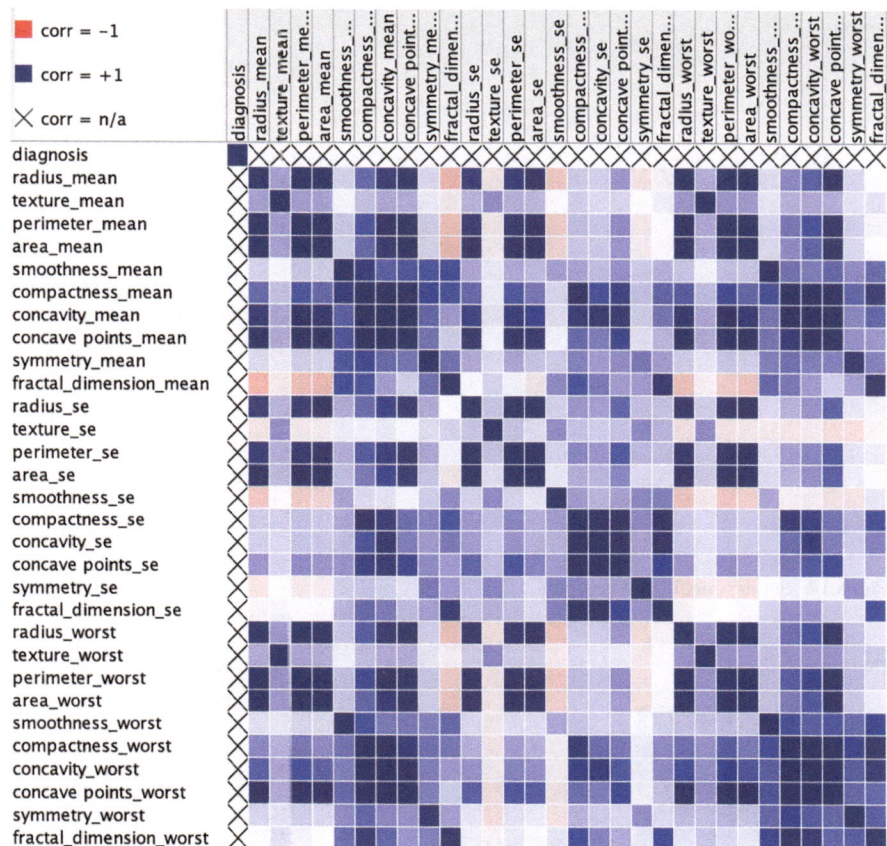

Fig. 6.13 Correlation matrix

6.2.4 Train and Test the Model

Double-click on the "SVM Learner" node and keep the default settings as shown in Fig. 6.16.

Double-click on the "SVM Predictor" node, and check the option "Append columns with normalized class distribution" (Fig. 6.17).

6.2.5 View the Result

We can view the result with the "ROC Curve" node and the "Scorer" node. The settings of the nodes are shown in Figs. 6.18 and 6.19.

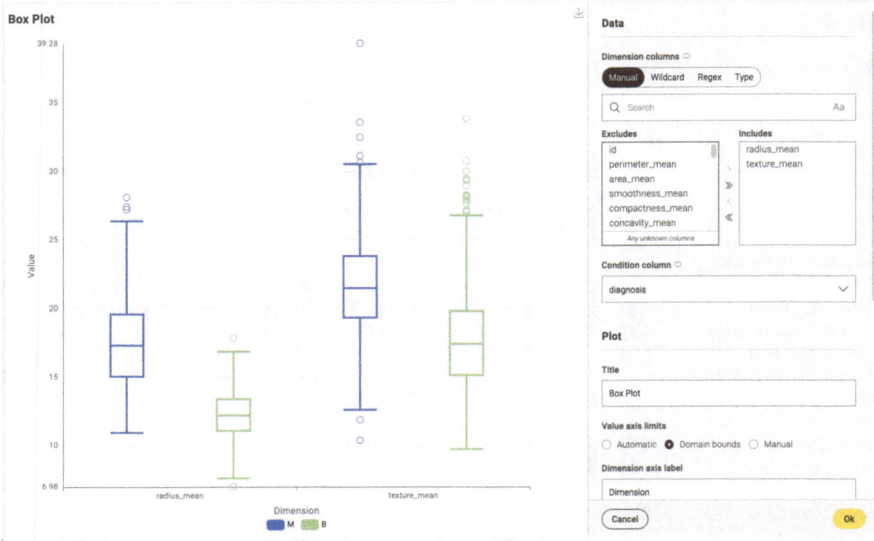

Fig. 6.14 Set the conditional box plot

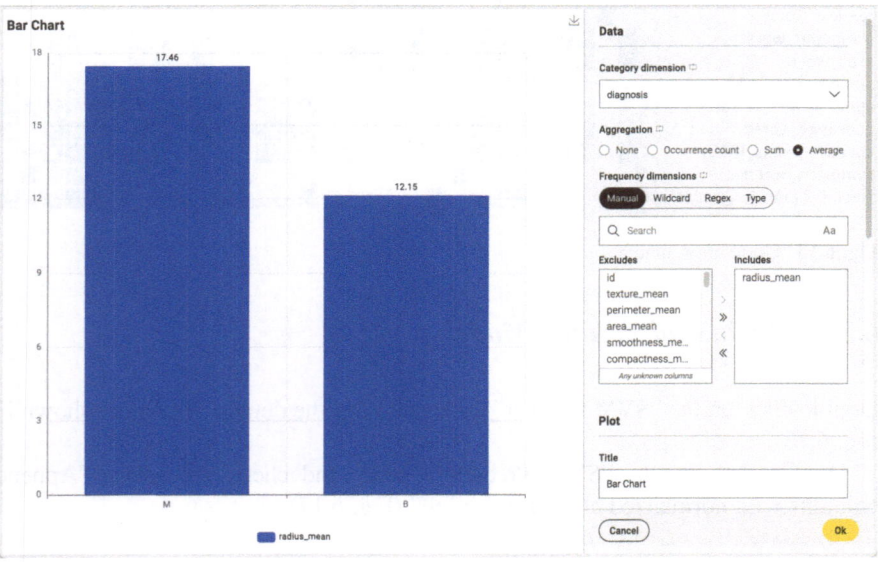

Fig. 6.15 Bar chart

As shown in Figs. 6.20 and 6.21, both the ROC curve and the confusion matrix indicate that our model is quite good.

In this section, we have successfully tackled the SVM problem with our prior knowledge and skills in linear regression and logistic regression.

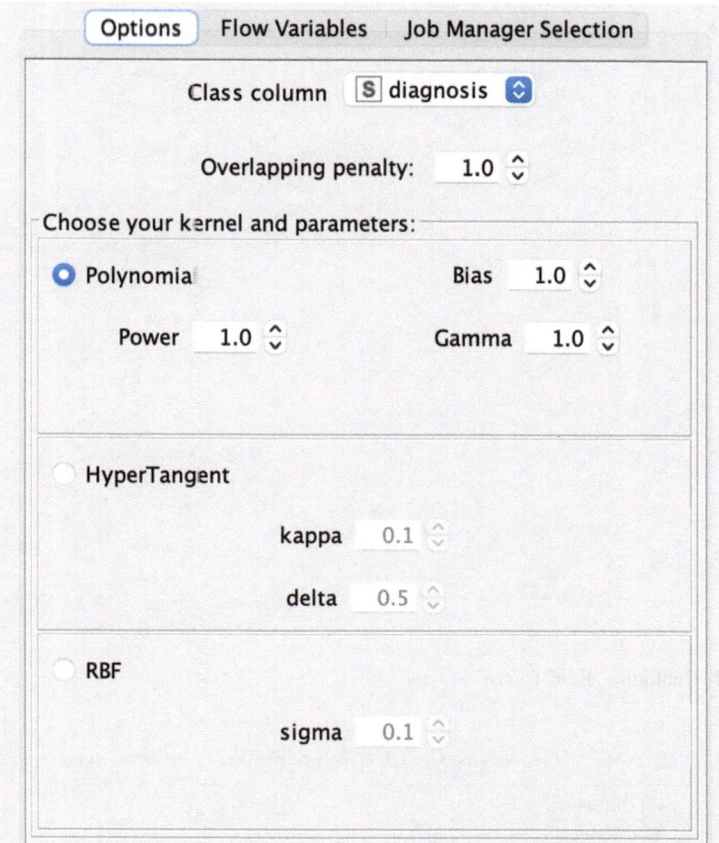

Fig. 6.16 Model configuration

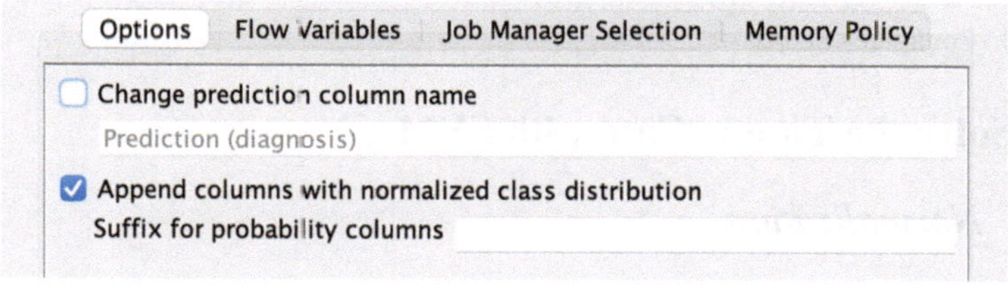

Fig. 6.17 Configure "SVM Predictor"

Give It a Try

• Try tuning hyperparameters to get a better model.

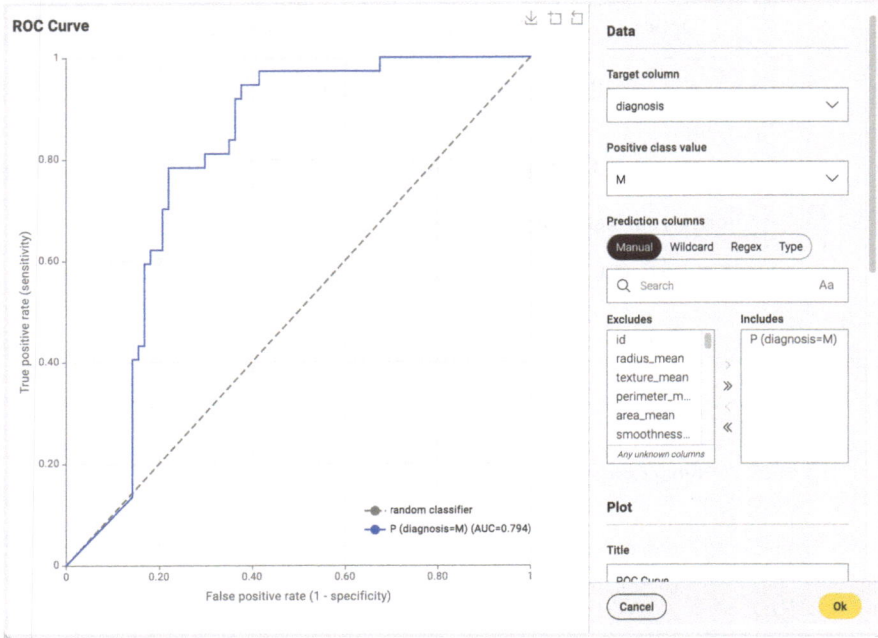

Fig. 6.18 Configure "ROC Curve"

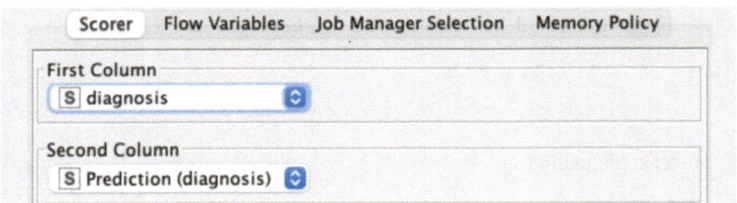

Fig. 6.19 Configure "Scorer"

6.3   Solve the Titanic Case with SVM

6.3.1   *Normalization*

Normalization plays a crucial role in SVM. As discussed above, SVM classifies data based on the geometric distribution. To illustrate this, let's consider a two-dimensional plane, as shown in Fig. 6.22. If the range of variation in the X axis is large while the range of variation in the Y axis is small, the margin will be primarily determined by the Y component, and the X component will have minimal impact. This sensitivity arises because even a slight change in Y corresponds to a significant range change in X. To address this issue, we can leverage the normalization techniques discussed in Chap. 3.

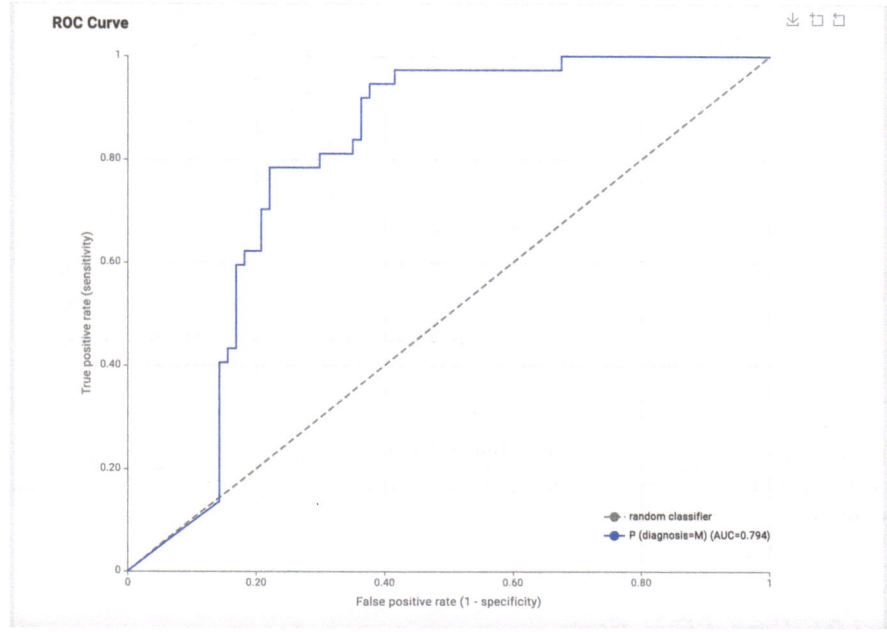

ROC Curve
True positive rate (sensitivity)
False positive rate (1 - specificity)
random classifier
P (diagnosis=M) (AUC=0.794)
0
0.20
0.40
0.60
0.80
1

Fig. 6.20 ROC

File Hilite
diagnosis ... M B
M 25 20
B 1 68
Correct classified: 93
Wrong classified: 21
Accuracy: 81.579%
Error: 18.421%
Cohen's kappa (κ): 0.584%

Fig. 6.21 Confusion matrix

y
0
x

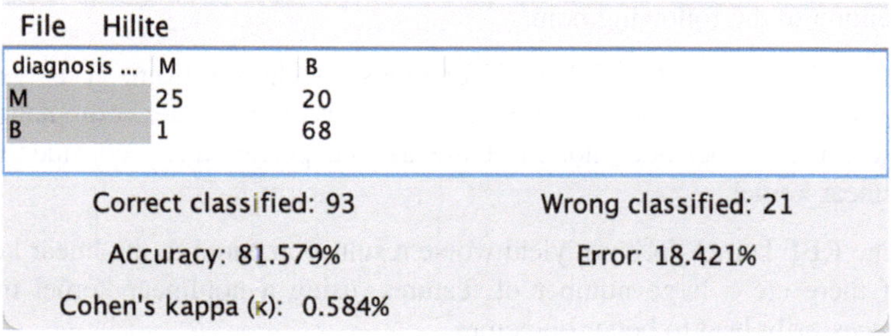

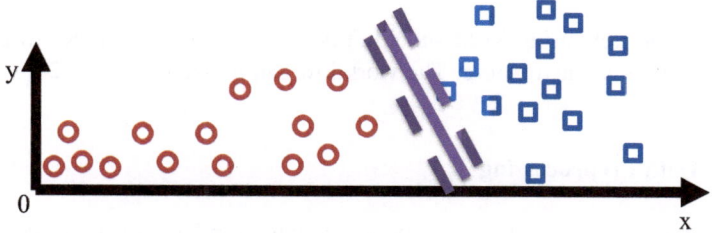

Fig. 6.22 *X* is much larger than *Y*

Fig. 6.23 *X* and *Y* are
similar after normalization

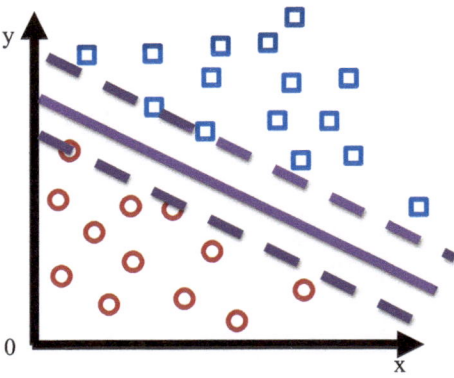

After normalization, as shown in Fig. 6.23, *X* and *Y* are similar in size. Generally, normalization improves the speed of gradient descent algorithm in finding optimal solutions.

6.3.2   Kernel Function

When considering kernel functions from an application perspective, we may need to pay attention to the following points:

- SVM offers a variety of kernel functions to choose from. It is generally recommended to start with a linear kernel as it provides faster computation.
- If the linear kernel does not meet the desired performance, consider using a nonlinear kernel.
  - The RBF kernel does not yield worse results compared to the linear kernel.
  - If there are a large number of features, using a nonlinear kernel may not necessarily lead to better outcomes.

6.3.3   Create a Workflow

Creating a workflow using SVM shouldn't be too difficult with the KNIME knowledge that you've learned before. The workflow can be seen in Fig. 6.24.

6.3.3.1   Data Preprocessing

First, we should process data before model training. Similar to the logistic regression workflow mentioned earlier, as shown in Fig. 6.25, after the data type conversion, the input data will be split into a training set and a test set. Subsequently, both sets

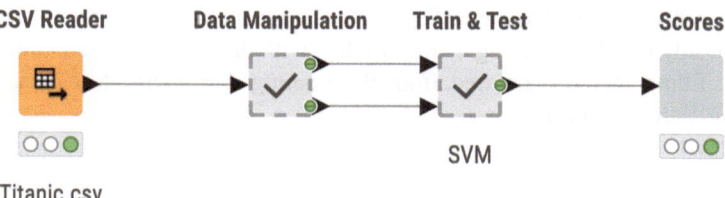

Fig. 6.24   Create the workflow

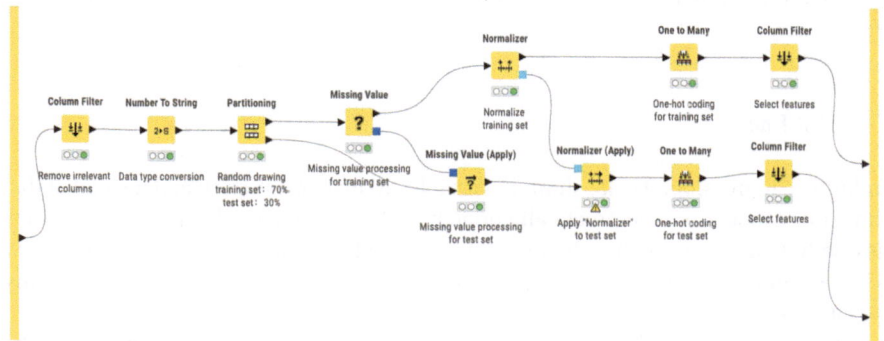

Fig. 6.25   Data processing

undergo the same processes: handling missing values, normalization, and one-hot encoding. The outputs from these processes serve for the next step in the workflow.

A large amount of content in this part has already been introduced earlier, and the same or similar parts will only be briefly explained.

Data Type Conversion

As we discussed in logistic regression, we need to convert the "Survived" variable to categorical data, which can be done by using the "Number To String" node in KNIME.

Missing Value Processing

As mentioned in logistic regression, this part of the work is done by the "Missing Value" node. Do you still remember the question we leave in the previous section when we talked about missing value? Here's the answer. The "Missing Value" and "Missing Value (Apply)" nodes are both activated. Why can't we just use "Missing Value"? The reason is that we need to ensure no information from the training data leaks into the test data. If "Missing Value" is used before the partitioning, it means that all the data is involved in data filling. This includes both future training data and test data, which can lead to the leakage of training data into the test data. Therefore, we must only use the "Missing Value" to deal with the training data and then apply the data filling method acquired from the training data to the test data, i.e., the "Missing Value (Apply)."

Data Normalization

This part has been introduced earlier; please note that the input for "Normalizer" must be training data, while the "Normalizer (Apply)" will apply the normalizer to the test set.

One-Hot Encoder

In logistic regression, we use dummy variables to handle categorical data, and at that time, this process was automatically done by KNIME. In fact, this process is known as one-hot encoding. In the SVM model of KNIME, the model cannot automatically generate dummy variables. We need to use one-hot encoder to generate dummy variables ourselves.

The one-hot encoding, simply put, is a vector representation where all the elements of the vector are set to 0 except 1, which has 1 as its value. It is commonly used for handling categorical data. For example, the division of gender can be encoded as 01 for male and 10 for female.

Now let's see how to do it using KNIME. Double-click to configure the node "One to Many" as shown in Fig. 6.26.

Here, "Sex" and "Embarked" are selected for data type conversion, and other categorical variables are not processed because we will filter them out. And it is

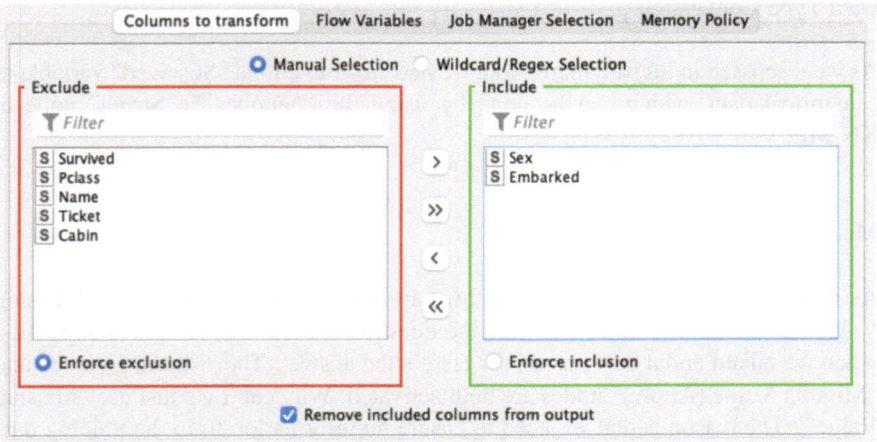

Fig. 6.26 Configure the one-hot encoder

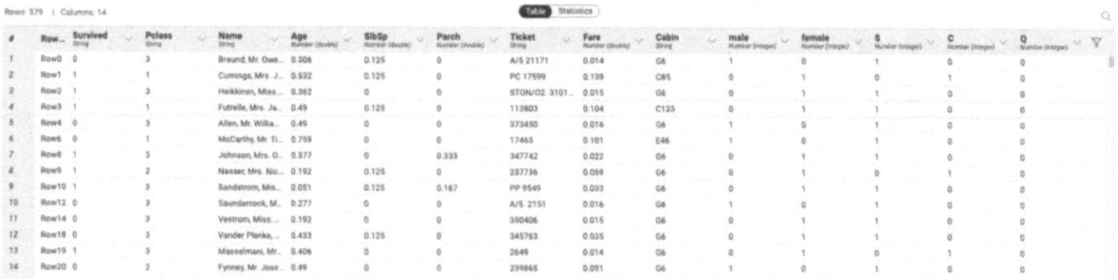

Rows: 579 | Columns: 14 · Table Statistics

#	Row..	Survived	Pclass	Name	Age	SibSp	Parch	Ticket	Fare	Cabin	male	female	S	C	Q
1	Row0	0	3	Braund, Mr. Owe..	0.306	0.125	0	A/5 21171	0.014	G6	1	0	1	0	0
2	Row1	1	1	Cumings, Mrs. J..	0.532	0.125	0	PC 17599	0.139	C85	0	1	0	1	0
3	Row2	1	3	Heikkinen, Miss..	0.362	0	0	STON/O2 3101..	0.015	G6	0	1	1	0	0
4	Row3	1	1	Futrelle, Mrs. Ja..	0.49	0.125	0	113803	0.104	C123	0	1	1	0	0
5	Row4	0	3	Allen, Mr. Willia..	0.49	0	0	373450	0.016	G6	1	0	1	0	0
6	Row6	0	1	McCarthy, Mr. Ti..	0.709	0	0	17463	0.101	E46	1	0	1	0	0
7	Row8	1	3	Johnson, Mrs. O..	0.377	0	0.333	347742	0.022	G6	0	1	1	0	0
8	Row9	1	2	Nasser, Mrs. Nic..	0.192	0.125	0	237736	0.059	G6	0	1	0	1	0
9	Row10	1	3	Sandstrom, Mis..	0.051	0.125	0.167	PP 9549	0.033	G6	0	1	1	0	0
10	Row12	0	3	Saundercock, M..	0.277	0	0	A/5. 2151	0.016	G6	1	0	1	0	0
11	Row14	0	3	Vestrom, Miss..	0.192	0	0	350406	0.015	G6	0	1	1	0	0
12	Row18	0	3	Vander Planke,..	0.433	0.125	0	345763	0.035	G6	0	1	1	0	0
13	Row19	1	3	Masselmani, Mr..	0.406	0	0	2649	0.014	G6	0	1	0	1	0
14	Row20	0	2	Fynney, Mr. Jose..	0.49	0	0	239865	0.051	G6	1	0	1	0	0

Fig. 6.27 Encoded data

necessary to select the "Remove included columns from output" option to delete the columns before the conversion. In the monitor, a new data table can be seen as shown in Fig. 6.27.

You may find that several columns have been added to the data table, i.e., male, female, and S, C, Q. You can observe the number of 1s in each set of data to determine if it meets the requirements of one-hot encoding.

Do you feel that this result is similar to the previously introduced concept of dummy variable?

When dealing with categorical data such as occupation, gender, etc., it cannot be quantitatively processed and we need to find out a way to quantify it. This "quantization" is typically accomplished by introducing "dummy variables." We can construct artificial variables that only take "0" or "1" based on the attribute types of these factors.

Filter Data

The last step in data processing is to filter out unnecessary data columns. Double-click on the settings of "Column Filter," as shown in Fig. 6.28. According to the requirements for dummy variables, if a variable conversion can have a total of n possible values, the number of dummy variables corresponding to this variable would be $n - 1$. Therefore, in this case, only one value can be chosen for "Sex," and one value needs to be removed for the "Embarked."

6.3.3.2 Training and Testing

The workflow for this part is shown in Fig. 6.29.

You may see that this workflow is only slightly different from the previous logistic regression workflow, as the learner has changed from "Logistics Regression Learner" to "SVM Learner." This difference is easily understood.

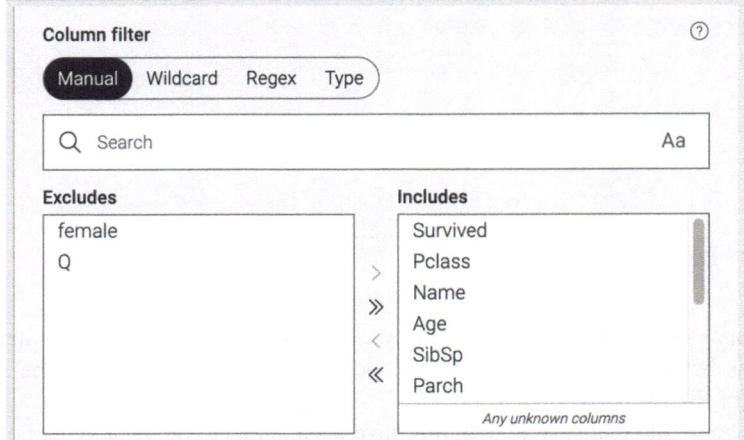

Fig. 6.28 Filtering redundant dummy variables

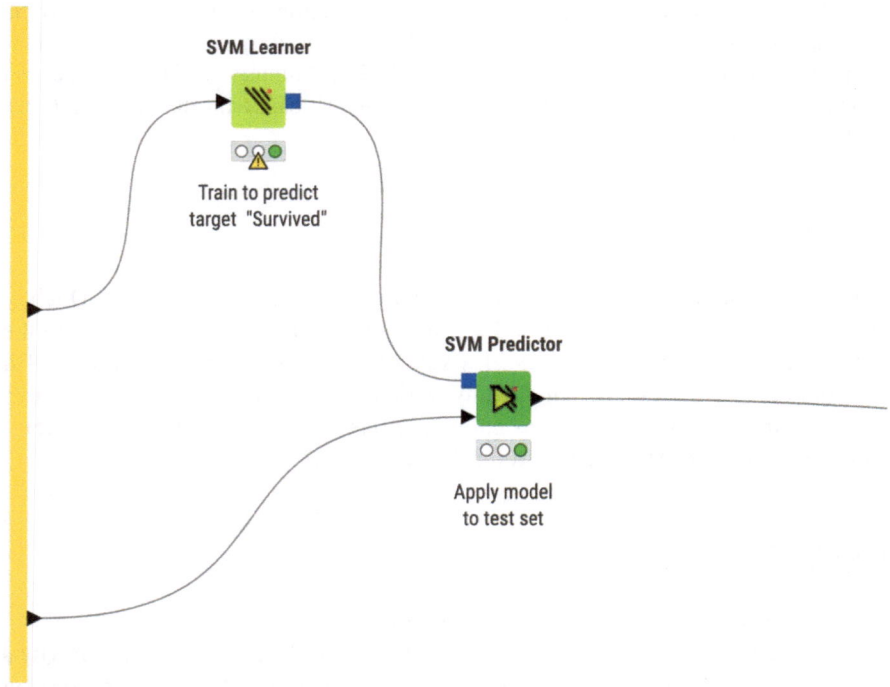

Fig. 6.29 Train and test the model

6.3.3.3   Scoring

The evaluation of model can be through the "Scores" component, as shown in Fig. 6.30. Unlike the previous "Scores" component we used, there is an additional "Table View" node as the output of the "Scorer" node, which allows tables such as "Confusion matrix" and "Accuracy statistic" to be displayed in the Component like a graph.

In Chap. 5, when optimizing the model, we always start by modifying the hyperparameters in the model, then execute the workflow, and finally open the "Scores" component to view the "Confusion matrix" and "Accuracy statistic" of the "Scorer" node. This process can be quite cumbersome. To simplify the process, we can add a "Table View" node to the output of the "Accuracy statistic" in the "Scorer" node. In this way, after adjusting the hyperparameters, we only need to execute the "Open view" command in the "Scores" component to view the accuracy and other information of the model.

6.3.4   C Parameter

Parameter tuning is the most important step. Open the settings of SVM node, as shown in Fig. 6.31.

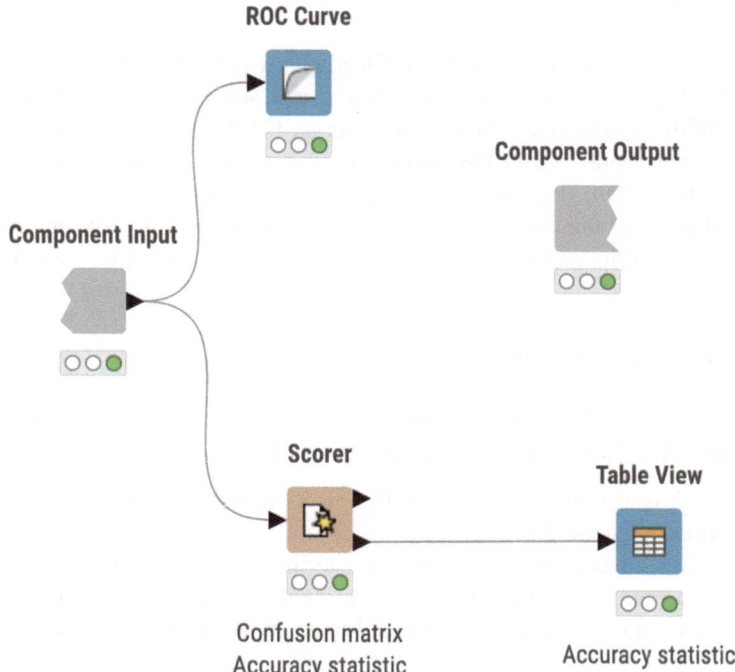

Fig. 6.30 Score the model

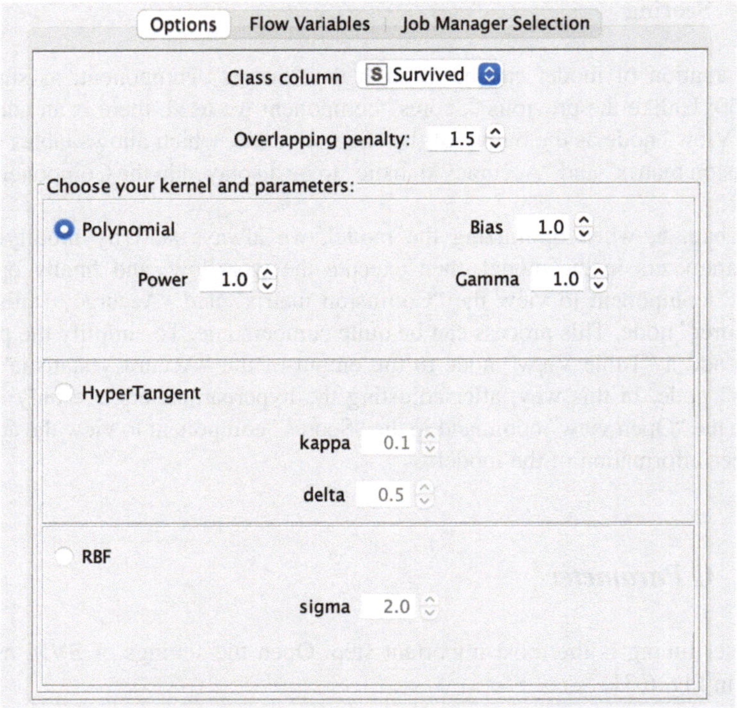

Fig. 6.31 Configure the model

The C parameter used in KNIME is "Overlapping penalty," which represents the model. The larger the value, the more severe it is, resulting in narrower margin and easier overfitting. Here, I set it to 1.5. We can try 0.1, 0.5, 1, 1.5, and 2 to see how they affect the results. From the results, the accuracy is between 0.77 and 0.79. Changing the C here does not make a difference in the outcome. This situation suggests that under the use of a linear kernel, the model is not easily improved by changing the margin, and we should consider using a more complex kernel.

6.3.4.1   Polynomial Kernel

As shown in Fig. 6.32, when the power is greater than 1, it is polynomial.
Here are what you should know about "Polynomial":

- Power = 1 uses a linear kernel, and Power >1 represents the use of a nonlinear high-order polynomial kernel.
- Bias is a parameter that can be adjusted to make the data symmetrical around zero. If the data is not naturally centered around zero, we can apply a bias to shift the values and make them symmetrical. For example, if we have a range of [0, 2],

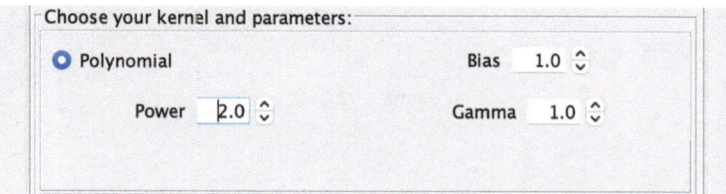

Fig. 6.32 Configure polynomial kernel

we can apply a bias of −1, which will shift the values to [−1, 1], making the data symmetrical around zero.
- Gamma, as mentioned earlier, represents the "hesitance" of the data. The larger the gamma, the more "hesitate" the data is, resulting in a smaller range of influence and a higher likelihood of overfitting.

In Fig. 6.32, we tried to maintain a constant value of *C* at 1.5 and Power at 2. The resulting accuracy was found to be 0.801. To avoid overfitting, the Power value was temporarily kept at 2, and different values of *C* were tested. The *C* values tested were 0.1, 0.5, 1, 1.5, 2, and 5, and the accuracy rates were found to be 0.80447, 0.80006, 0.79888, 0.79333, 0.79333,and 0.7912, respectively. From this trend, we can conclude that as the value of *C* increases, the model becomes more stringent in its demands on the training data. This increased strictness can make the model more prone to overfitting, which can negatively impact its performance on unseen data.

Next, let's keep *C* at 0.1 and Power at 2 and try changing "Gamma" to 0.1, 0.5, 1, 1.5, 2, and 3. The accuracy rates are 0.803, 0.813, 0.818, 0.817, 0.816, and 0.815, respectively. From these results, we can observe that as the value of gamma increases, the model's accuracy initially starts low, then improves, and eventually decreases again. This indicates that the model transitions from underfitting (low accuracy) to an appropriate fit (high accuracy) and then starts overfitting (decreasing accuracy) again. Based on these findings, the best choice is to select the highest accuracy rate, which corresponds to a gamma value of 2.

Similar methods can be used to tune "Bias." And we chose 1 after several rounds of trial.

Through the hyperparameter tuning process, we can gradually establish an understanding of the model. Next, based on the above analysis and methods, we continue to optimize RBF.

6.3.4.2 RBF Kernel

The RBF, also known as Gaussian kernel, has only one parameter to tune, i.e., sigma. And sigma is inversely proportional to gamma, as shown in Fig. 6.33. We don't need to elaborate it here again. You can try different values for sigma and see what impact it has on the model.

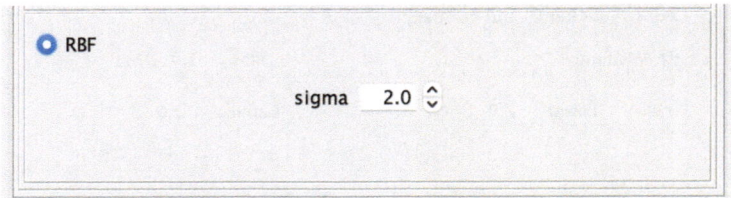

Fig. 6.33 Configure RBF

The parameters *I* set here are definitely not optimal. You can try to tune it yourself and challenge the Titanic Kaggle Competition.

The model scoring is exactly the same as logistic regression and will not be elaborated further.

Points for Discussion

Coming all the way to Chap. 6, there are some details that need to be reviewed.

- First, in the section of "Underfitting and Overfitting" in Chap. 5, do the "training data" and "other data" we use equal to the "test data"?
- Second, is it proper to use test data to optimize parameters like C, gamma, and other model parameters in this chapter?
- Starting from the first question, you may proceed from the perspective of overfitting and think over the second one.

We will analyze it in the next chapter.

6.4   Practice

1. Briefly describe the difference between support vector machine and logistic regression.
2. Briefly describe the role of kernels.
3. Please raise an example of one-hot encoding.

Reference

https://www.deeplearning.ai/courses/machine-learning-specialization/

Chapter 7
Tree-Based Algorithms

> Two heads are better than one.
> —*Idiom*

Abstract This chapter delves into tree-based algorithms, exploring their fundamental concepts, practical applications, and techniques to optimize their performance. The discussion begins with an introduction to decision trees, highlighting their characteristics, visualization, and the common problem of overfitting. Strategies to prevent overfitting, such as early stopping and pruning, are thoroughly examined. The concept of ensemble learning is introduced, emphasizing how it combines multiple weak learners to create a strong learner. Specific focus is given to various ensemble learning techniques like bagging, stacking, random forest, and boosting. The chapter also includes practical applications of these algorithms using KNIME, a data analysis tool, demonstrating their effectiveness in real-world scenarios like the Titanic case and bank customer classification. The importance of feature engineering, handling of abnormal data, and comparison of different tree-based models like bagging, random forest, and boosting is thoroughly explored. The chapter concludes with practice exercises to reinforce the concepts covered.

Keywords Tree-based algorithms · Decision trees · Overfitting · Early stopping · Pruning · Ensemble learning · Bagging · Stacking · Random forest · Boosting · Hyperparameter tuning

This chapter covers

- **Tree-based algorithms.**
- **Concept of decision tree.**
- **Early stopping.**
- **Ensemble learning.**

© The Author(s), under exclusive license to Springer Nature Singapore Pte Ltd. 2024
Y. Geng et al., *Practical Machine Learning Illustrated with KNIME*,
https://doi.org/10.1007/978-981-97-3954-7_7

Tree-based algorithms are a set of popular approaches in many fields because they work in a way that is similar to how humans make decisions. They follow a step-by-step process, just like we do when we think through a problem. For example, we often use "if ... then ... " statements to make decisions, covering a wide range of different scenarios to make the decision better.

7.1   Introduction to Decision Trees

Depending on the task, the decision tree is also referred to as classification tree or regression tree. Leaf nodes provide classification, internal nodes represent features, and branches represent decision rules. The construction of a decision tree usually follows a top-down approach, choosing the best attribute to split at each step. The definition of "the best" is to make the training set in the child nodes as pure as possible. Different algorithms use different metrics to define "the best."

7.1.1   Characteristics of Decision Trees

There are several advantages of decision trees compared to other algorithms:

- Easy to understand and explain.
- The features do not need to be normalized.
- Easy to visualize.

For illustration purposes, the decision tree for survival on the familiar Titanic case could be depicted as shown in Fig. 7.1.

Fig. 7.1 Decision tree

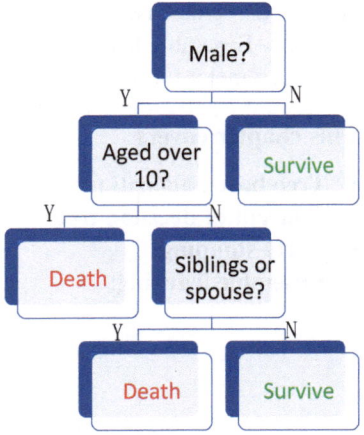

Even readers with no prior knowledge of decision trees can easily grasp the meaning conveyed by this diagram. It is evident at a glance. If the individuals are not male, they survive. However, if they are male, their survival depends on their age. If they are over 10 years old, they will be determined as dead; or otherwise .... That's it, a picture is worth a thousand words, and Fig. 7.1 effectively captures the essence of the above three advantages. Of course, decision trees do have their limitations, such as the tendency to overfit.

7.1.2 Overfitting Prevention

Imagine you have a book with pictures of lots of different dogs. You start learning about dogs by looking at every tiny detail in the pictures. It happens that all dogs in this book have a spot on its nose and a curly tail and are holding a red ball.

If you then decide that all dogs must have a spot on their nose and a curly tail or carry a red ball to be a dog, this is like overfitting. You're paying too much attention to the small, specific details in your book.

Now, suppose you have the same book of dogs. But this time, you only look at the first page, which has a picture of a big, brown dog. You decide that all dogs must be big and brown because that's what you saw on the first page.

If you see a small, white dog, you might not think it's a dog because it doesn't look like the big, brown one in your book. This is like underfitting. You're not learning enough about dogs because you're only using one simple example to understand all dogs.

Points for Discussion

- Can you think of other examples in daily life that are similar to overfitting?

One of the questions that arises in a decision tree algorithm is the optimal size of the final tree.

- A small tree might not capture important structural information about the sample space. For example, in the Titanic case, we can simply assume that all men died and all women survived. And the tree would be very simple, with only two branches, but it does not use more features for prediction, compromising the accuracy. This situation will lead to underfitting.
- A tree that is too large risks overfitting the training data and poorly generalizing to new samples.

We will mainly discuss about overfitting and how we can solve it.

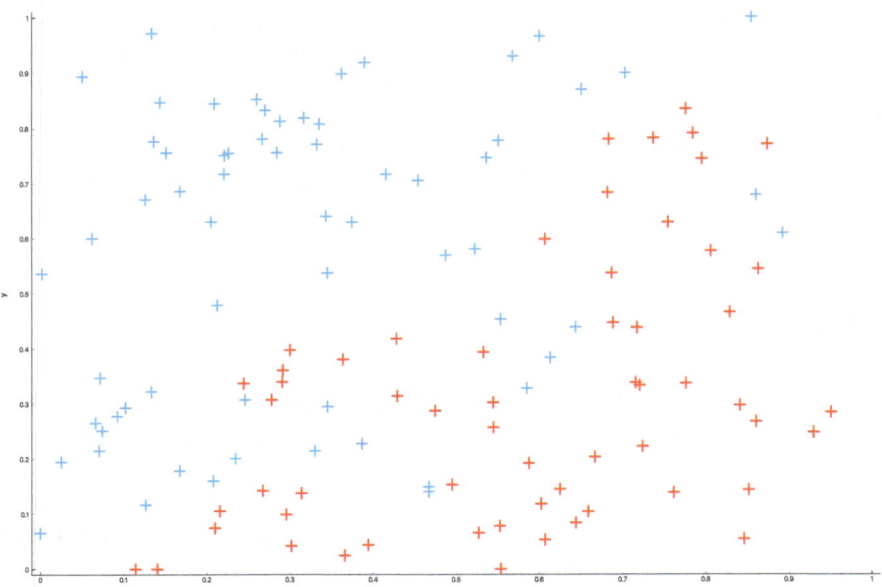

Fig. 7.2 Data distribution

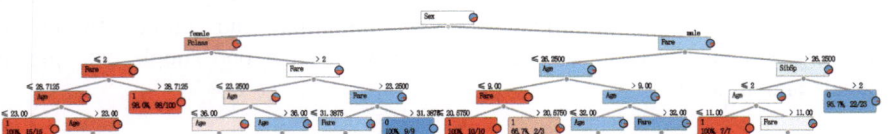

Fig. 7.3 Decision tree structure

7.1.3 Problem Analysis

Suppose we want to classify the data as shown in Fig. 7.2.

One possible construction of its classification tree is shown in Fig. 7.3.
Observe this tree and ask yourself three questions:

- Do we need such a "tall" tree?
- Does the error rate decrease quickly with increasing node depth?
- Are the sample sizes at some nodes too small?

These three questions bring us two solutions:

- Prevent problems before they occur.
- Solve the problem after it arises.

These two solutions are.

- Early stopping.
- Pruning.

7.1.4   Occam's Razor

The common feature of these two methods is to reduce the size of the tree, and their philosophical basis can be seen as the Occam's razor. It was a principle proposed by William of Occam, a fourteenth-century English logician and Franciscan friar, which means "entities should not be multiplied beyond necessity" or "the simplest solution is usually the best one."

7.1.5   Early Stopping

First, let's explore the concept of early stopping in the context of decision trees. As the name suggests, it involves halting the construction of the tree before reaching the deepest layer, thus terminating the process in advance. There are three methods commonly used for early stopping: limiting the tree's depth, determining whether to stop based on the error rate, and deciding whether to stop based on the amount of data in the node (https://www.deeplearning.ai/courses/machine-learning-specialization/).

Points for Discussion

- How do you think the three indicators above can be used?

7.1.5.1   Limit Tree Depth

As the depth of the decision tree increases, the risk of overfitting becomes more pronounced. To mitigate this, we typically split the data into training and other data and monitor the changes in the loss function for both as shown in Fig. 7.4.

Fig. 7.4 Tree depth and model

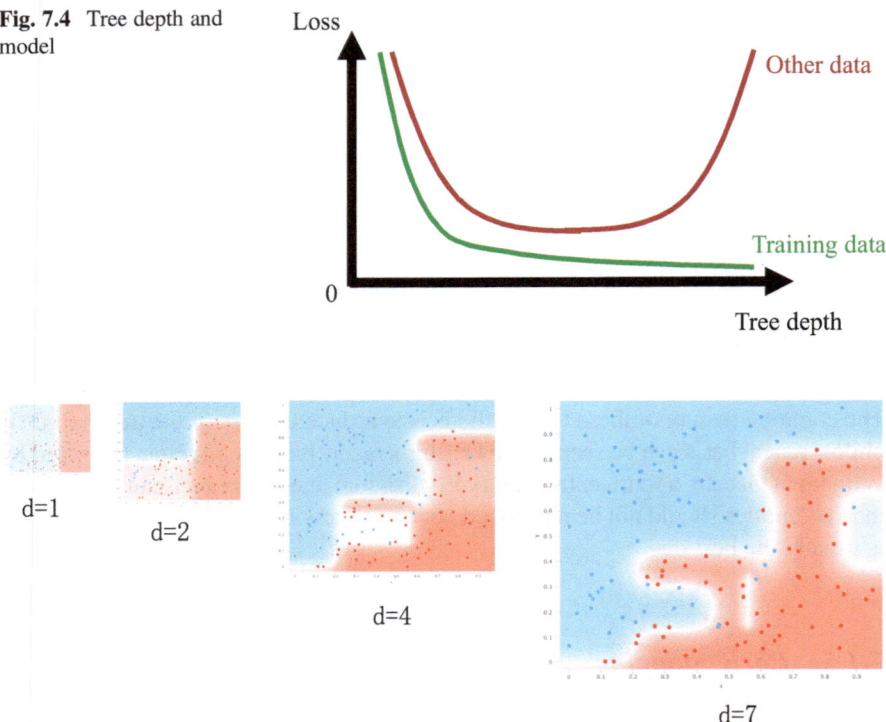

Fig. 7.5 Decision boundary and model complexity

We can observe a gradual decrease in the loss function of the training data as the depth of the decision tree increases. However, the loss function of other data may exhibit a pattern of initially decreasing and then increasing. This increasing trend indicates that the model is overfitting.

We can assess the risk of overfitting by examining the boundaries determined by the decision tree. Figure 7.5 illustrates that as the depth increases, the boundaries become increasingly intricate. For instance, when $d = 1$, the boundary is simply a straight line. However, when $d = 7$, the complexity of the boundary is significantly heightened, indicating a high possibility of overfitting.

When plotting the decision boundary graph for data classification alongside the graph depicting the change in loss values, we can see their relationship as shown in Fig. 7.6.

To reduce the risk of overfitting, we can choose to stop the growth of the tree at the appropriate depth position, and the best position is of course to select the position where the loss value is the smallest in other data. In the above figure, it corresponds to the tree depth of 4.

7.1.5.2 Hyperparameters

The tree depth is one of the hyperparameters that need to be selected based on our own experience. Let's take a closer look at Fig. 7.6: what does "Other data" refer to? Can we use test data instead?

It is crucial to adhere to the requirement for testing data, which states that the testing data should never be involved in the training of the model. The testing data should remain completely unaware of the entire training process to ensure unbiased evaluation. Using the testing set to select the optimal depth would violate this requirement, as it would reveal the training content to the test data, which is considered cheating. Therefore, it is strictly prohibited to utilize the testing set for such purposes.

At this point, a serious issue arises: what should be done when training data cannot be used and testing data is also unavailable? The answer is that we need to validate the dataset. For example, we can divide the entire dataset into 60% for training, 20% for validation, and 20% for testing (Fig. 7.7). The training set is used to train the model, the validation set is used to determine proper hyperparameters, and the test set is used to test the model (https://www.deeplearning.ai/courses/machine-learning-specialization/).

In this way, our model training takes the following steps:

- Split the dataset into training set, validation set, and test set.
- Use the training set to tune hyperparameters and train the model.
- Validate a set of hyperparameters using a validation dataset to ensure that the model achieves satisfactory performance and avoids overfitting to the validation dataset.
- Test the model using the test set to ensure that overfitting and other problems do not occur.

Fig. 7.6 Decision boundary and loss function

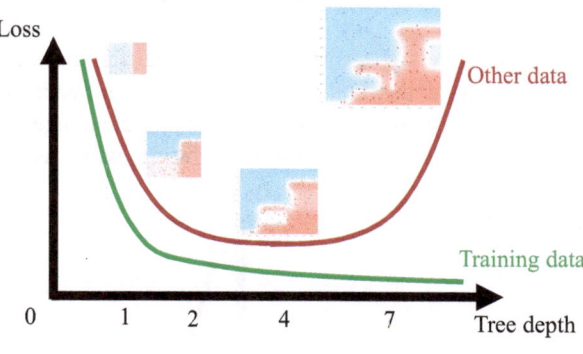

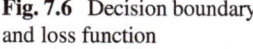

Fig. 7.7 Dataset partition

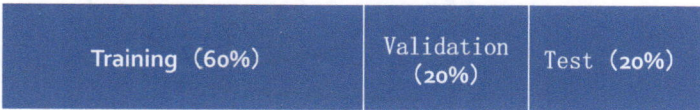

7.1.5.3   Threshold Error Rate

Limiting the depth of a decision tree can present a challenge as certain branches may benefit from deeper depths while others may be more effective with shallower depths. However, the traditional approach of using a fixed depth limit fails to address this issue adequately. An alternative method involves determining when to stop growing the tree based on the classification error rate. The idea is to cease tree growth when the error rate falls below a predefined threshold.

As depicted in Fig. 7.8, when the error rate increases, the complexity of the model decreases, resulting in a decrease in tree depth and an increased risk of underfitting. Setting a low error rate threshold can easily lead to overfitting, where the model performs well on the training set but poorly on the validation set. It is advisable to select hyperparameters that correspond to the lowest error rate on the validation set, indicating the optimal range for the model's performance.

7.1.5.4   Threshold Sample Size

Another crucial method for early stopping is the threshold sample size. When the data is classified to a point where the volume becomes small and lacks sufficient representativeness, it is advisable to early stop the growth of the decision tree. This method is often enforced and serves as the default option in many machine learning tools. For instance, if the sample size falls below 5, it becomes questionable how representative the differences exhibited by these samples are at this stage. As illustrated in Fig. 7.9, as the threshold sample size increases from low to high, the model transitions from overfitting to underfitting.

7.1.5.5   Pros and Cons of Early Stopping

The advantages of early stopping in tree training are apparent as it eliminates the need to fully expand all the data, resulting in faster training speed. However, there are also evident disadvantages. One major challenge is determining when to stop, as stopping too early can lead to underfitting, which is the opposite of overfitting.

Fig. 7.8 Threshold error rate

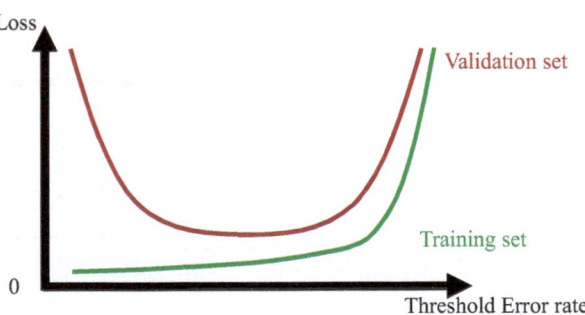

Fig. 7.9 Threshold sample size

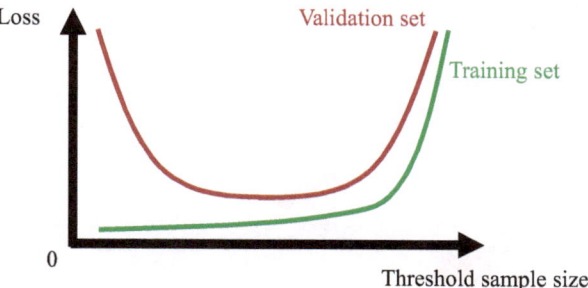

Fig. 7.10 Much more complicated real situation

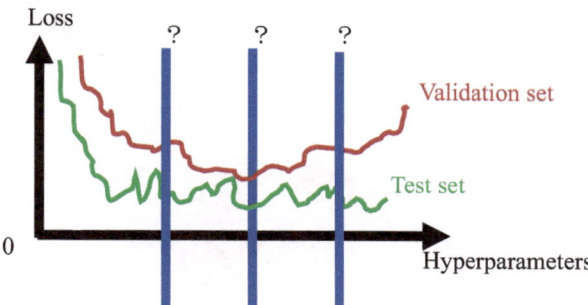

You might be wondering, how could we stop it too early? Isn't it clear from the previous figure? But the figures presented earlier are too idealistic. The real situation may be even worse than Fig. 7.10.

To mitigate the issue of too early stopping, pruning can be employed to address the problem of underfitting. In fact, early stopping is often referred to as pre-pruning, while pruning itself is commonly known as post-pruning.

7.1.5.6 Pruning

Pruning, as the name implies, is to prune off the branches and twigs to prevent the model from getting caught up in details. The most basic method is to first grow the entire tree and then prune unnecessary branches based on the error rate as shown in Fig. 7.11.

Pruning removes the branches to reduce the size of the decision tree, reducing the complexity of the final classifier and improving prediction accuracy by reducing overfitting.

In this context, a post-pruning technique is introduced. If the error rate of the parent node is lower than that of the child nodes, we prune the child nodes. This means that if the child nodes are unable to reduce the error rate, they should be discarded.

As depicted in Fig. 7.12, the red numbers within the nodes indicate the error rates, while the blue nodes represent the average error rates of the child nodes if the node

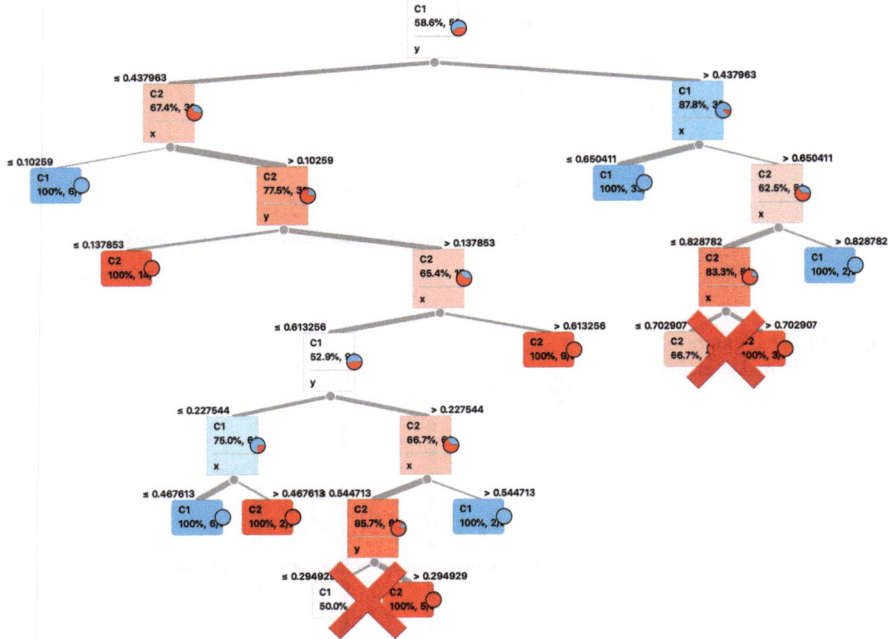

Fig. 7.11 Pruning

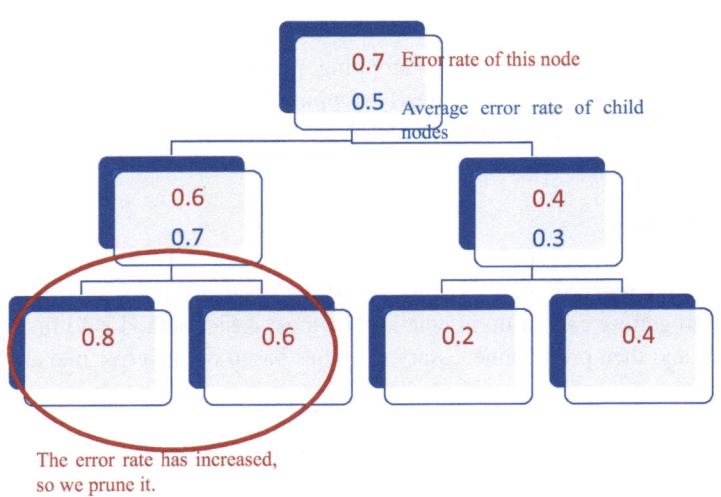

Fig. 7.12 Pruning based on the error rate

continues to branch out. Let's start with the root node. Its error rate is 0.4, and the average error rate of its child nodes is 0.3. Furthermore, the error rate of this node is higher than the error rate after branching, indicating that branching can proceed. Moving on to the right branch, its error rate is also higher than the average error rate of the child nodes, allowing for further branching. Now, let's examine the left branch. After branching, the error rate increases from 0.6 to 0.7. Therefore, in this scenario, there is no need for additional branching, and the two branches of this node should be pruned and removed.

Pruning can effectively prevent overfitting, so it is recommended to perform pruning during model training.

7.1.6 Ensemble Learning

If a model performs poorly in classification or regression tasks, it is considered a **weak learner**. Conversely, if it performs well, it is considered a **strong learner**. However, there is a trade-off between model complexity and performance. If a model is too simple, it may underfit the data and fail to capture important patterns. On the other hand, if it is too complex, it may overfit the data and perform poorly on unseen examples. To address this challenge, we can employ ensemble learning techniques. Ensemble learning combines multiple weak learners to create a strong learner, with the goal of surpassing the performance of any individual model. Two heads are better than one.

Ensemble learning primarily utilizes four methods:

- Bagging (Bootstrap AGGregatING).
- Stacking.
- Random forest.
- Boosting.

The above algorithms are all derived from decision tree algorithms, and they all belong to ensemble learning.

7.1.6.1 Bootstrap

Before delving into the various algorithms of ensemble learning, let's first understand the concept of **bootstrap**. Imagine there is a bag filled with balls of different colors. The process of bootstrapping involves randomly selecting a ball from the bag, noting its color, and then returning it back to the bag. This process is repeated multiple times, with each selection being independent and with replacement (Fig. 7.13).

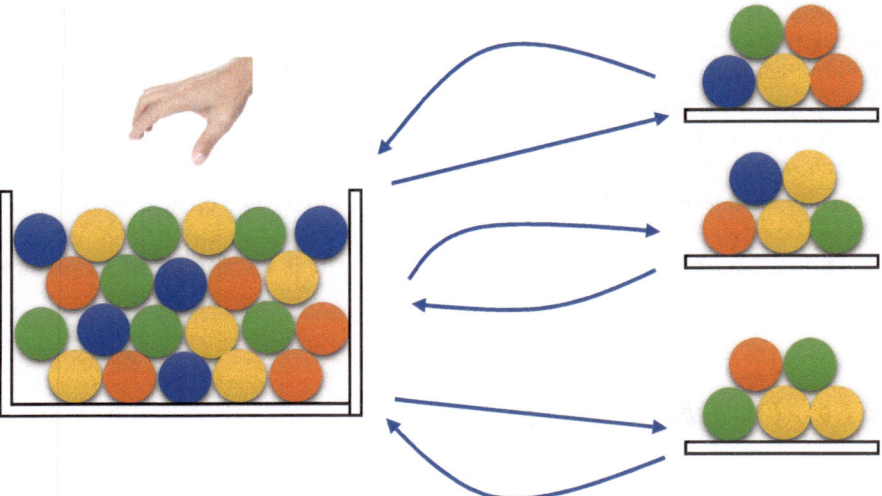

Fig. 7.13 Bootstrap

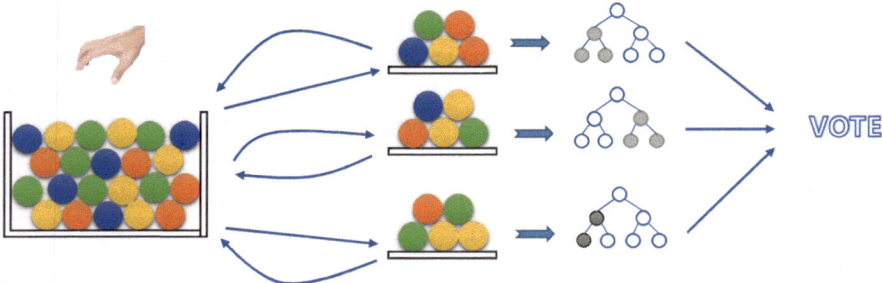

Fig. 7.14 Bagging using the voting method

7.1.6.2 Bagging

Bagging involves randomly sampling n instances with replacement from the training set, building n decision tree models, and then combining their predictions through voting to determine the final result (Fig. 7.14). Let's take the customer churn analysis model as an example. Suppose we perform bootstrapping on the data three times and use three learners to build three models. And these models predict "churn," "churn," and "not churn," respectively, for a specific record. By applying the principle of majority rule, the final classification result is determined as "churn" since it receives the majority of votes. Voting is a commonly employed method in bagging, where the predictions of the individual models are combined. To avoid ties, it is recommended to use an odd number of models if we use voting.

7.1.6.3 Stacking

Stacking is a technique similar to bagging, but with a notable difference in the voting stage. Unlike bagging, where a simple majority vote is used, stacking involves using the prediction results of each individual model as input. These predictions are then passed into an "integrator" model, which can be any available machine learning model, such as logistic regression (Fig. 7.15). The integrator model combines the predictions from the individual models to determine the final result.

7.1.6.4 Random Forest

Random forest is a technique that shares similarities with bagging, but it goes beyond just bootstrapping samples. In addition to sampling samples, random forest also samples features during the training process. Specifically, it randomly selects a subset of features, typically the square root of the total number of features, at each split. This feature sampling is performed to mitigate the impact of feature correlation on the model.

For instance, when Jack seeks rental advice from his friends, each friend may inquire about different aspects such as Jack's personal preferences, budget, location preferences, and other relevant factors. Based on their individual responses, Jack will receive various pieces of advice, resembling the decision tree algorithm. As more friends contribute with different questions and advice, Jack accumulates a comprehensive set of suggestions and makes a decision by considering the collective advice received from his friends. And this is the random forest algorithm.

Points for Discussion

- Would you solve using similar algorithms such as decision trees, random forests, or stacking to solve problems in your life? Please raise some examples.

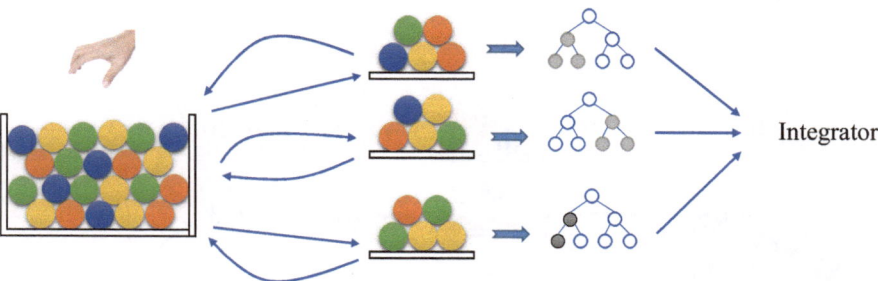

Fig. 7.15 Stacking

7.1.6.5   Boosting

Similar to bagging, boosting algorithms aim to combine multiple weak learners to create a strong learner. However, there is a fundamental difference between the two approaches. In bagging, each step is independent and samples are selected randomly. In contrast, boosting builds on the previous iterations and focuses on selecting misclassified samples to improve performance in subsequent iterations. To illustrate this concept, imagine a student who compiles a book of mistakes from every practice and exam. The student then learns from these mistakes. After going through the entire set of incorrect exercises, the student can summarize a new set of incorrect exercises based on the mistakes made. This process continues, constantly improving performance by learning from the new set of incorrect exercises. This iterative approach is analogous to boosting (Fig. 7.16).

For example, Jack seeks advice from friends about renting a house. He asks the next friend based on the previous friend's response, and each friend's question and response are also based on the previous friend's answer. Eventually, Jack decides to follow the advice of the last friend to make his final decision. This sequential decision-making process resembles the boosting algorithm.

In boosting, the final result is determined solely by the last learner, while each previous learner is making silent contributions.

Give It a Try

- Try using the above algorithms in KNIME.

7.1.6.6   Adaptive Boosting

Boosting algorithm is a very popular algorithm in data analysis. Here we introduce the basic algorithm AdaBoosting (adaptive boosting) in boosting algorithm. The adaptiveness lies in the fact that the misclassified samples by the previous learner will be used to train the next learner. Suppose we want to classify the points of two colors in Fig. 7.17.

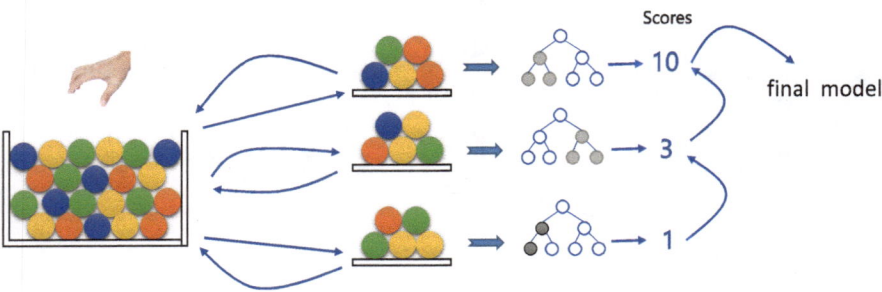

Fig. 7.16 Boosting

Fig. 7.17 Classification of
points of two colors

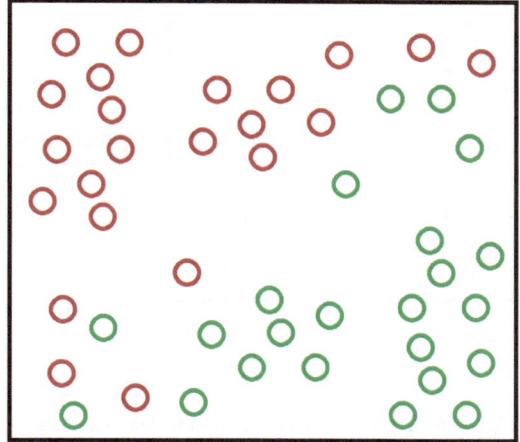

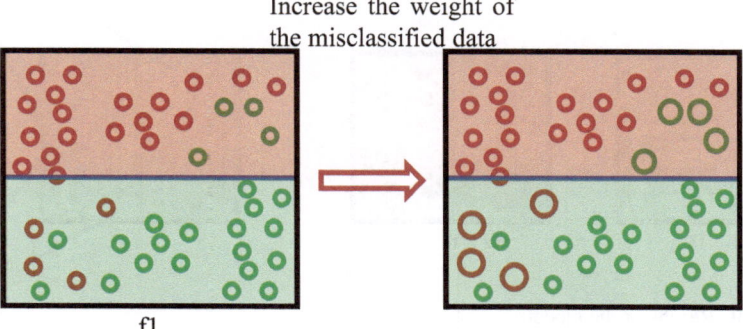

Fig. 7.18 Increase the weight of misclassified data

Here, the weight of each data is the same, and model f1 simply classifies these points as shown in Fig. 7.18.

It is evident that this simple classification exhibits a significant number of errors. Subsequently, the algorithm assigns higher weights to the misclassified data, as depicted in Fig. 7.18 on the right, where the size of the points is enlarged. As the weight of the incorrect data in model f1 increases, model f2 becomes more focused on accurately classifying the points that f1 misclassified, as illustrated in Fig. 7.19.

The models are assigned weights based on their error rates, with higher weights given to models with lower error rates and lower weights assigned to models with higher error rates. This indicates that the algorithm places greater value on the predictions of models that exhibit better classification performance. The weighted predictions of the models are then aggregated to obtain the final result of adaptive boosting. As depicted in Fig. 7.20, model f1 is assigned a weight w1, model f2 is assigned a weight w2, and the final model is $f = w1 \times f1 + w2 \times f2$. In this context,

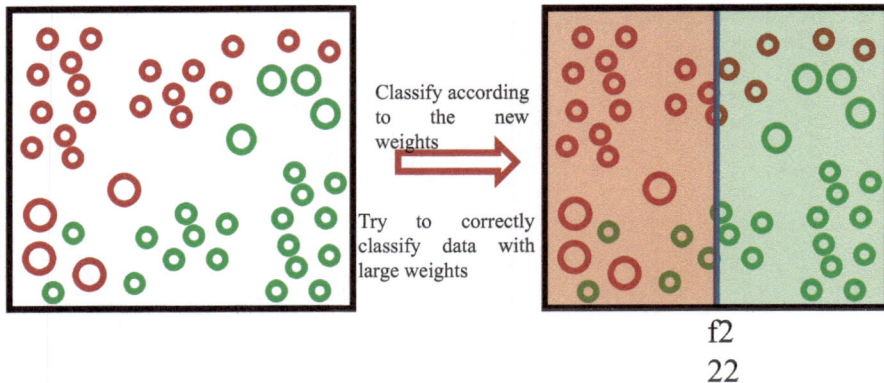

Fig. 7.19 Try to correctly classify the data with large weights

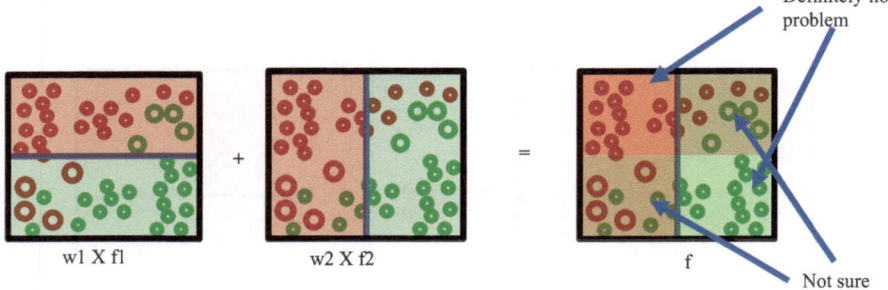

Fig. 7.20 Results of the model

the areas where both models make consistent predictions are considered certain, while other locations are deemed uncertain and may require further improvement.

Generally, increasing the number of weak learners can improve the accuracy of the final model. As shown in Fig. 7.21, more models work together to produce a strong learner.

Points for Discussion

- Have you used the idea of adaptive boosting in your daily work and life?

7.2   Solve the Titanic Case with Decision Tree

You may use decision tree to solve the Titanic case. The important issue to note is how to use the validation dataset in KNIME. A hint is given here as shown in Fig. 7.22.

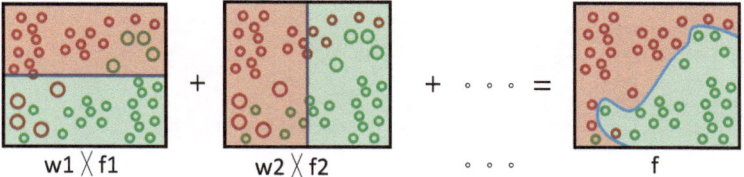

Fig. 7.21 More models yield better results

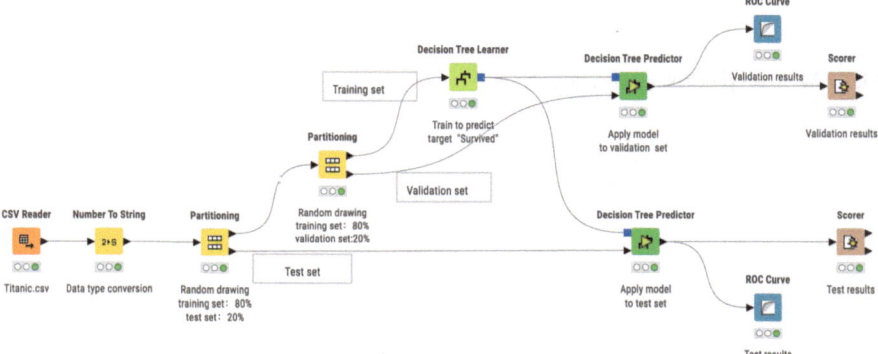

Fig. 7.22 Set the training set, validation set, and test set

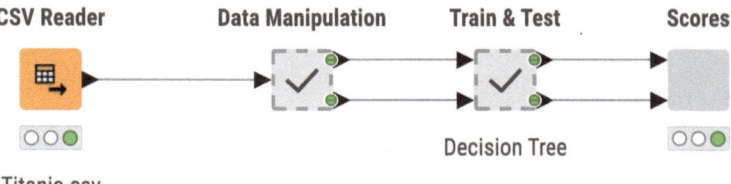

Fig. 7.23 KNIME workflow

Give It a Try

• Think about how to build the entire workflow on your own before moving on the next paragraph.

Now let me briefly introduce this workflow. As always, our workflow remains as shown in Fig. 7.23. Regardless of the problem we are solving, the workflow itself will not change significantly; only the specific content inside needs to be modified.

First, this workflow does not require handling categorical data, as decision tree models can directly use both numeric and categorical data. Therefore, there is no need to use the "One to Many" node, and the model does not require the conversion of dummy variables. Therefore, the data processing Metanode can be seen in

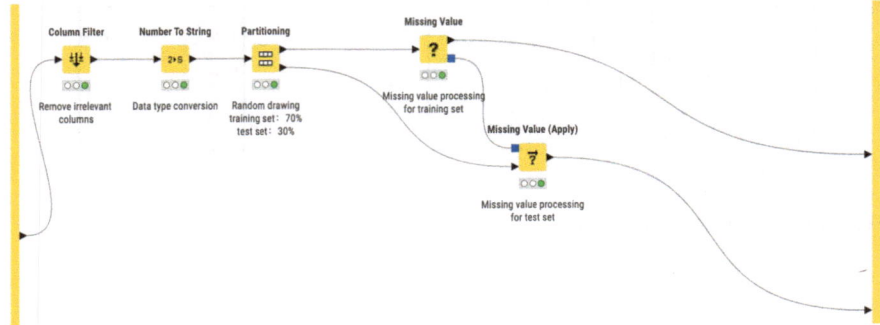

Fig. 7.24 Data processing Metanode

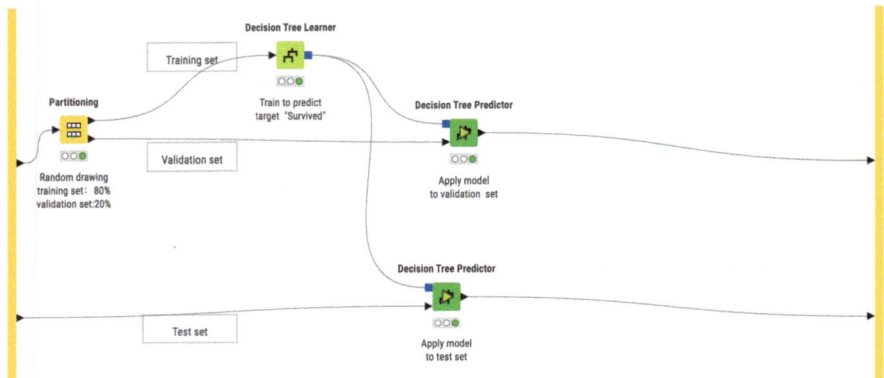

Fig. 7.25 "Train & Test"

Fig. 7.24. Essentially, this is the data processing Metanode used by the logistic regression model.

Secondly, except for the model to be used, the "Train & Test" Metanode remains unchanged, as shown in Fig. 7.25.

It is important to note the data partitioning process in this workflow. Firstly, the data is divided into the test set and other data. Then, the "other data" is further split into the training set and the validation set. The training data is utilized to train the model, while the trained model and the validation data are fed into a decision tree predictor for making predictions. Subsequently, a scorer assesses the performance of the model. If overfitting is detected, adjustments to the hyperparameters are necessary. Once it is confirmed that there is no overfitting on the validation set, the model is tested using the test set to ensure that no issues such as overfitting are present.

The model evaluation is conducted within the "Scores" component, which is composed of "ROC Curve" and "Scorer" nodes, and it evaluates both the validation set and the test set, respectively.

7.3   Decision Tree Advanced: Feature Engineering

This is a Kaggle competition for bank customer classification (https://www.kaggle.com/c/santander-customer-satisfaction). You will receive a dataset containing a large number of anonymous numeric variables. The "TARGET" column is the variable to be predicted. For dissatisfied customers, TARGET equals 1, and for satisfied customers, it equals 0. Our task is to predict the probability of each customer being a dissatisfied customer. The data for this project can be found in the "customer train.csv" in the above kaggle competition link.

The challenge lies in the fact that the dissatisfied group is relatively small compared to other groups, and the meaning of the data features is unknown. Additionally, there are a large number of other data points with a value of 0 and related features. Therefore, this problem requires extensive feature engineering to address these complexities.

For us beginners, let's stick to a simple processing method:

Columns with only zeros or the same values? Delete them! Linearly correlated columns? Delete them! Columns have almost no changes? Delete them!

Finally, we will use the previously introduced decision tree techniques to optimize the model.

The workflow of this project is shown in Fig. 7.26. In this workflow, we first perform data type conversion of "Survived," followed by exploratory data analysis (EDA) and feature engineering. Due to the massive workload of feature engineering, we will create a separate Metanode for it. After the partition, data will separately

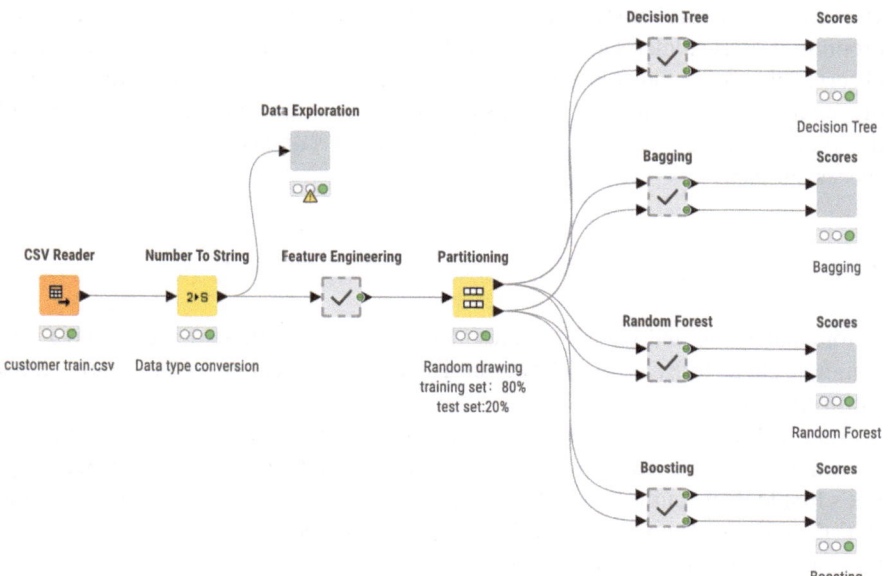

Fig. 7.26 Decision tree workflow

input into four modules—decision tree, random forest, boosting, and bagging—for training, testing, and scoring.

7.3.1 Exploratory Data Analysis

In the node monitor of the "CSV reader" node, it can be found that the dataset consists of 76,020 samples, 370 features, and 1 binary target. There are no missing values in the entire dataset, but it cannot be ruled out that the data provider has encoded missing data.

The dataset is partially anonymous, so it is unclear what each feature represents. Our only clue is the feature name, which is not random but also vague in meanings. So we need to guess what it is. For example, the names of the first few features are ID, var3, var15, and imp_ent_var16_ult1. Except for being sure that the ID column should be a kind of ID, nothing else can be confirmed.

In order to understand the meaning of each feature, we need to conduct EDA on the data. The composition nodes of the EDA component are shown in Fig. 7.27. Before that, we need to use the "Number To String" node to convert the "TARGET" column from numerical type to character type (as shown in Fig. 7.26). This section is based on the analysis of van den Elsen (2017).

7.3.1.1 Data Balance

Use the "Pie Chart" node with the settings shown in Fig. 7.28 to explore the proportion of customer satisfaction for "TARGET." From the pie chart, it can be observed that 0 (customer is not dissatisfied) accounts for 96.04% and 1 (customer is dissatisfied) accounts for 3.96%. This dataset has a significant class imbalance, and this proportion should be in line with the expectations of customer satisfaction for a successful bank.

7.3.1.2 Feature Group

Before analyzing individual features, let's first see if these features have any common characteristics. As shown in Table 7.1, all feature names start with a substring from the table.

Start with what we can guess better. "delta" and "num" should be the simplest, as we usually use these two to represent the difference and numeric value, respectively. Let's first look at "num." Add a "Statistics" node in the main workflow and view the statistical data of "num" in Fig. 7.29. It can be observed that a significant portion of this type of data is 0 and concentrated at 0.

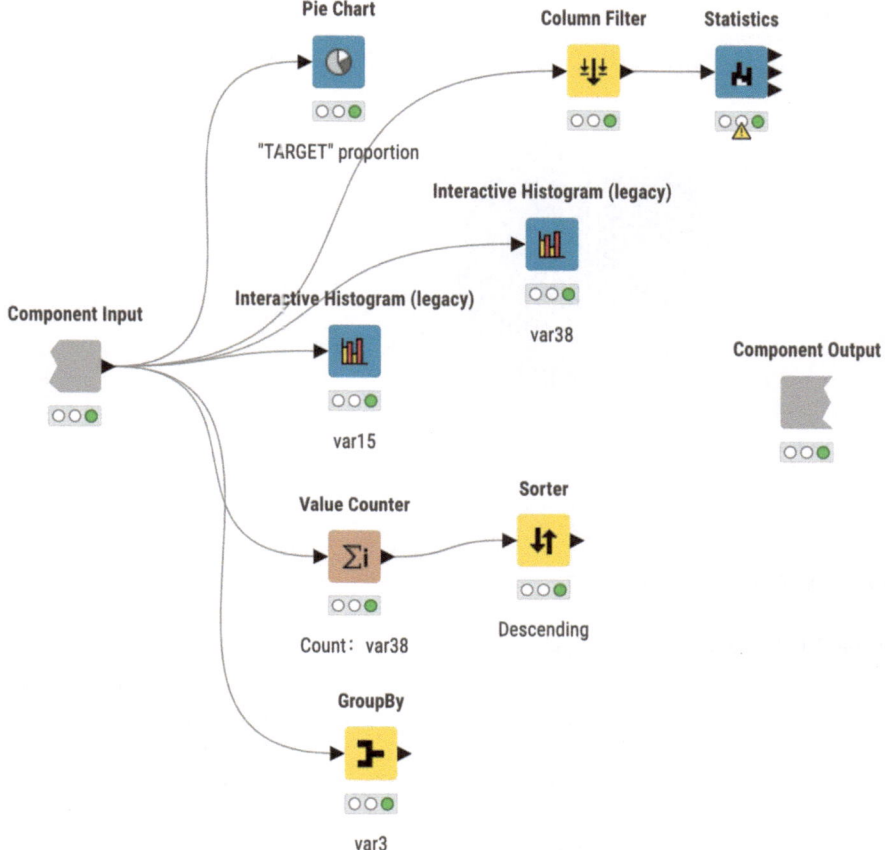

Fig. 7.27 EDA component

Based on Fig. 7.30, it can be quantitatively observed that 0 occupies a significant proportion in terms of numerical data volume ranking. This type of data exists in integer form and is likely to be some kind of categorical data.

Let's take a look at "delta." Similar methods can reveal their concentration around a specific value. Guess what difference is it?

If we look at "imp," there are also large number of 0s, but the values are very large. Some people say that "imp" corresponds to the Spanish word of "importe," which means quantity. So this part of the data may be related to the amount of money or other quantities (Fig. 7.31).

Next, take a look at "ind": could it be an abbreviation for the English word "indicator"? Observe the distribution of this type of data in Fig. 7.32, and it is found that the distribution only consists of two values: "0" and "1." Therefore, it should be an indicator.

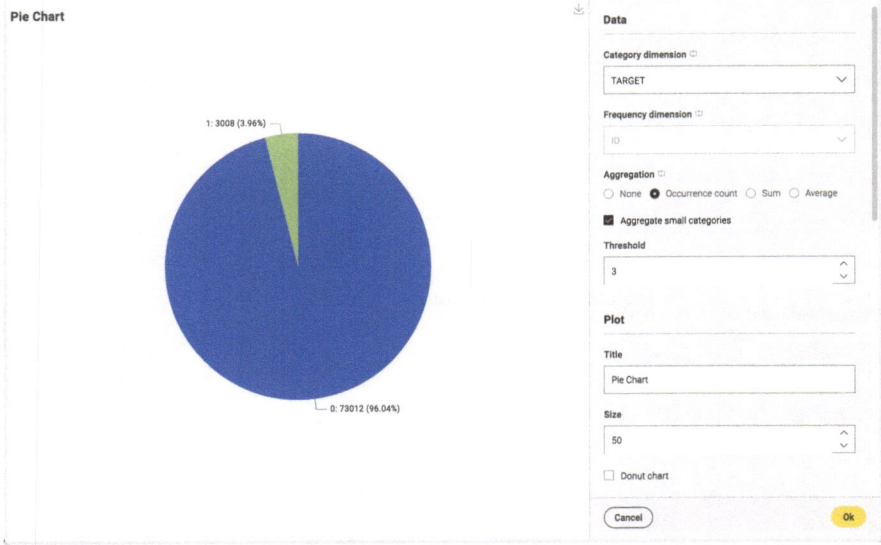

Fig. 7.28 Proportion of "TARGET"

Table 7.1 String features

Substrings	Features
delta	delta_imp_amort_var18_1y3
imp	imp_aport_var17_hace3
ind	ind_var1
num	num_var12
saldo	saldo_var28
var	var15

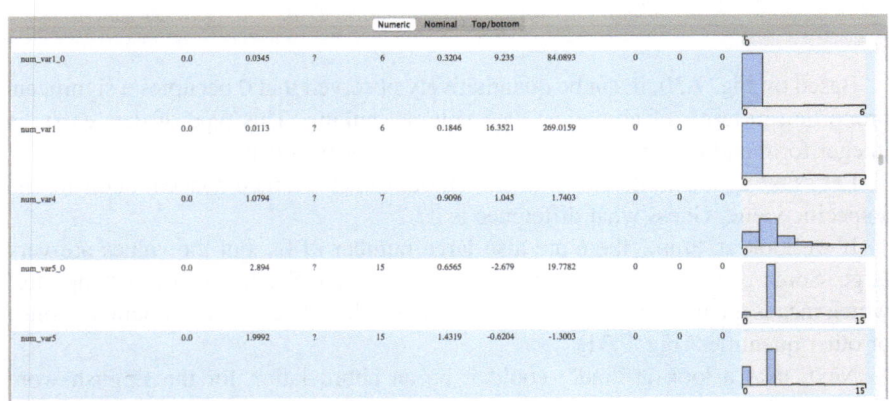

Fig. 7.29 Distribution of "num"

	num_var26	num_var25_0	num_var25	num_op_var40_hace2	num_op_var40_hace3	num_op_var40_ult1	num_op_var40_ult3	num_op_var41_hace2	num_op_var41_hace3	num_op_var41_ult1	num_op_var41_ult3
0	No. missings: 0	No. missings: 0	No. missings: 0	No. missings: 0	No. missings: 0	No. missings: 0	No. missings: 0	No. missings: 0	No. missings: 0	No. missings: 0	No. missings: 0
	Top 20: 0 : 74147 3 : 1574 6 : 240 9 : 42 12 : 12 15 : 2 33 : 1 21 : 1 27 : 1	Top 20: 0 : 74223 3 : 1525 6 : 216 9 : 40 12 : 11 15 : 2 33 : 1 21 : 1 27 : 1	Top 20: 0 : 74223 3 : 1525 6 : 216 9 : 40 12 : 11 15 : 2 33 : 1 21 : 1 27 : 1	Top 20: 0 : 75918 3 : 26 6 : 18 9 : 16 12 : 10 15 : 8 18 : 4 24 : 4 21 : 3 45 : 2 51 : 2 27 : 2 117 : 1 30 : 1 72 : 1 42 : 1 96 : 1 54 : 1 81 : 1	Top 20: 0 : 76013 3 : 3 6 : 2 9 : 1 48 : 1	Top 20: 0 : 75791 3 : 56 6 : 34 9 : 22 12 : 27 15 : 17 18 : 13 24 : 13 21 : 9 33 : 6 27 : 5 36 : 5 30 : 4 60 : 3 78 : 2 45 : 2 39 : 2 57 : 2 69 : 1 234 : 1	Top 20: 0 : 75772 3 : 48 6 : 39 12 : 27 24 : 17 15 : 16 9 : 15 21 : 12 18 : 9 30 : 8 33 : 8 42 : 6 27 : 5 36 : 5 45 : 5 54 : 3 39 : 3 69 : 2 75 : 2 60 : 2	Top 20: 0 : 67679 3 : 2617 6 : 1405 9 : 865 12 : 624 15 : 467 18 : 377 21 : 313 24 : 255 27 : 242 30 : 180 33 : 142 36 : 130 39 : 107 42 : 84 45 : 81 48 : 62 51 : 47 54 : 43 57 : 38	Top 20: 0 : 75044 3 : 491 6 : 206 9 : 107 12 : 55 15 : 36 21 : 19 18 : 19 30 : 10 24 : 8 27 : 6 33 : 6 39 : 3 48 : 2 60 : 1 81 : 1 36 : 1 42 : 1 57 : 1 51 : 1	Top 20: 0 : 64388 3 : 2929 6 : 1742 9 : 1090 12 : 811 15 : 733 18 : 619 21 : 503 24 : 400 27 : 365 30 : 315 33 : 272 36 : 214 39 : 204 42 : 162 45 : 157 48 : 146 51 : 120 54 : 112 57 : 102	Top 20: 0 : 62355 3 : 3024 6 : 1816 9 : 1237 12 : 896 15 : 745 18 : 638 21 : 550 24 : 447 27 : 408 30 : 346 33 : 324 36 : 282 39 : 273 42 : 253 48 : 196 45 : 175 51 : 155 54 : 154 57 : 102
	Bottom 20:	Bottom 20:	Bottom 20:	Bottom 20:	Bottom 20:	Bottom 20: 135 : 1 75 : 1 81 : 1 66 : 1 42 : 1 63 : 1 51 : 1 105 : 1 168 : 1 171 : 1	Bottom 20: 63 : 2 351 : 1 180 : 1 129 : 1 84 : 1 66 : 1 51 : 1 177 : 1 90 : 1 264 : 1 48 : 1 57 : 1 87 : 1 171 : 1 189 : 1	Bottom 20: 93 : 9 99 : 8 96 : 7 102 : 6 87 : 6 114 : 4 129 : 3 120 : 2 108 : 2 117 : 2 171 : 2 123 : 1 156 : 1 201 : 1 162 : 1 153 : 1 249 : 1 165 : 1 144 : 1 186 : 1	Bottom 20: 66 : 1 69 : 1	Bottom 20: 129 : 2 210 : 2 159 : 2 186 : 2 156 : 2 162 : 2 147 : 2 174 : 1 141 : 1 213 : 1 171 : 1 144 : 1 177 : 1 192 : 1 240 : 1 168 : 1 204 : 1 270 : 1 216 : 1 468 : 1	Bottom 20: 306 : 2 267 : 2 363 : 1 258 : 1 291 : 1 225 : 1 318 : 1 279 : 1 288 : 1 276 : 1 231 : 1 249 : 1 252 : 1 312 : 1 399 : 1 321 : 1 219 : 1 270 : 1 240 : 1 468 : 1

Fig. 7.30 Ranking of data volume for "num"

imp_amort_var18_ult1	0.0	0.2312	?	15,691.8	57.321	270.3322	73,891.7966	0	0	0	0 … 15.692
imp_amort_var34_hace3	0.0	0.0	?	0.0	0.0	0.0	0.0	0	0	0	
imp_amort_var34_ult1	0.0	0.0181	?	1,096.02	4.0995	255.4113	67,480.7015	0	0	0	0 … 1.096
imp_aport_var13_hace3	0.0	2,823.9491	?	840,000	25,334.4681	12.4604	193.6248	0	0	0	0 … 840.000
imp_aport_var13_ult1	0.0	619.585	?	450,000	11,252.9951	23.9059	670.2091	0	0	0	0 … 450.000
imp_aport_var17_hace3	0.0	98.7888	?	6,083,691.87	22,120.7192	273.6739	75,255.8869	0	0	0	0 … 6.083.692

Fig. 7.31 Distribution of "imp"

ind_var31_0	ind_var31	ind_var32_cte	ind_var32_0	ind_var32	ind_var33_0	ind_var33	ind_var34_0	ind_var34	ind_var37_cte	ind_var37_0	ind_var37	ind_var39_0	ind_var40_0
No. missings: 0	No. missings: 0	No. missings: 0	No. missings: 0	No. missings: 0	No. missings: 0	No. missings: 0	No. missings: 0	No. missings: 0	No. missings: 0	No. missings: 0	No. missings: 0	No. missings: 0	No. missings: 0
Top 20: 0 : 75695 1 : 325	Top 20: 0 : 75741 1 : 279	Top 20: 0 : 75928 1 : 92	Top 20: 0 : 75938 1 : 82	Top 20: 0 : 75963 1 : 57	Top 20: 0 : 75972 1 : 48	Top 20: 0 : 76018 1 : 2	Top 20: 0 : 76018 1 : 2	Top 20: 0 : 76018 1 : 2	Top 20: 0 : 70524 1 : 5496	Top 20: 0 : 71059 1 : 4961	Top 20: 0 : 71059 1 : 4961	Top 20: 0 : 66955 0 : 9065	Top 20: 0 : 75152 1 : 868

Fig. 7.32 Distribution of "ind"

Fig. 7.33 Distribution of "saldo"

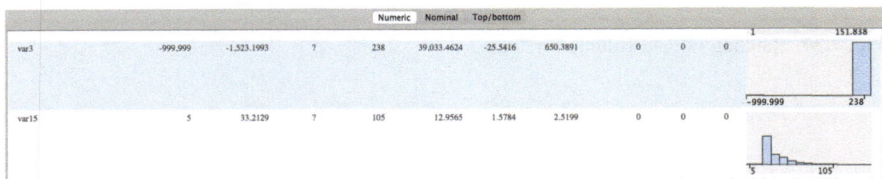

Fig. 7.34 Distribution of "var"

Let's take a look at "saldo" again. As can be seen in Fig. 7.33, this type of data is somewhat similar to "imp," with a large number of 0s, but the data value is much larger than "imp," so it is speculated that this type of data should also be related to money, but in a larger quantity.

Finally, let's take a look at "var" as shown in Fig. 7.34. It seems that this type of data does not have any obvious common characteristics, so let's proceed to observe each of these features individually.

7.3.1.3 Var

Add the "Interactive Histogram (legacy)" node and the "GroupBy" node to the main workflow for future use.

If we view "var3," it can be seen from Fig. 7.35 that most of the data is 2, with very few other values, as indicated by the "Statistics" node.

We can use the "GroupBy" node to see how many different values there are in total. Set up this node as shown in Fig. 7.36, and then view the results in the node monitor after executing it. There are a total of 208 different values as shown in Fig. 7.37, which is close to the total number of countries and regions in the world. Therefore, the analysis suggests that this data likely represents the country or region from where the customers are coming.

Fig. 7.35 Ranking of data volume in "var3"

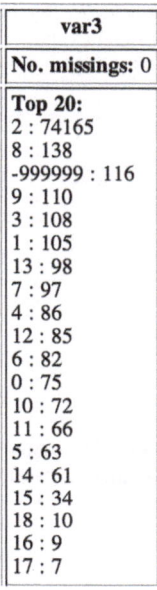

var3
No. missings: 0

Top 20:
2 : 74165
8 : 138
-999999 : 116
9 : 110
3 : 108
1 : 105
13 : 98
7 : 97
4 : 86
12 : 85
6 : 82
0 : 75
10 : 72
11 : 66
5 : 63
14 : 61
15 : 34
18 : 10
16 : 9
17 : 7

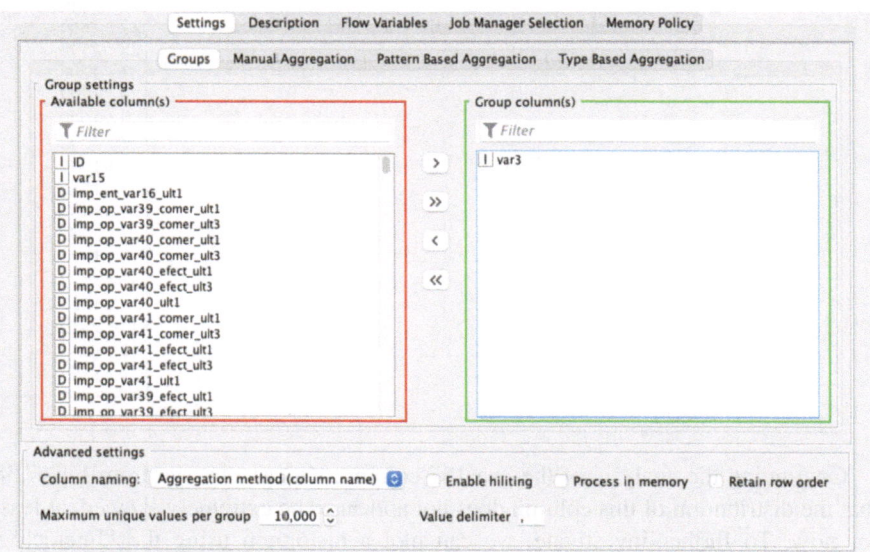

Fig. 7.36 View "var3" with "GroupBy" node

Based on the data and our assumption, it appears that the column represents countries or regions, with a possibility of 2 indicating the United States, which has the highest population. However, there is one outlier with a value of −999,999. This value seems unusual and could potentially represent a missing data point.

Fig. 7.37 Number of
different values

Rows: 208 | Columns: 1

#	RowID	var3 Number (integer)
203	Row202	225
204	Row203	228
205	Row204	229
206	Row205	231
207	Row206	235
208	Row207	238

Fig. 7.38 Ranking of data
volume for "var15"

var15
No. missings: 0
Top 20: 23 : 20170 24 : 6232 25 : 4217 26 : 3270 27 : 2861 28 : 2173 31 : 1798 29 : 1727 30 : 1640 32 : 1592 36 : 1546 35 : 1536 34 : 1489 37 : 1401 38 : 1390 33 : 1377 39 : 1329 40 : 1310 41 : 1242 42 : 1199

Continuing the analysis of the "var15" column, we can observe from Fig. 7.38 that the distribution of this column does not appear to be extremely skewed, at least for now. To further investigate, we can plot a histogram using the "Interactive Histogram (legacy)" node. By clicking on the Configure button, as shown in Fig. 7.39, we can set it to display all nodes. Then, by clicking on "Open View," we can observe and analyze the data, as shown in Fig. 7.40. From the histogram, we can see that the data volume suddenly increases to its maximum value starting from 18 and then gradually decreases. There is not much data remaining around the age of 60. Based on this observation, we can make a bold assumption that this column represents age. It is reasonable to assume that a person starts working at the age of 18 and retires at 60, so their interactions with the bank would likely begin at the age

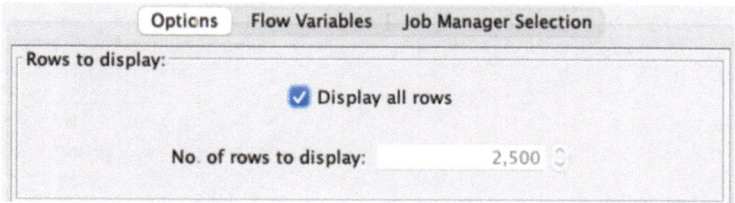

Fig. 7.39 Ensure all data is displayed

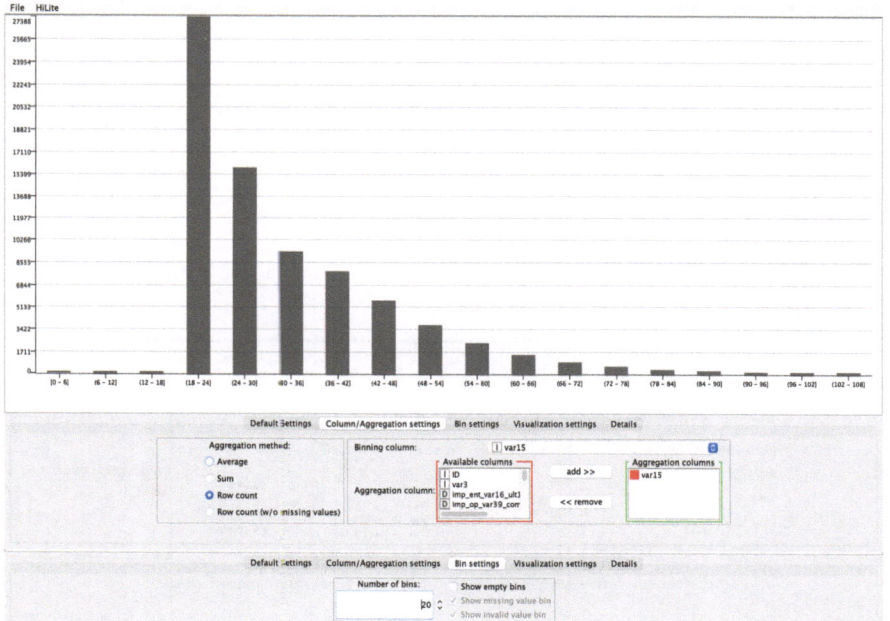

Fig. 7.40 View the histogram

of 18 and gradually decrease toward retirement. Therefore, it is plausible to conclude that this column represents age.

Next, let's take a look at "var21" and "var36." We are currently unsure of their representations merely by observation.

Lastly, let's examine the "var38" column. Similar to the previous method, by observing Fig. 7.41, we can see that the values are more concentrated in the lower range, around 100,000. However, there is an abnormal peak that stands out. Upon closer examination of the data statistics table in Fig. 7.41, we can observe that the average value of this data is 117,235.8094. Interestingly, this value falls precisely within the outlier range shown in Figs. 7.40 and 7.42.

To further evaluate the "var38" column, we can add the "Value Counter" and "Sorter" nodes to the main workflow in the following sequence: firstly, we use the

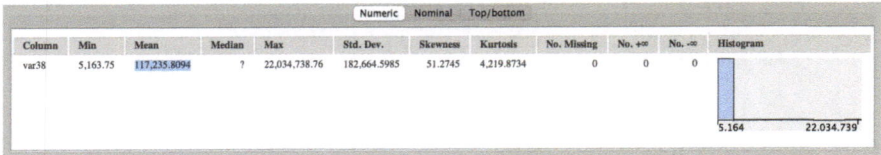

Fig. 7.41 Mean value

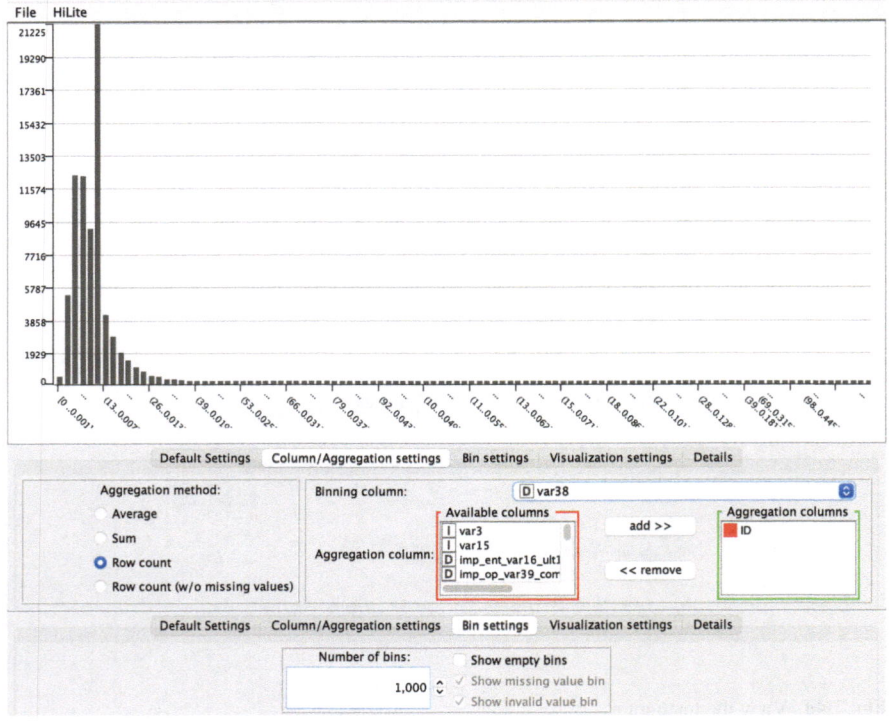

Fig. 7.42 Histogram

"Value Counter" node to count the occurrences of each value of "var38," as depicted in Fig. 7.43. This calculation yields the results shown in Fig. 7.44. Next, we employ the "Sorter" node to arrange the occurrences of each value of "var38" in descending order, as illustrated in Fig. 7.45. The sorted results are displayed in Fig. 7.46. Upon analysis, it becomes evident that the most frequent value is 117310.979016494, which appears significantly more frequently than the other values. This raises suspicions about other values. Discussions on the Kaggle competition forum suggest that, during the data processing stage, these values were replaced with the average value of the data.

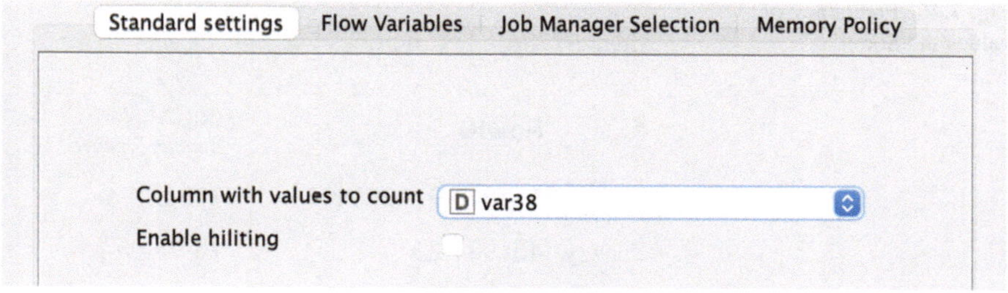

| Standard settings | Flow Variables | Job Manager Selection | Memory Policy |

Column with values to count   D var38

Enable hiliting

Fig. 7.43 Configure the "Value Counter"

Fig. 7.44 Results of "Value Counter"

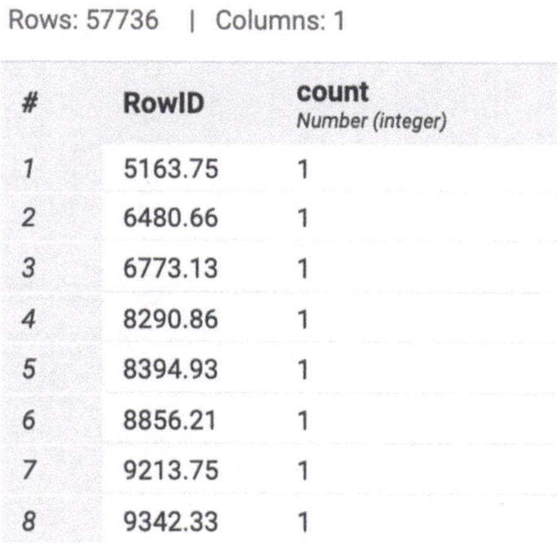

Rows: 57736 | Columns: 1

#	RowID	count Number (integer)
1	5163.75	1
2	6480.66	1
3	6773.13	1
4	8290.86	1
5	8394.93	1
6	8856.21	1
7	9213.75	1
8	9342.33	1

| Sorting Filter | Advanced Settings | Flow Variables | Job Manager Selection | Memory Policy |

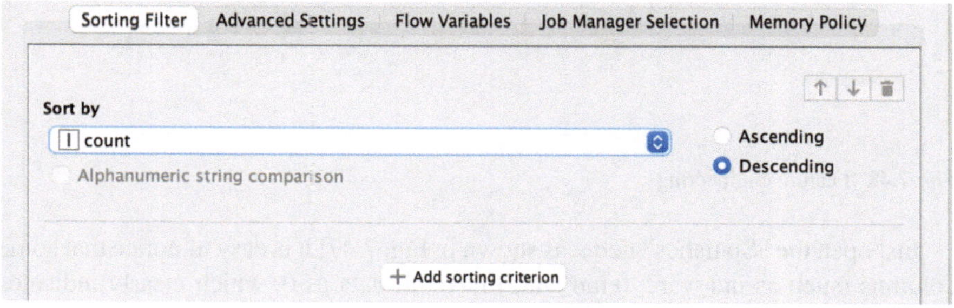

Sort by

I count   ○ Ascending
                                          ● Descending

Alphanumeric string comparison

+ Add sorting criterion

Fig. 7.45 Configure the "Sorter"

7.3.1.4   Data Quality

In the general analysis above, we found that there are many problems with the data, but they have been cleverly concealed. Now let's take a brief look at the problems that may arise to prepare for the subsequent analysis.

Fig. 7.46 Frequency of values in "Sorter"

Rows: 57736 | Columns: 1

#	RowID	count ↓ Number (integer)
1	117310.979016494	14868
2	451931.22	16
3	463625.16	12
4	104563.8	11
5	288997.44	11
6	236690.34	8
7	67088.31	7
8	104644.41	7

Numeric Nominal Top/bottom

ind_var2_0	ind_var2	ind_var5_0	ind_var5	ind_var6_0	ind_var6	ind_var8_0	ind_var8	ind_var12_0	ind_var12	ind_var13_0	ind_var13_corto_0	ind_var13_corto	ind_va
No. missings: 0	No. missings: 0	No. missings: 0	No. missings: 0	No. missings: 0	No. missings: 0	No. missings: 0	No. missings: 0	No. missings: 0	No. missings: 0	No. missings: 0	No. missings: 0	No. missings: 0	No. mi
Top 20: 0 : 76020	Top 20: 0 : 76020	Top 20: 1 : 72829 0 : 3191	Top 20: 1 : 50459 0 : 25561	Top 20: 0 : 76012 1 : 8	Top 20: 0 : 76018 1 : 2	Top 20: 0 : 73524 1 : 2496	Top 20: 0 : 73846 1 : 2174	Top 20: 0 : 70887 1 : 5133	Top 20: 0 : 72564 1 : 3456	Top 20: 0 : 72048 1 : 3972	Top 20: 0 : 72756 1 : 3264	Top 20: 0 : 72867 1 : 3153	Top 20 0 : 752 1 : 773

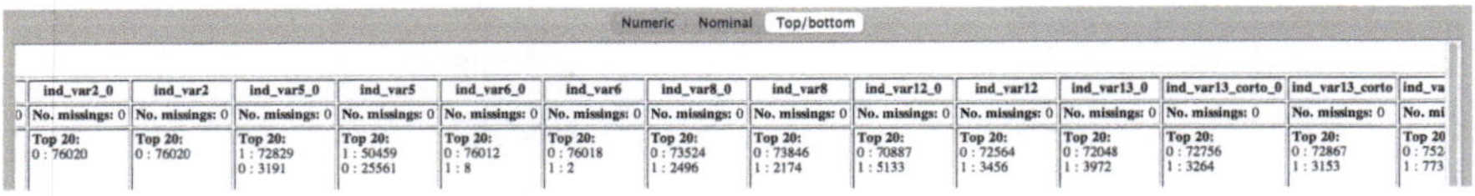

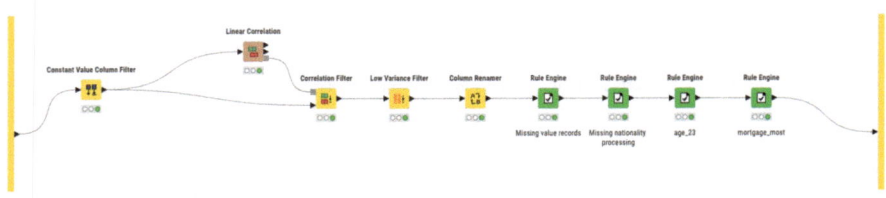

Fig. 7.47 Further view the data

Fig. 7.48 Feature engineering

Just open the "Statistics" node, as shown in Fig. 7.47; it is easy to notice that some columns (such as ind_var2_0,ind_var2) have all data as 0, which clearly indicates that such data is not useful.

Continue to view the data and you may find more problems. Below, we will address these problematic data through feature engineering.

7.3.2 *Feature Engineering*

Open "Feature Engineering" Metanode. As depicted in Fig. 7.48, the feature engineering of this project involves a multitude of nodes, many of which are unfamiliar

to us. However, the overall concept remains consistent with previous feature engineering approaches, which involve the combination of data visualization and processing. Now, let's delve into each step of the process for a closer examination.

7.3.2.1 Remove Constant Value Column

Use the "Constant Value Column Filter" node to remove constant columns. Constant columns do not provide any information and are not necessary, so they can be deleted directly.

7.3.2.2 Filter by Relevance

We may then filter out unnecessary columns based on the correlation of the data. First, use the "Linear Correlation" node to calculate the correlation of each column, and then use the "Correlation Filter" node to filter out highly correlated nodes.

The correlation matrix is presented in Fig. 7.49, illustrating a significant number of highly correlated features. To determine which features to retain and which to discard, a threshold for the correlation coefficient should be set. The specific threshold value depends on the specific problem at hand and requires analysis. The "Correlation Filter" node is employed to filter out features with excessively high correlation. By examining Fig. 7.48, it is evident that the inputs to this node consist of the data and the correlation model, derived from the outputs of preceding nodes.

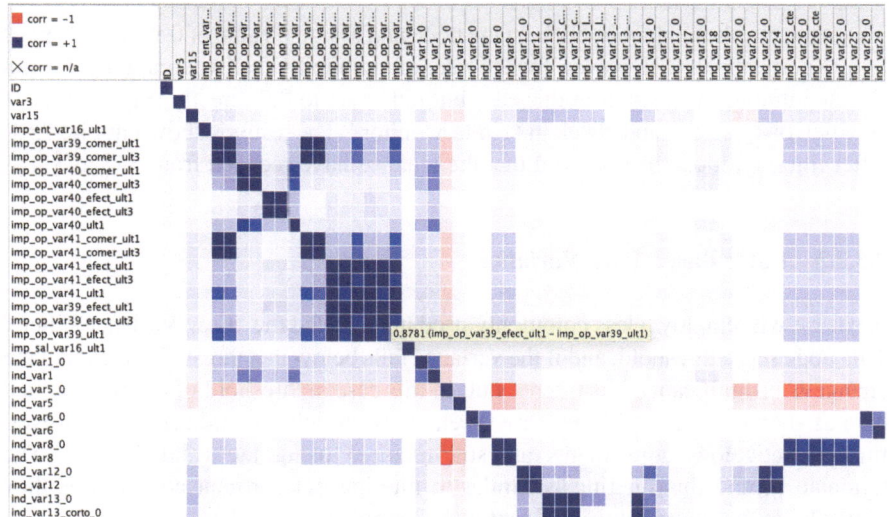

Fig. 7.49 Graphic matrix of data correlations

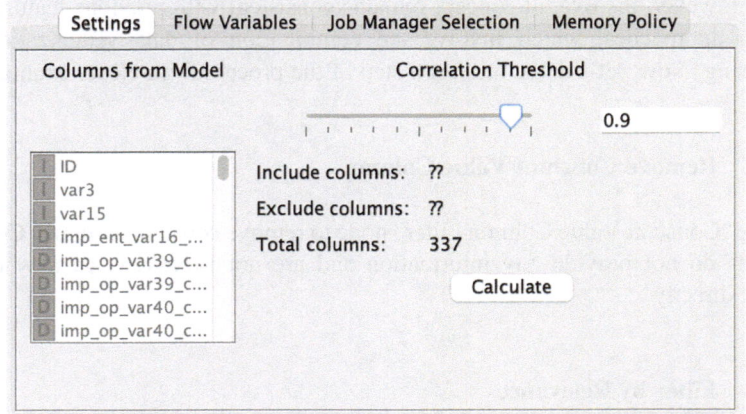

Fig. 7.50 Select all nodes

#	Row...	ID Number (inte...	var3 Number (inte...	var15 Number (inte...	imp_ent... Number (dou...	imp_op... Number (dou...	imp_op... Number (dou...	imp_op... Number (dou...	imp_op... Number (dou...	imp_op... Number (dou...	imp_op... Number (dou...	imp_op... Number (dou...	imp_op... Number (dou...	imp Num
1	Row0	1	2	23	0	0	0	0	0	0	0	0	0	0
2	Row1	3	2	34	0	0	0	0	0	0	0	0	0	0
3	Row2	4	2	23	0	0	0	0	0	0	0	0	0	0
4	Row3	8	2	37	0	195	195	0	0	0	0	0	0	0
5	Row4	10	2	39	0	0	0	0	0	0	0	0	0	0
6	Row5	13	2	23	0	0	0	0	0	0	0	0	0	0
7	Row6	14	2	27	0	0	0	0	0	0	0	0	0	0
8	Row7	18	2	26	0	0	0	0	0	0	0	0	0	0
9	Row8	20	2	45	0	0	0	0	0	0	0	0	0	0
10	Row9	23	2	25	0	0	0	0	0	0	0	0	0	0

Rows: 76020 | Columns: 171

Fig. 7.51 Results after filtering

To set the threshold, double-click on the node and assign a value of 0.9, which serves as a hyperparameter. However, for the sake of simplicity, the hyperparameter is not selected through validation in this case but rather set to 0.9 directly (Fig. 7.50).

Click on this node and check in "Node Monitor." The result is shown in Fig. 7.51. After filtering, it can be observed that the features have reduced from 337 to 171.

7.3.2.3 Filter Based Data Variance

Next, we will employ a less commonly used node called the "Low Variance Filter." This node sets a threshold, and if the variance falls below this threshold (meaning the change is insignificant), it is filtered out. In this case, a threshold of 0.3 was applied here as shown in Fig. 7.52, which is a relatively straightforward setting. However, further meticulous adjustments are still required using the validation set. The rationale behind this filtering is similar to filtering out constant columns, as they typically do not contribute significantly to the model.

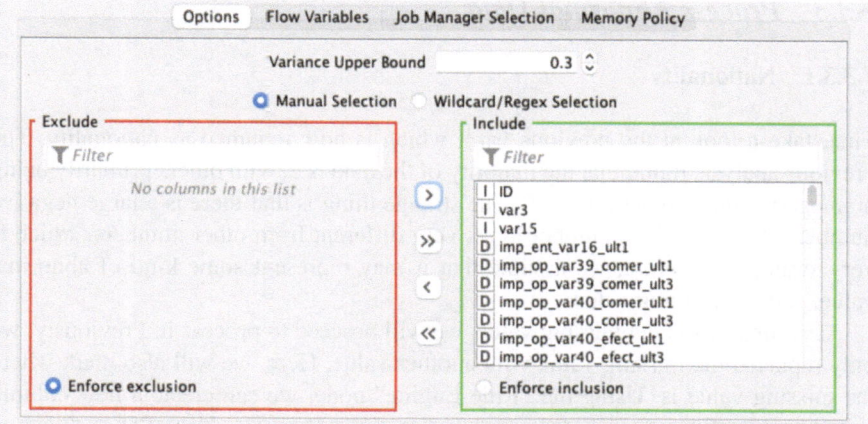

Fig. 7.52 Filter out data with small changes

Column		New name	
var3	⌄	nationality	🗑

Column		New name	
var15	⌄	age	🗑

Column		New name	
var38	⌄	mortgage	🗑

Column		New name	
num_var4	⌄	number_products	🗑

⊕ Add column

Fig. 7.53 Rename the characteristics based on the assumptions

7.3.2.4   Feature Assumption

This step will be quite complex and will require a significant amount of domain knowledge. Here, only a brief explanation will be provided. In general, a well-educated guess will be made based on the range and distribution of these data, combined with the bank's business. "Column Renamer" node is configured according to Fig. 7.53.

7.3.3   Process Abnormal Data

7.3.3.1   Nationality

First, take a look at the previous var3, which is now renamed as nationality. The previous analysis found that the majority of the data is 2, with others primarily being nonnegative integers less than 20. The strange thing is that there is a large negative number $-999,999$. This number looks very different from other numbers, which is very strange. Therefore, we assume that it may represent some kind of abnormal value, such as missing value.

Assuming this is the missing value, we will proceed to process it. Previously, we only replaced the missing value with another value. Here, we will also mark where the missing value is. Using the "Rule Engine" node, we can create a new column called "nationality_missing" to record the locations of missing data. Set as shown in Fig. 7.54, be sure to select "Append Column" and enter "nationality_missing" in the text box.

But here we need to "program" for the first time. But don't worry, this programming is relatively simple.

```
$nationality$ = -999999 => 1
TRUE => 0
```

The program terminates upon execution of the conditional statement. The variables within the two "$" symbols are column names. The "=" symbol is used to

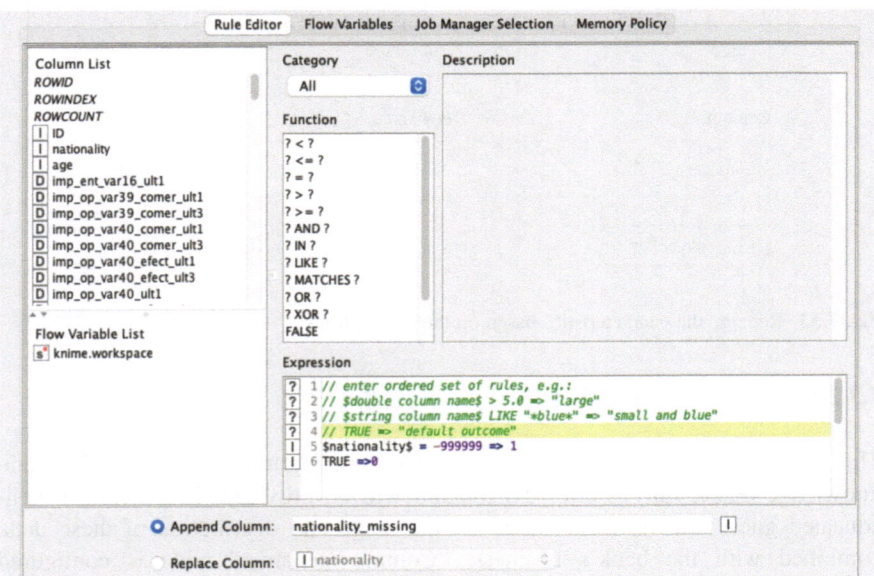

Fig. 7.54 Mark the missing value

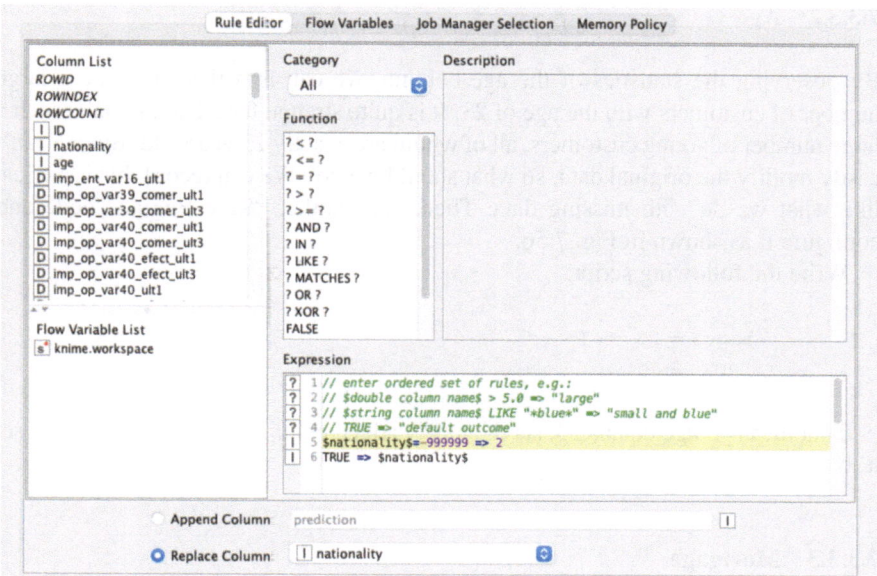

Fig. 7.55 Process missing data

determine if the left and right items are equal, while the "=>" symbol indicates assignment. The first line means that if the nationality column is equal to −999,999, then set its value to 1. If a row of data meets the condition in the first line, the value of the added column, "nationality_missing," will be set to 1 and the program ends. If a row of data does not meet the condition in the first line, the second line will be executed. The condition in the second line is "TRUE," which means it should always be true. Therefore, the value of the column "nationality_missing" that does not meet the condition "$nationality$ = −999,999" is set to 0, and the program ends.

Then, replace the value −999,999 with 2, which is the mode of the column. For the missing values, fill them with more meaningful values using the "Rule Engine" node, as shown in Fig. 7.55. Select the "Replace Column" option, and enter "nationality" in the text box to specify that the "nationality" column will be updated based on the rules for filling missing values.

The script filled in is

$$\$nationality\$ = -999999 \implies 2$$
$$TRUE \implies \$nationality\$$$

The first line is "Replace -999999 with 2," and the second line is "Keep the original value if it is not -999999."

7.3.3.2 Age

By observing the statistics of the age column, we can find that there are a large number of customers with the age of 23. It is quite strange for a bank to have such a large number of young customers, all of whom are exactly 23 years old. But we can't easily modify the original data, so what should we do? We can record this issue just like what we do with missing data. Then, use another "Rule Engine" node and configure it as shown in Fig. 7.56.

Write the following script:

```
$age$ = 23 => 1
TRUE  => 0
```

That means if age equals 23, set the newly added column data to 1; otherwise, set it to 0.

7.3.3.3 Mortgage

When viewing the data, you may notice that there are so many 117310! How is it possible for everyone to agree on this mortgage amount? Similar to the handling of age, we also add a column to record this issue. Use a "Rule Engine" node to record item 117310 as shown in Fig. 7.57, and use "mortgage_most" to record this result.

The script is as follows:

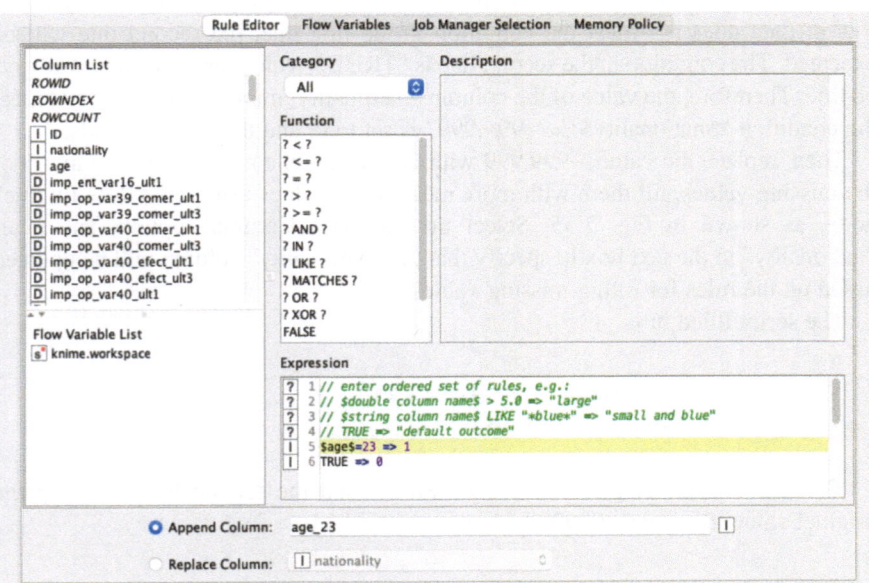

Fig. 7.56 Add a node to record age issues

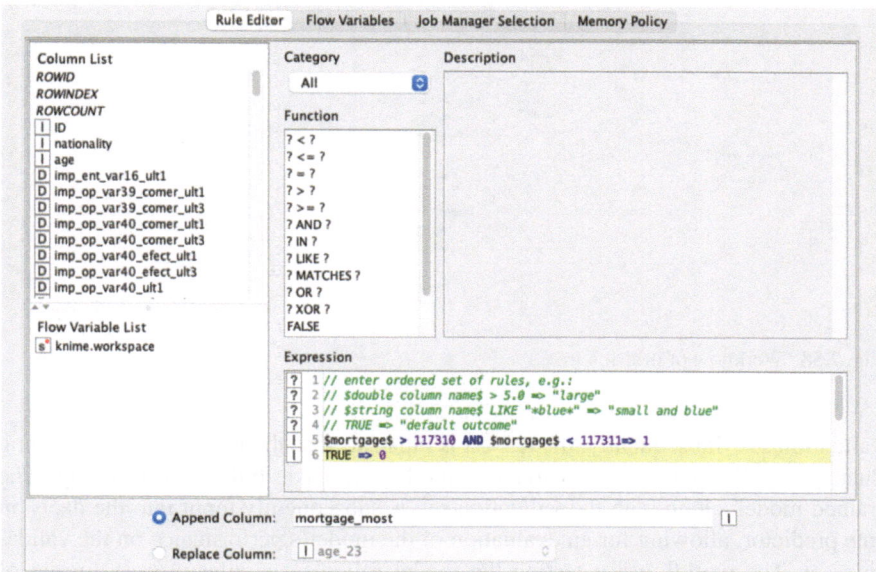

Fig. 7.57   Record the most frequent item

$mortgage$ > 117310 AND $mortgage$ < 117311=> 1
TRUE=> 0

Because the value is approximately 117310, a conditional statement is used here: if the value of mortgage is between 117310 and 117311, it is 1; otherwise, it is 0.

At this point, we have completed the most basic feature processing, and the next step is to start building the model.

7.4   Advanced Application of Decision Tree: Model Building and Comparison

First, we will split the data into a test set and other data and then train with different models.

7.4.1   Decision Tree

We can start with the most basic decision tree, and the workflow shown in Fig. 7.58 involves several steps. Initially, the input data is divided into a training set and a

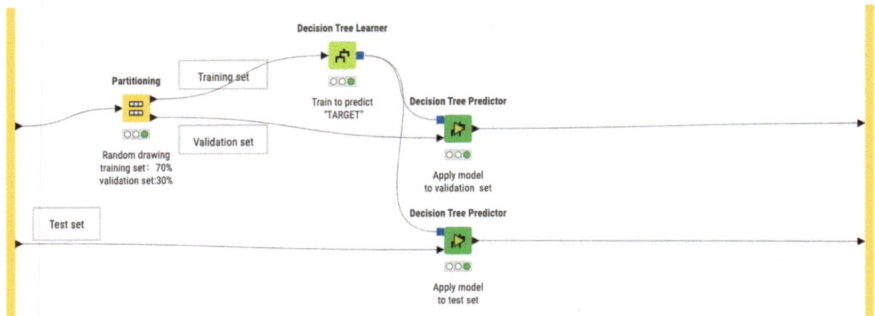

Fig. 7.58 Workflow of decision tree

validation set. Subsequently, the test set is introduced as the next input. The training data is then fed into the decision tree model, which is trained accordingly. The trained model, along with the validation set, is subsequently input into the decision tree predictor, allowing for an evaluation of the model's performance on the validation set. The workflow for testing the model follows a similar process to that of validating the model.

Double-click on the "Decision Tree Learner" node to make settings as shown in Fig. 7.59.

You may configure the node as follows:

- Pruning method: set to "MDL" and check "Reduced Error Pruning."
- Records per node: "Min number records per node" is set to 2.

After training the model, you can click on this node and select the "Open View" to view the decision tree, as illustrated in Fig. 7.60. On the left side of the figure, you will find specific details about the decision tree, while the right side displays a thumbnail of the tree. By clicking on the "+" or "−" symbols, you can expand or collapse branches of a node, respectively.

From this decision tree, it can be seen that initially, 96.0% of the data is 0, totaling 40,873 in number; and 4.0% of the data is 1, totaling 1698.

Next, we can use the comparison of "Imp_op_var40_effect_ult1" with 975 to determine the branching: when "s Imp_op_var40_effect_ult1" ≤ 975, there are 42,566 data points, 96.0% of them with 0 and 4.0% of them with 1; when "Imp_op_var40_effect_ult1" >975, there are only 5 data, 1 with 0 and 4 with 1.

When observing the validation set results of the model, we can refer to the ROC curve depicted in Fig. 7.61. The AUC is slightly larger than 0.5, indicating that the model's performance is relatively average.

By viewing the confusion matrix and model accuracy results (Fig. 7.62), it can be seen that the model accuracy of 96.202% and the percentage of "TARGET" being 0 at 96.04% are very close. What does this indicate?

This is caused by data imbalance, which will be discussed in detail in Chap. 8.

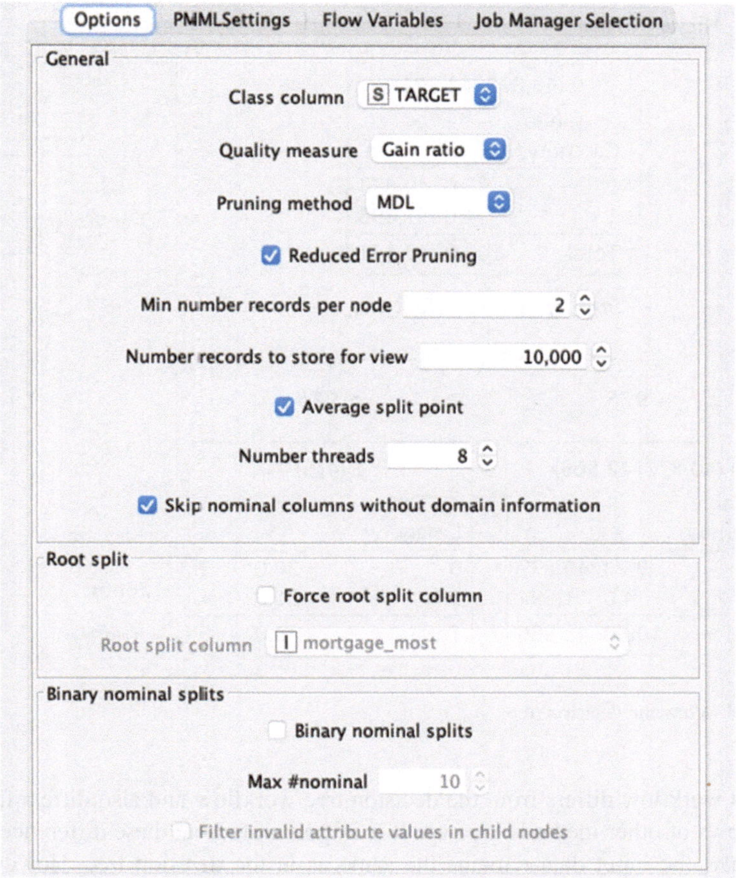

Fig. 7.59 Configure the decision tree

The decision tree we trained this time is quite simple, consisting of only two branches without any additional features for making branching decisions. As mentioned earlier, this type of decision tree tends to lack sufficient accuracy and may lead to underfitting. It is generally recommended to address the issue of data imbalance before training the model, as this can help in achieving a relatively reasonable decision tree model.

7.4.2   Bagging

The next step is to see how bagging is done. Enter the "Bagging" Metanode (created by the author, not built-in KNIME) in the main workflow, as shown in Fig. 7.63.

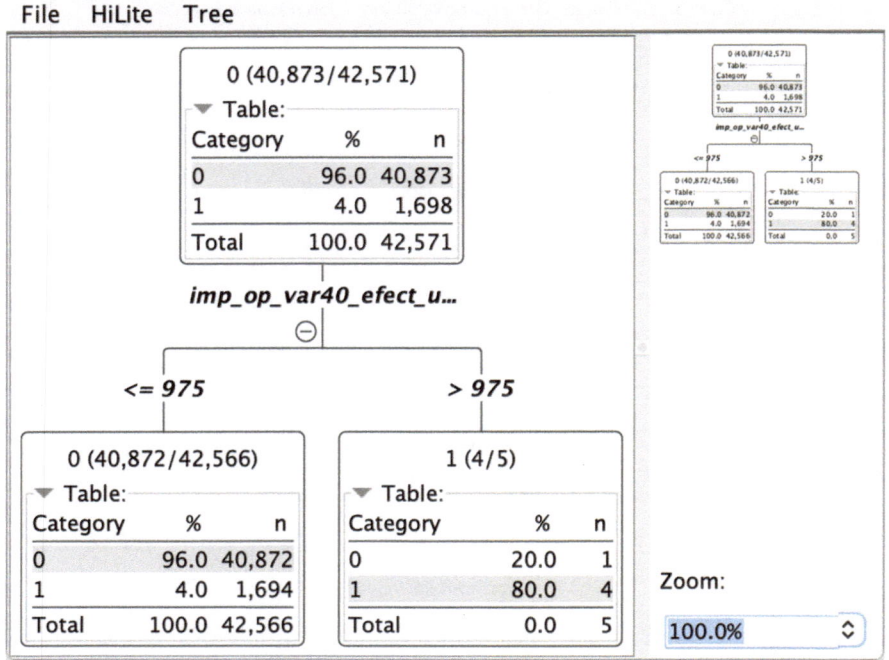

Fig. 7.60 View the decision tree

This workflow differs from the decision tree workflow and also differs from the workflows of other methods later on. It is important to note these differences.

Firstly, the input data remains the same as in the decision tree. However, we encounter unfamiliar nodes such as "Bagging" (built-in node of classic KNIME UI) and "Joiner."

Let's delve into the "Bagging" node first, where we can observe the workflow depicted in Fig. 7.64.

The "Bagging" created directly from KNIME is shown in Fig. 7.65. In Fig. 7.64, the original workflow was modified because validation data was used.

You can compare Fig. 7.65 and Fig. 7.64, and after modification, an input and an output have been added. Modifying the workflow is not difficult, but the process of adding input and output can be a bit confusing. In Fig. 7.66, you can click on the system's built-in "Bagging" node, which will display two ⊕ buttons below the Metanode. The left ⊕ button is for adding input configuration, and the right ⊕ button is for adding output configuration. By clicking on ▶ **Table** as shown in Fig. 7.67, you can add the necessary input and output. After that, you can proceed to add the corresponding workflow.

The workflow is divided into two parts: one part uses training data and validation data, and the other part uses test data to test the model. Now we don't focus on the test part, the remaining part can be divided into two sections:

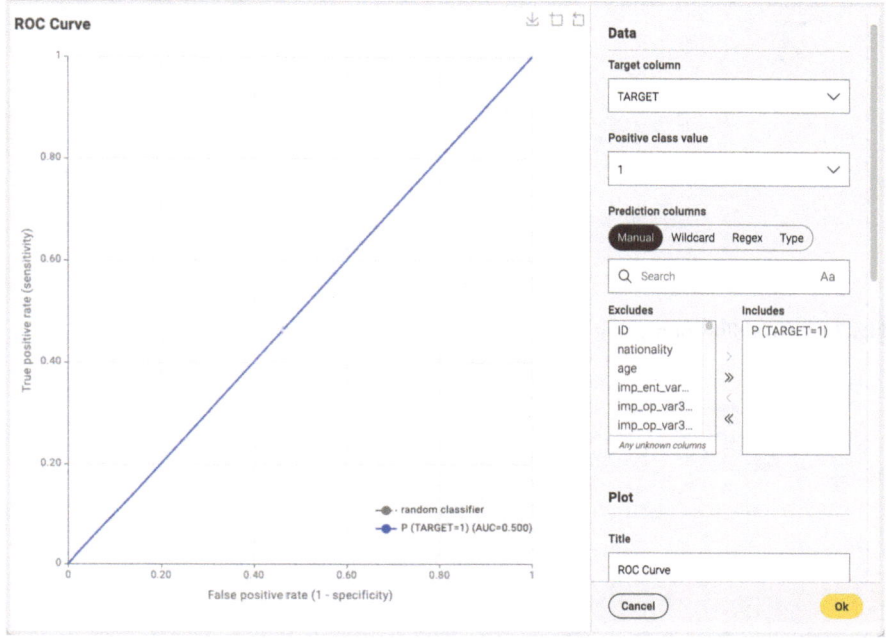

Fig. 7.61 ROC of decision tree validation set

File	Hilite		
TARGET \ ...	0	1	
0	17550	1	
1	694	0	

Correct classified: 17,550   Wrong classified: 695

Accuracy: 96.191%   Error: 3.809%

Cohen's kappa (κ): −0%

Fig. 7.62 Accuracy of decision tree validation set

- Train multiple models using training data.

  - Data shuffling.
  - Train multiple models separately.

- Vote for the model results using validation data.

  - Predict using the trained model with validation data separately.
  - Vote for the results.

To begin with, "Shuffle" node is responsible for randomly shuffling the order of the data. Then, at the "Chunk Loop Start" node, you can open the settings interface,

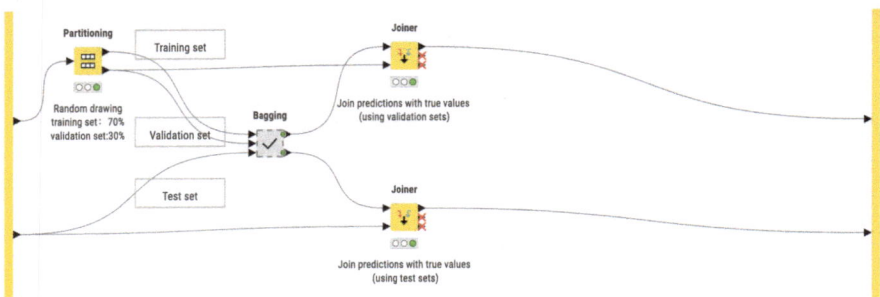

Fig. 7.63 Workflow of bagging

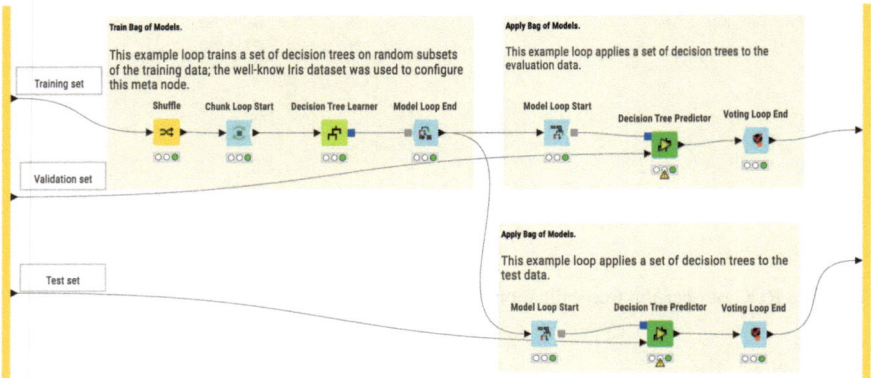

Fig. 7.64 Node of bagging

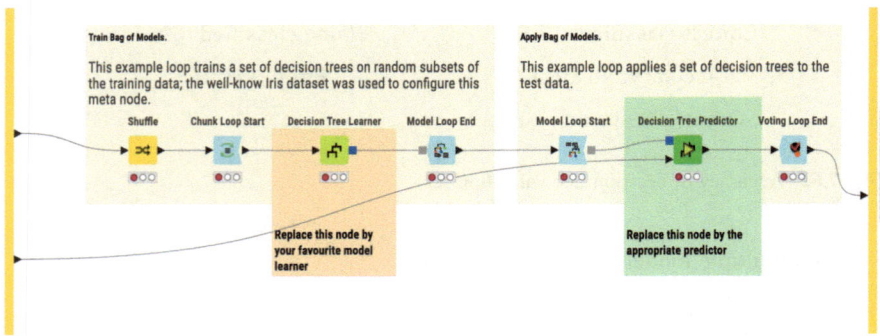

Fig. 7.65 "Bagging" in the classic KNIME UI

as shown in Fig. 7.68, and divide the data into ten chunks for ten iterations. This allows for multiple epochs of training.

Then configure the "Decision Tree Predictor" node as shown in Fig. 7.69. Please rename the prediction column. (The new name does not matter, but it must be

Fig. 7.66 Configure the
Metanode

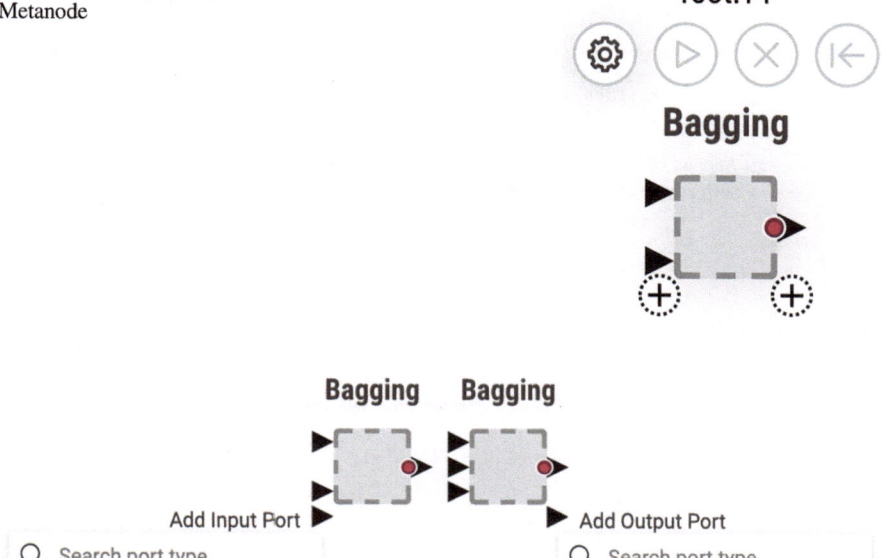

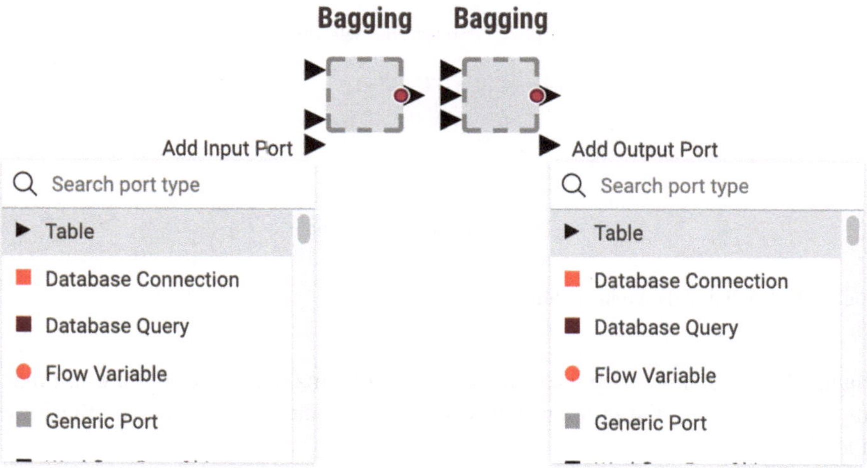

Fig. 7.67 Add input and output in Metanode

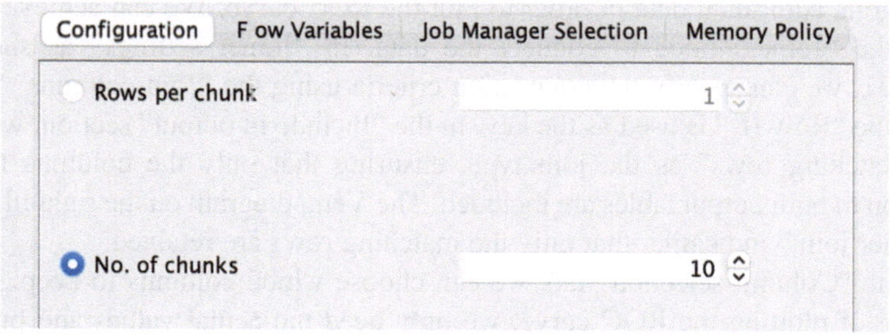

Fig. 7.68 Divide the data into ten chunks

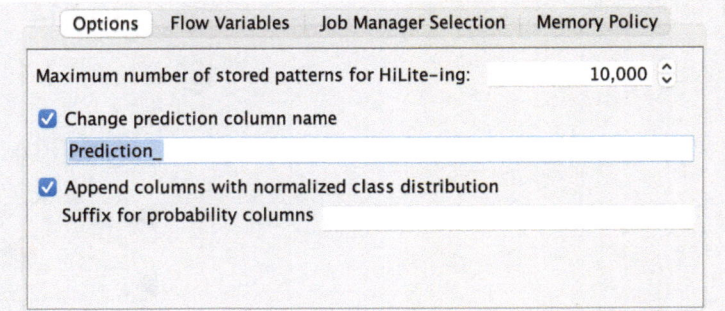

Fig. 7.69 Configure the "Decision Tree Predictor" node

| Options | Flow Variables | Job Manager Selection | Memory Policy |

Select the winner column   D P (TARGET=1)

Remove individual predictions

Fig. 7.70 Configure the voting column

changed. You can try it yourself to see what problems may arise if you don't make any changes. Hint: the problem will occur in the data joining part, i.e., the Joiner node below.)

Next, we can configure the "Voting Loop End" node as shown in Fig. 7.70 and select the voting column as "$P(\text{TARGET} = 1)$."

Since this Metanode only outputs voting results without any other data, we need to integrate additional data in order to plot the ROC curve. We can achieve this by using the "Joiner" node to connect the data. In "Joiner settings" as shown in Fig. 7.71, we can specify the connection criteria using the "Join columns" feature, where the "Row ID" is used as the key. In the "Include in output" section, we select the "Matching rows" as the join type, ensuring that only the columns that are common to both output tables are included. The Venn diagram on the right illustrates an "Inner join," indicating that only the matching rows are retained.

In the "Column selection" tab, we can choose which columns to keep. For the purpose of plotting the ROC curve, we only need the actual values and predicted values. Therefore, we will select the columns "P (TARGET = 1)" and "TARGET" to be included in the output (Fig. 7.72).

Finally, let's examine the results of the ROC curve. Although we didn't extensively optimize the model, we can see a slight improvement compared to the decision tree as shown in Fig. 7.73.

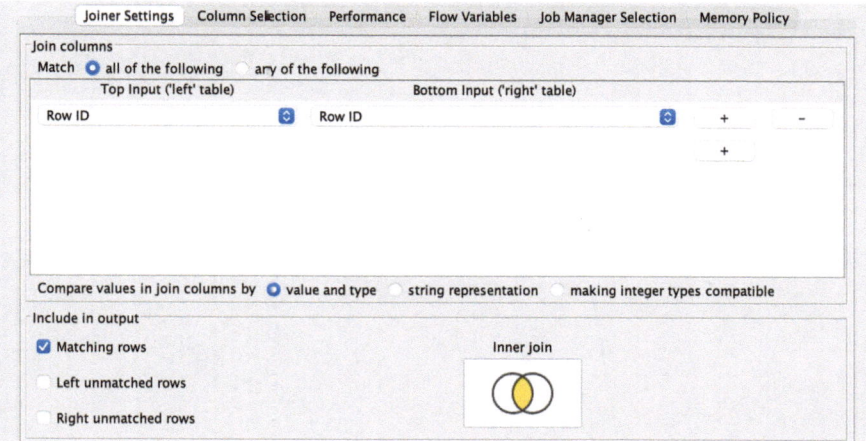

Fig. 7.71 Configure the join columns

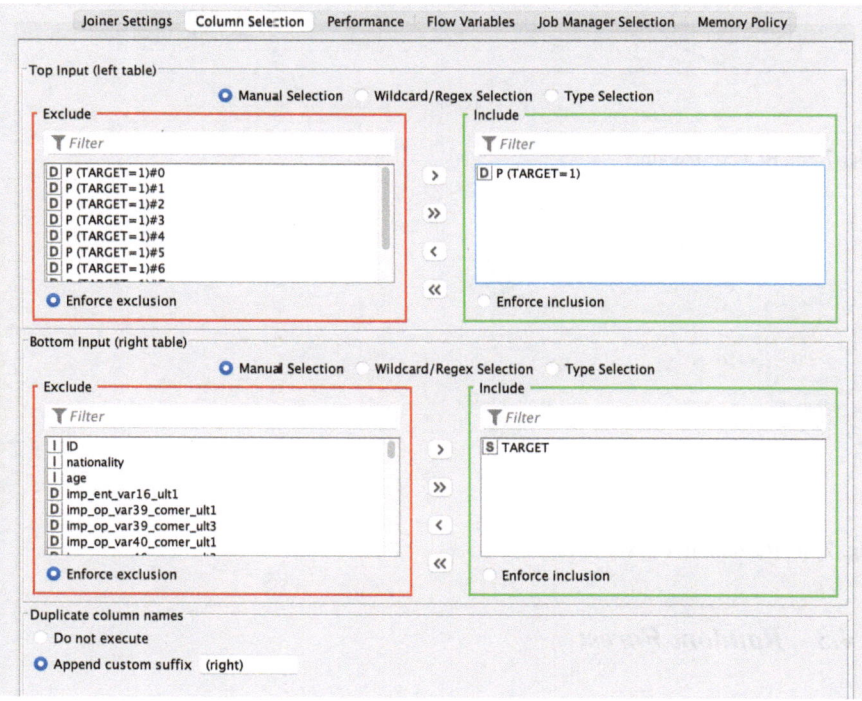

Fig. 7.72 Choose which column should be kept

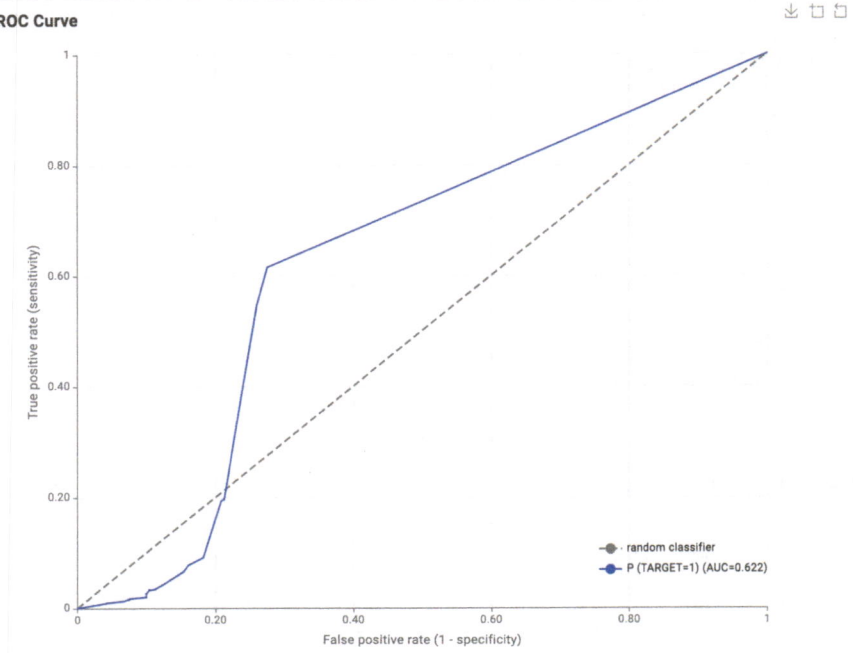

Fig. 7.73 ROC of bagging

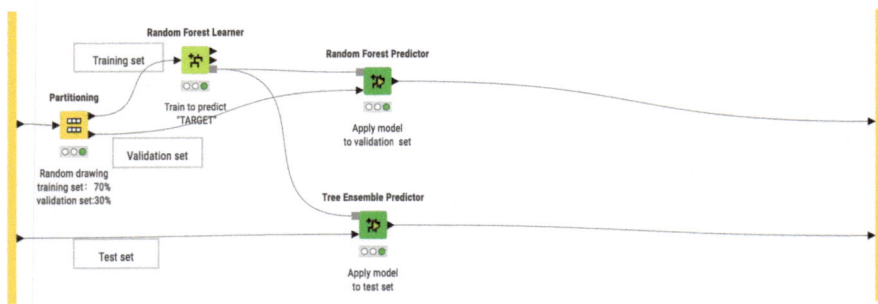

Fig. 7.74 Random forest workflow

7.4.3 *Random Forest*

The workflow of random forest is basically the same as the workflow of decision tree as shown in Fig. 7.74. The only difference is that we use "Random Forest Learner" instead.

In Fig. 7.75, the main settings here are that each node has a minimum of 5 data points ("Minimum node size") and 400 models ("Number of models").

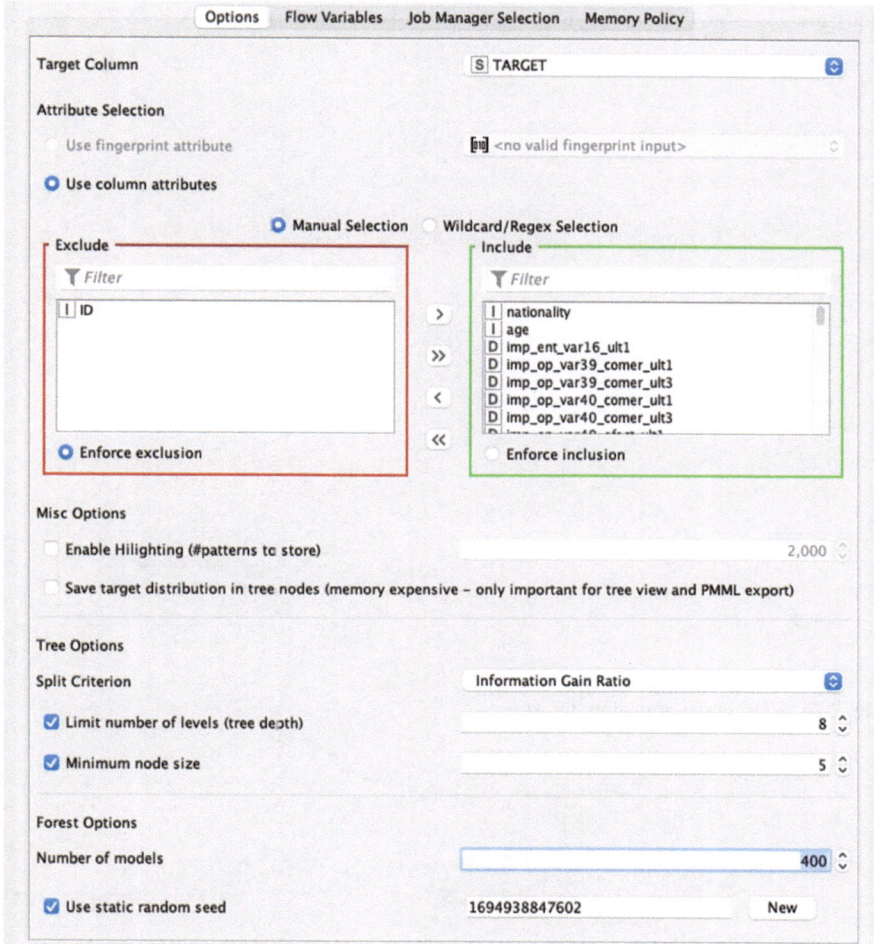

Fig. 7.75 Configure the random forest

After initiating the execution, it may take a few minutes for the process to complete. During this time, you may notice an increase in the noise generated by the computer fan. This indicates that the random forest algorithm is utilizing multiple models and consuming a significant amount of system resources.

If we look at the ROC results in Fig. 7.76, the model is still not very good.

7.4.4   Boosting

Finally, let's take a look at boosting, with its workflow shown in Fig. 7.77.

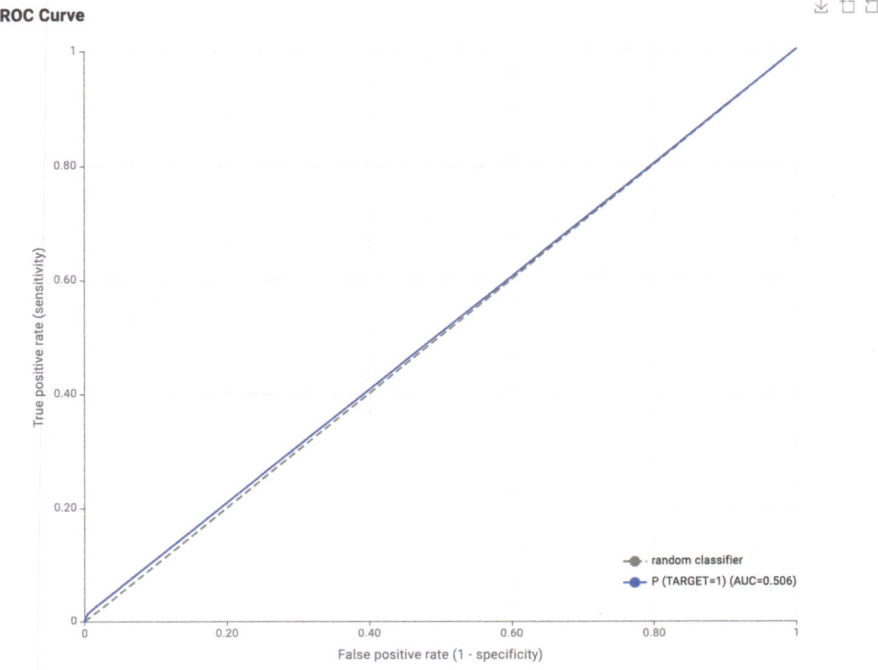

Fig. 7.76   ROC curve of random forest

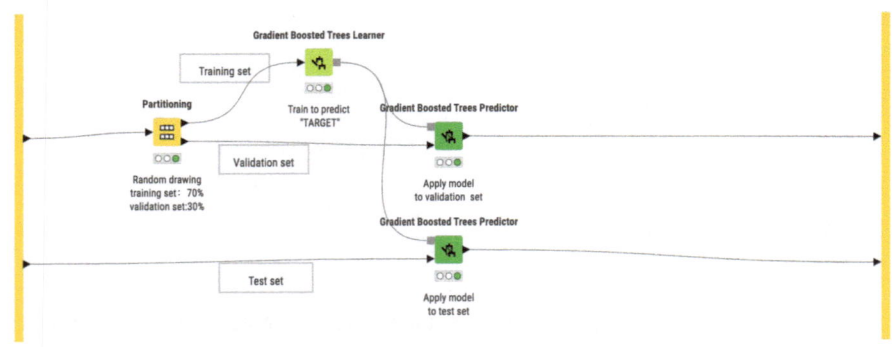

Fig. 7.77   Boosting workflow

In this case, we used the "Gradient Boosted Trees Learner" node and its corresponding "Gradient Boosted Trees Predictor" node built into KNIME, with other nodes being familiar to us.

As shown in Fig. 7.78, the main setting here is to use 100 models ("Number of models"). This algorithm is also quite resource-intensive, so it may take a few minutes.

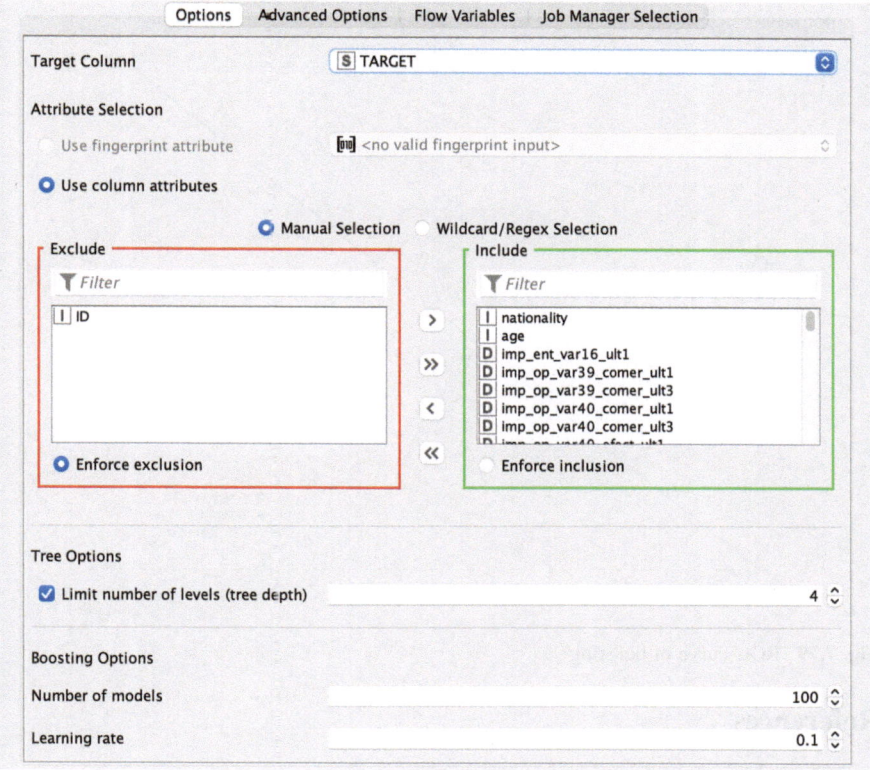

Fig. 7.78 Configure boosting

Finally, let's check the results. As seen in Fig. 7.79, the boosting effect is really good, living up to its name.

Give It a Try

- You can try these different methods and debug the model and see if a better model can be obtained.

7.5 Practice

1. Can the validation set and the test set be the same group of data?
2. Why can pruning reduce overfitting risks?
3. Please briefly describe the difference between bagging and boosting.

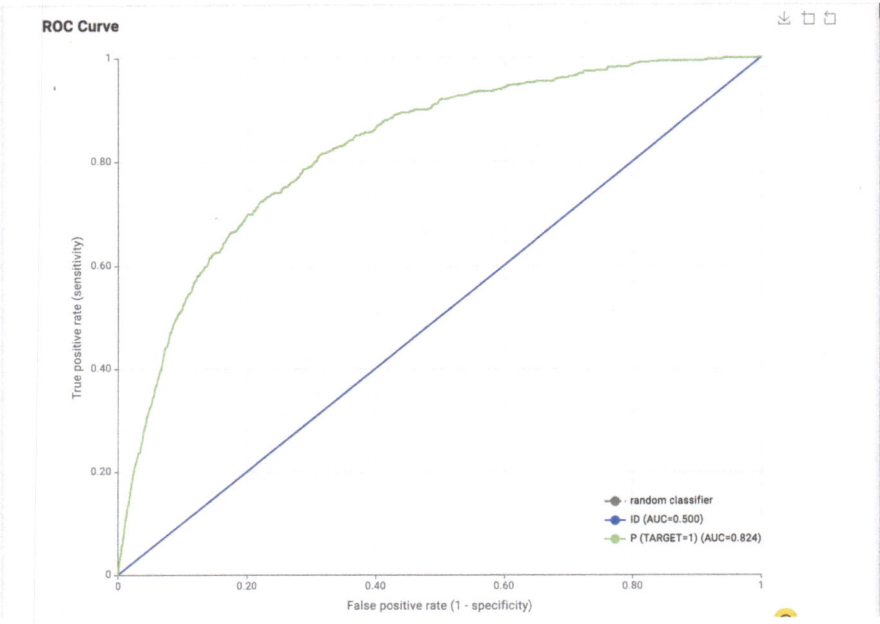

Fig. 7.79 ROC curve of boosting

References

van den Elsen, D. (2017). Santander Customer Satisfaction, Vrije Universiteit Amsterdam, Faculty of Economics and Business Administration. https://vu-business-analytics.github.io/internship-office/papers/paper-elsen.pdf
https://www.deeplearning.ai/courses/machine-learning-specialization/

Chapter 8
Understanding Decision Trees In-Depth

Surprisal is used to describe the unpredictability of a single digit or letter.

—*Myron Tribus*

Abstract This chapter delves into advanced concepts and techniques associated with decision trees in machine learning. The focus is on understanding the construction of decision trees, with an emphasis on identifying optimal features for data classification. The chapter introduces two key algorithms: ID3, which uses information entropy and information gain, and CART, which utilizes Gini impurity. These algorithms are explored in-depth to explain their methodologies in determining the optimal feature for tree construction. The concept of information entropy is elaborated upon, providing insights into its calculation and significance in data prediction. The chapter also addresses the challenge of imbalanced data and presents several strategies for optimization, including downsampling majority data, oversampling minority data, and using synthetic data generation techniques like SMOTE. Practical applications and settings in KNIME, a popular data analysis tool, are discussed to demonstrate the implementation of these concepts. The chapter concludes with a practice section, encouraging readers to explore related concepts and the implications of data imbalance.

Keywords Decision trees · ID3 · CART · Information entropy · Gini impurity · Imbalanced data · Information gain · Downsampling · Oversampling

This chapter covers

- **Information entropy.**
- **Gini impurity.**
- **Imbalanced data and its processing.**

8.1   Decision Trees Advanced

8.1.1   How to Construct a Decision Tree

The process of constructing a decision tree involves identifying the optimal features for data classification, utilizing these features as the root nodes, applying specific rules to classify the data, and finally iteratively repeating this process to build sub-trees until the tree is fully built. There are two crucial questions to be answered here:

- What is the optimal feature?
- When should we complete the tree?

The second question has already been addressed earlier, but what does "optimal" mean? In this section, we will explore the answer together. There are many good metrics, and common ones include information entropy and Gini impurity. They are used, respectively, in the ID3 algorithm and the CART algorithm (https://en.wikipedia.org/wiki/Decision_tree_learning).

8.1.1.1   ID3 Algorithm

The ID3 (Iterative Dichotomiser 3) algorithm is based on information theory and uses information entropy (Entropy) and information gain as metrics to answer what is the optimal feature.

8.1.1.2   CART Algorithm

CART (classification and regression trees) algorithm is a binary classification recursive algorithm, which only performs binary splits of either "yes" or "no" and uses Gini impurity as the criterion to determine the optimal feature.

Now let's take a look at what these two algorithms mean separately.

8.1.2   Determine the Optimal Feature with ID3

First, let's understand what the ID3 algorithm is designed to do. Because ID3 uses information entropy and information gain as the measurement criteria, let's first take a look at these two concepts (https://en.wikipedia.org/wiki/Entropy_(information_theory)).

8.1.2.1   Entropy

Entropy was originally a thermodynamic concept used to measure the degree of disorder in a system. For instance, in Fig. 8.1, there are three distinct types of eggs.

Fig. 8.1 Which picture will you choose to count the egg

Now, the question arises: which picture would be preferable for determining the total count of eggs?

Most people would probably choose to count the eggs in the first picture without any surprises, because it is the least "chaotic" one. The more chaotic, the greater the entropy, and accordingly, the first picture has the smallest entropy, while the third picture has the largest entropy.

8.1.2.2   Information Entropy

In the realm of information theory, entropy, also referred to as "information entropy," represents the average amount of information contained within each received message. Let's consider a scenario where three individuals have a picture of Fig. 8.1 in front of them, and their task is to determine the total count of eggs. Now, imagine that a cheat sheet is placed on the table, revealing the exact number of eggs in front of each person. In this situation, the person experiencing the highest level of "surprise" would be the one facing the most "chaotic" or uncertain circumstances. This is because the information conveyed by the sheet carries a significant amount of information content.

Let's consider another example to delve deeper into our understanding of information entropy and explore its calculation method. Imagine yourself in a wilderness, as depicted in Fig. 8.2, with no Internet access on your phone and limited battery life. Your only option is to make a single phone call for help. Then here comes the question: what sentence would you choose to convey the most crucial information within this constraint?

- I'm in the wilderness.
- I'm at the boundary of the forest and the grassland.
- I'm at the entrance of Hogwarts.

Assume that in the eyes of the person searching, you are wandering around, which means that the probability of you appearing in any location in the wilderness is equal.

Fig. 8.2 Lost in the wilderness

- Everyone knows that you are in the wilderness, so clearly this statement offers no additional information.
- "I am at the boundary between the forest and the grassland" can be helpful. In such a vast wilderness, the probability of being right at the boundary is not high. You're on a line in a plane. So at least we know the point I'm looking for is on this line.
- "I am at the entrance of Hogwarts" is telling us your location, so we can go directly to find you. Also note that the probability of you appearing exactly at the entrance of Hogwarts is extremely low, making the information in this sentence greatest.

From this example, you may have observed that as the amount of information increases, the probability of the event occurring decreases. To ensure the additivity of information, we define the information of an event X as the logarithm of the reciprocal of its probability:

$$I(X) = \log\left(\frac{1}{P(X)}\right) = -\log(P(X))$$

In a computational system, there exists a multitude of events, and the anticipated quantity of information associated with these events (average) is the system's entropy.

$$H(X) = E[(I(X))] = \sum_{i=1}^{n} P(x_i) \log_b \frac{1}{P(x_i)}$$

Here, n represents the number of possible outcomes for an event, and b denotes the base of the logarithm, typically set to 2, the natural constant e, or 10. When $b = 2$

(as in the ID3 algorithm), the entropy is measured in bits; when $b = e$, the entropy is measured in nats; and when $b = 10$, the entropy is measured in Harts.

If we view each message as a system and each bit within the message as an event, information entropy can be defined as the average amount of information contained in each received message. It quantifies the level of uncertainty associated with a message. Drawing from the earlier example of survival in the wilderness, when an event with lower probability actually occurs, it contains a greater amount of information. Another perspective is to imagine that you're asked to predict the next event with utmost effort. Entropy then represents the degree of surprise you experience upon making a correct guess. In essence, entropy serves as a measure of how unexpected the correct guess is. The greater the surprise, the more valuable the information.

8.1.2.3   Example of Information Entropy Calculation

Let's continue with a simpler example of flipping a coin to explain the value of information entropy. If you toss a fair coin, what is the probability that you will guess it correctly? It's one-half of course. What if the two sides of this coin are exactly the same? You will never be wrong. Which experiment of tossing a coin do you think would give you more surprises? The one with higher information entropy? Let's look at formula:

- When tossing a fair coin, we are faced with only two possibilities; thus, $n = 2$. In the case of a fair coin, the probability of obtaining heads or tails is equal, with each having a probability of one-half.

$$H(X) = \sum_{i=1}^{2} \frac{1}{2} \log_2 2 = 1$$

- Tossing a coin with the same faces on both sides has only one possibility, so the probability is definitely 1.

$$H(X) = \log_2 1 = 0$$

In reality, it is definitely between the two extreme cases.

8.1.2.4   Information Gain

Information gain refers to the extent by which information entropy decreases under specific conditions. Put simply, it signifies the amount of surprise or uncertainty reduction that occurs when I disclose a secret to you. It quantifies the level of doubt

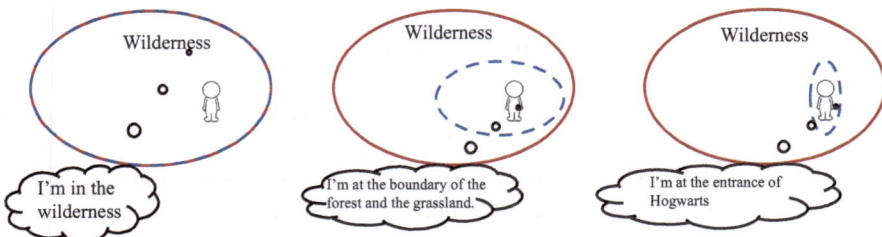

Fig. 8.3 The amount of information for each answer is different

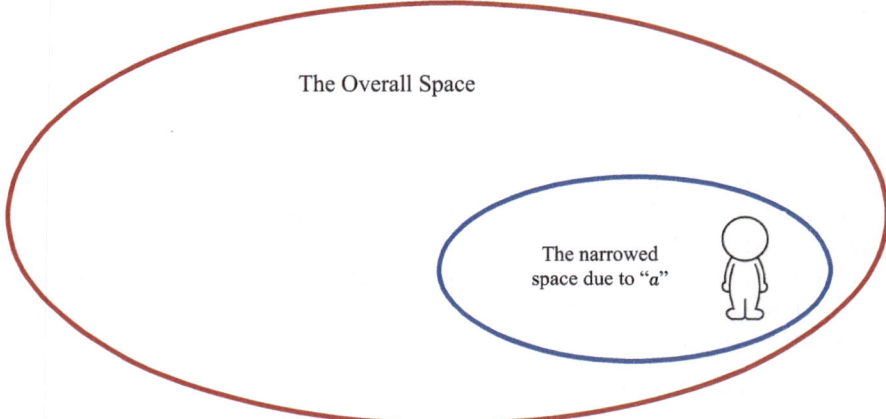

Fig. 8.4 The reduction of space due to information gain

that can be resolved or the degree of surprise that can be experienced through the revealed information.

In the example of survival in the wilderness, each answer is treated as a secret. Information gain is certainly the highest for the third response "I'm at the entrance of Hogwarts." As shown in Fig. 8.3, the red circle represents all possibilities, and the dashed blue circle represents the possible location range after obtaining information. Different information can lead to different ranges. The larger the information, the more accurate the positioning.

Let's look at the formula below:

$$IG(T, a) = H(T) - H(T|a)$$

The "a" is the call you make in the wilderness, which narrows down the search scope. In the given formula, "$|a$" can be seen as the space narrowed down due to "a." And information gain refers to the amount of space reduction as shown in Fig. 8.4.

With this understanding, the above formula quantifies how much information can be increased by limiting you to a smaller range.

8.1.2.5 Build a Tree Using Information Entropy

The steps involved in building a tree using information entropy are as follows:

- Calculate the information entropy of the given dataset.
- Calculate the information entropy after each feature is classified.

  - Calculate the information entropy of this feature for each value.
  - Calculate the weighted average (expected value) of these information entropies.
  - Calculate the information gain of this feature.

- Select the feature that yields the highest information gain as the root node, with its distinct values serving as branches of the tree.
- Continue to build branches using other features.
- If further division is not possible due to identical nodes or exhaustion of features, terminate the process.

8.1.3 Determine the Optimal Feature with CART

8.1.3.1 Gini Impurity

Everyone should have heard of the Gini coefficient or Gini index in the news. It is an indicator for assessing the fairness of income distribution. However, Gini impurity is not the same as the Gini coefficient. They are two different measures that are related but distinct from each other.

There is often confusion between the Gini coefficient and Gini impurity in various documents. It is crucial for everyone to be aware of this distinction. In fact, Wikipedia explicitly mentions in the "Gini impurity" section: "Not to be confused with Gini coefficient." This highlights the significance of the confusion that arises between these two concepts.

In essence, Gini impurity quantifies the level of surprise that arises when we make our best attempt to accurately guess the label of a randomly sampled element from a set. More precisely, it involves randomly selecting an element and assigning a label based on the distribution of labels. Gini impurity serves as a metric to gauge the degree of surprise associated with this process.

Does it sound similar to the concept of information entropy?

We may compare the formulas:

$$\text{Gini} : \text{Gini}(E) = 1 - \sum_{j=1}^{c} p_j^2$$

$$\text{Entropy} : H(E) = -\sum_{j=1}^{c} p_j \log p_j$$

The formulas are also basically the same, aren't they? In fact, it doesn't make much difference which one is used as a measure of error level.

8.1.3.2   Build a Tree Using Gini Impurity

The steps are the same as the way we use information entropy.

- Calculate the Gini impurity of the given dataset.
- Calculate the Gini impurity after each feature is classified.

  – Calculate the Gini impurity for each value of this feature.
  – Calculate the weighted average (expected value) of these Gini impurities.
  – Calculate the Gini gain of this feature.

- Select the feature that yields the highest Gini gain as the root node, with its distinct values serving as branches of the tree.
- Continue to build branches using other features.
- If further division is not possible due to identical nodes or exhaustion of features, terminate the process.

The biggest difference between CART and ID3 is the use of different criteria for classification. Another difference is that ID3 can only be used for classification problems, while CART, as the name implies, can be used for both classification and regression.

8.1.4   Settings in KNIME

In the previous chapter on bank customer classification, as shown in Fig. 8.5, in the decision tree training node settings, you can choose "Gain ratio" or "Gini index," where the former is using information gain to build the tree and the latter is using Gini impurity. You can try both to see which one produces a better model.

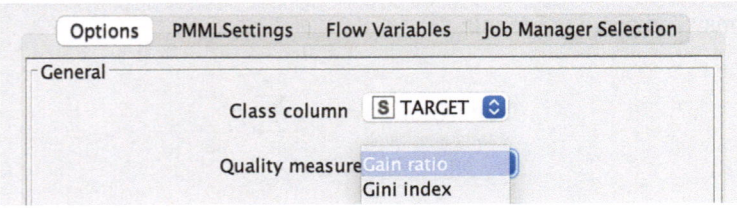

Fig. 8.5 Quality measure for a decision tree

8.2   Optimization of Imbalanced Data

Generally, imbalanced data can have a significant impact on the model.

In the previous chapter, we encountered a problem with bank customer classification. We needed to determine the TARGET, but this data only accounts for around 4%, with the majority of data being 0. Our goal is to identify those customers who have TARGET as 1 as accurately as possible. However, assuming all customers have TARGET as 0 will yield an overall accuracy of 96% (Fig. 8.6), which means the model cannot effectively solve our problem (https://machinelearningmastery.com/what-is-imbalanced-classification/, https://machinelearningmastery.com/tactics-to-combat-imbalanced-classes-in-your-machine-learning-dataset/).

To solve this problem, there are several algorithms that we can use:

- Downsample majority data.
- Oversample minority data.
- Adjust the loss function to give more weight to minority data.
- Generate more "minority data."

8.2.1   Downsample Majority Data

This approach involves randomly selecting a subset of the majority data in order to balance its quantity with that of the minority data in the original dataset. As depicted in Fig. 8.7, there is a significant imbalance between the number of red and blue data points. Initially, a random sampling of the majority data is performed, and subsequently, as illustrated in Fig. 8.8, only the sampled data is utilized during model training. However, it is evident that this method discards a substantial amount of data, which may not be advantageous for effective model training.

KNIME uses the "Equal Size Sampling" node to perform this operation. The workflow shown in Fig. 8.9 demonstrates downsampling of majority data. Note that "Equal Size Sampling" can only be used on the training set, never on the validation or test set.

The configuration of the "Equal Size Sampling" node is relatively simple, mainly choosing whom to operate on. In this case, "Nominal column" is set to "TARGET."

File Hilite		
TARGET \ ...	0	1
0	17553	0
1	692	0

Correct classified: 17,553	Wrong classified: 692
Accuracy: 96.207%	Error: 3.793%
Cohen's kappa (κ): 0%	

Fig. 8.6 All classified as "not dissatisfied"

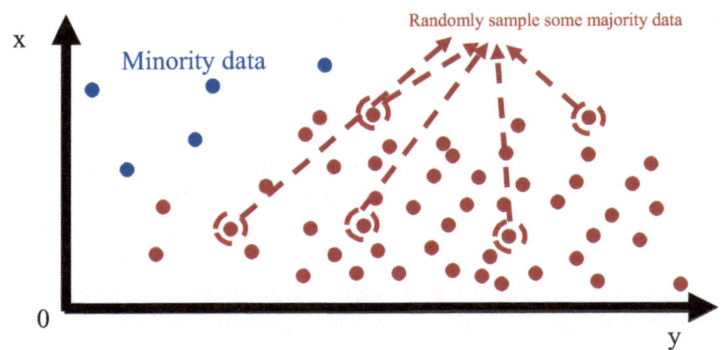

Fig. 8.7   Randomly sample some majority data

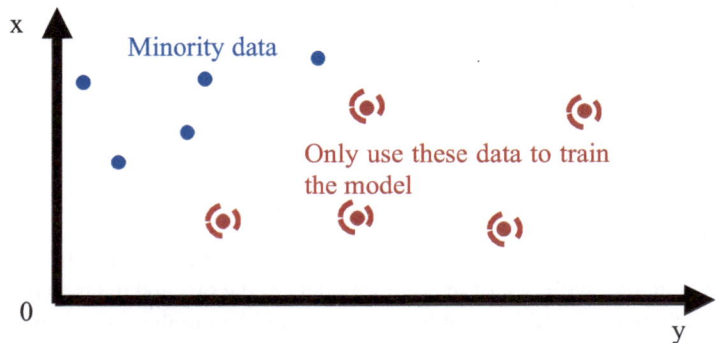

Fig. 8.8   Only use these sampled data to train the model

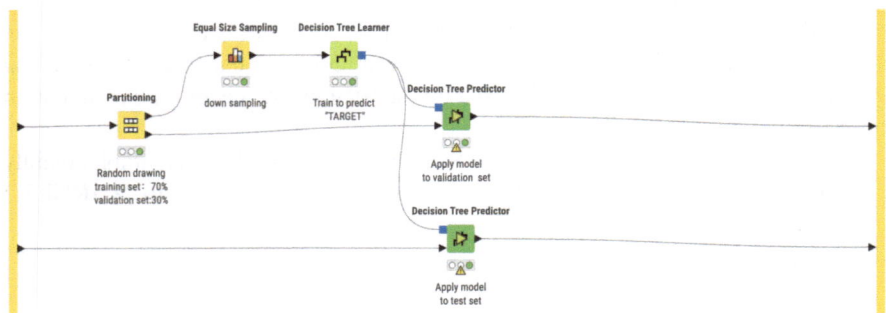

Fig. 8.9   Workflow for majority data downsampling

"Use exact sampling" requires an equal number of data for both classes and is resource-intensive, so we won't check it here. We only need roughly the same amount of two types of data, so we choose "Use approximate sampling" as shown in Fig. 8.10.

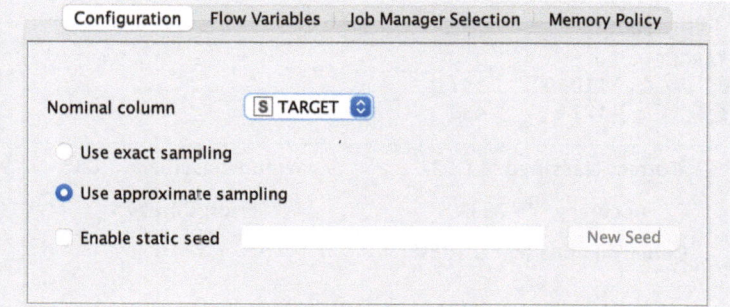

Fig. 8.10   Configure "Equal Size Sampling"

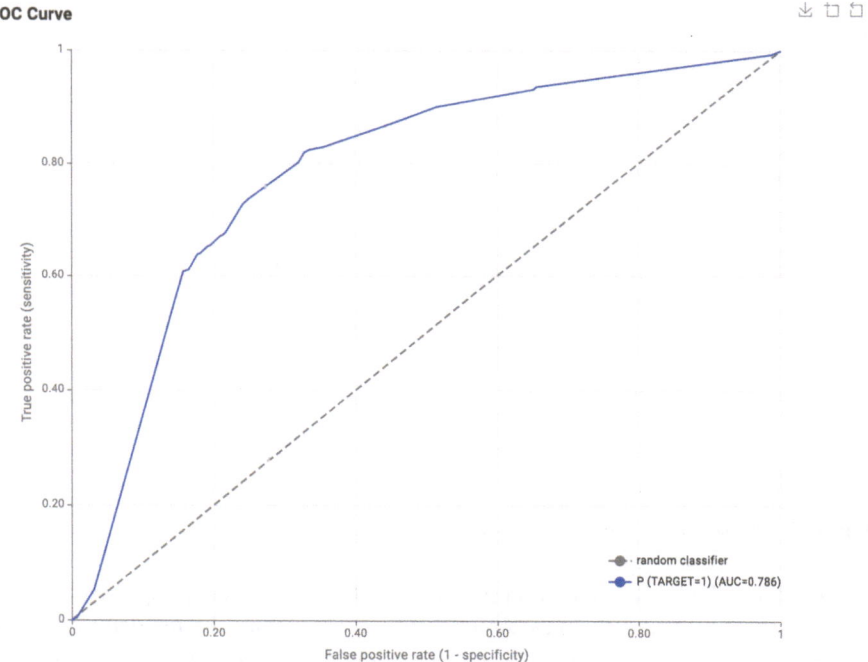

Fig. 8.11   Decision tree ROC

Let's see how it goes next. I had a total of 48,652 training data, but after downsampling, there are only 3937 left. Here, we directly observe the results of the test set. It can be seen from Fig. 8.11 that the ROC is much better than before.

Based on the results in Fig. 8.12, although the accuracy is not very high, we have captured more dissatisfied customers. The downsampling methods for other models such as bagging are similar. You can try them out for yourselves.

File	Hilite		
TARGET \ ...	0	1	
0	11069	3511	
1	171	453	

Correct classified: 11,522   Wrong classified: 3,682

Accuracy: 75.783%   Error: 24.217%

Cohen's kappa (κ): 0.136%

Fig. 8.12 Results of the "Scorer" with downsampling

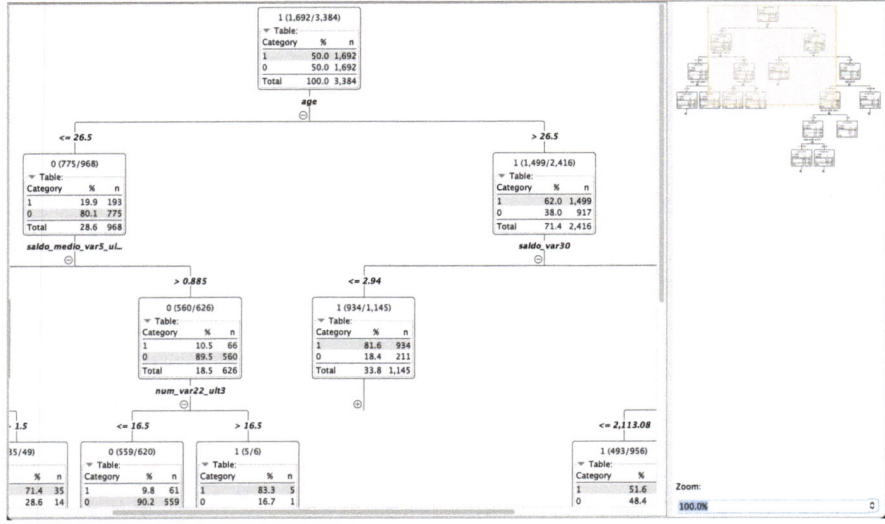

Fig. 8.13 Decision tree model trained with downsampling

Following Sect. 7.4.1, after training the model with downsampling, we may view the decision tree model again as shown in Fig. 8.13. Obviously, the decision tree is "taller" now, which indicates that after handling the imbalance of the data, more features are involved in the determination of the results.

At the beginning of the decision tree, 49.2% of the data is 0 and 50.8% is 1. The proportions of the two types of data samples are almost the same.

Next, we can compare the feature "age" with a threshold of 26.5 and "saldo_var30" with 2.565 to determine the branching direction. It becomes evident that as the branches diverge, the proportion of either branch 0 or 1 in the decision tree increases while the proportion of the other branch decreases. When the category proportion approaches 100%, it signifies the outcome determined by that particular branch. In Fig. 8.14, when "imp_op_var41_effect_ult3" is greater than 615, all 14 samples have the label of 1. At this point, these 14 data points no longer undergo

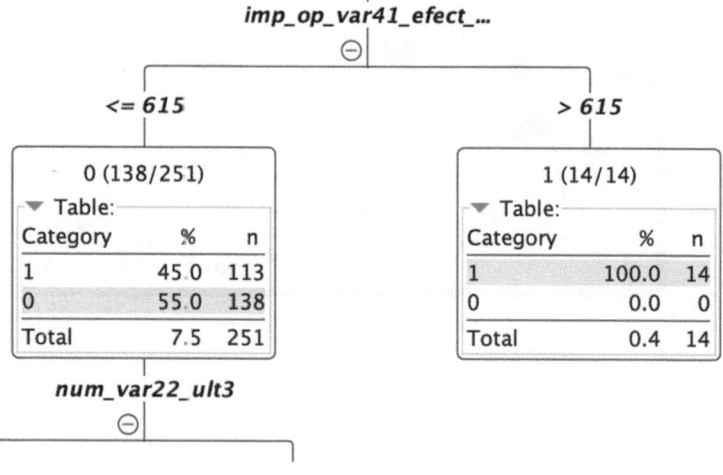

Fig. 8.14 Accuracy reaching 100%

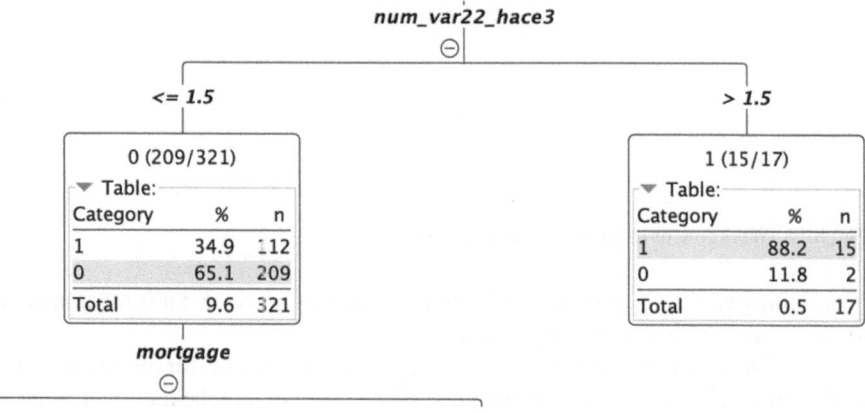

Fig. 8.15 Minimum number of records per node

further branching decisions. The evaluation for the condition "imp_op_var41_effect_ult3" >615 terminates, and the decision tree proceeds to evaluate the branching data for the condition "imp_op_var41_effect_ult3" ≤615.

Given that the "Min number records per node" is set to 2, it implies that the branching process should halt when the number of nodes reaches 2. Therefore, if the minimum number of nodes falls below 2, regardless of the accuracy of the results, the branching process must stop. In Fig. 8.15, there are 8 data points in the branch where "num_var22_hace3" >1.5. Among the 17 samples, 15 samples have a label of 1, while 2 sample has a label of 0. If further branching were to occur, there would be a node with only one data point, rendering it impossible to divide any further. The "Min number records per node" is a critical hyperparameter that plays a vital role in

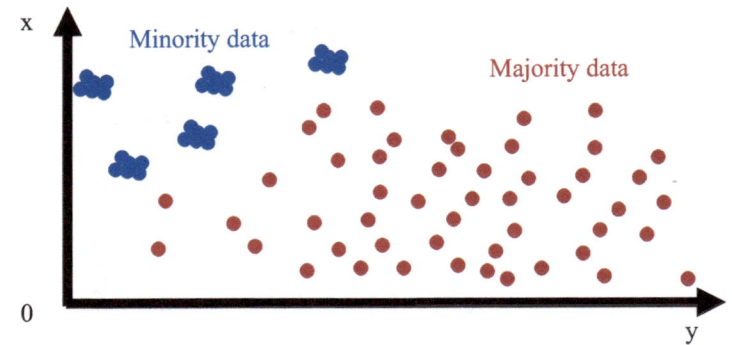

Fig. 8.16 Oversample minority data

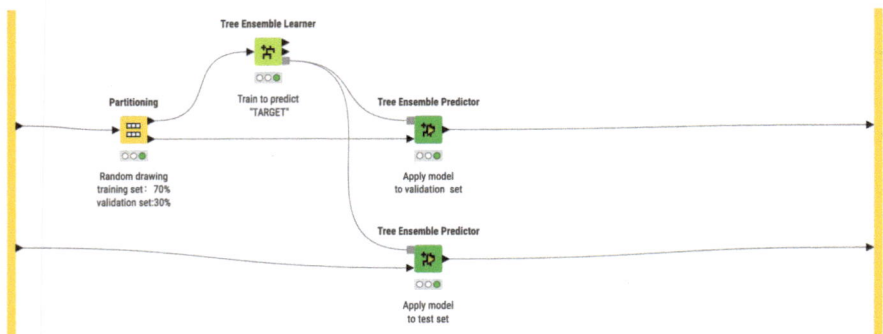

Fig. 8.17 Workflow of minority data oversampling

preventing overfitting. Setting it to 2 is not an optimal choice, and it is recommended to experiment with slightly larger values.

Note. The training results of the decision tree are not unique, for reference only. It is also reasonable if the setting is the same as the textbook but the training results are different.

8.2.2   Oversample Minority Data

If we employ minority data oversampling, we randomly resample the minority data to augment its quantity. As depicted in Fig. 8.16, we replicate a substantial amount of minority data and subsequently utilize the entire dataset for model training. While this approach avoids discarding any data, it is susceptible to overfitting due to the increased likelihood of obtaining duplicate instances.

KNIME does not provide this algorithm node directly but indirectly provides a similar function. The specific workflow is shown in Fig. 8.17, which is actually replacing the decision tree nodes with "Tree Ensemble Learner."

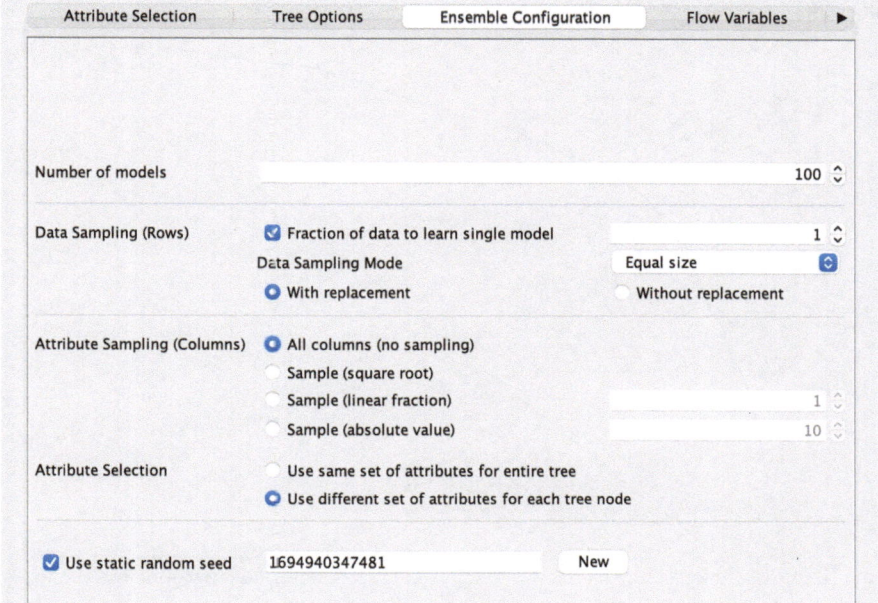

Fig. 8.18 Configure "Tree Ensemble Learner"

The "Tree Ensemble Learner" node can be considered as a random forest model with more options, as shown in Fig. 8.18. The settings under the "Ensemble Configuration" label are related to sampling. In "Data Sampling (Rows)," the "Fraction of data to learn single model" is set to 1, which means using the same amount of data as the actual data size. "Data Sampling mode" is set to "Equal size," which means using the same number of minority data. But how can the majority of data possibly be the same as the minority of data? So, here we use the method of sampling with replacement, which may result in oversampling of minority data.

By observing the ROC curve of the test results in Fig. 8.19, it can be seen that the performance of this algorithm has significantly improved.

As shown in Fig. 8.20, if we look at its confusion matrix, we'll notice that the accuracy is slightly lower than that of decision tree. However, the performance of identifying TARGET as 1 is much better than merely using the decision tree.

8.2.3   SMOTE Algorithm

In machine learning, there is a technique called SMOTE (synthetic minority oversampling technique) which involves increasing the representation of the minority class by synthesizing new samples. Let's see how it works.

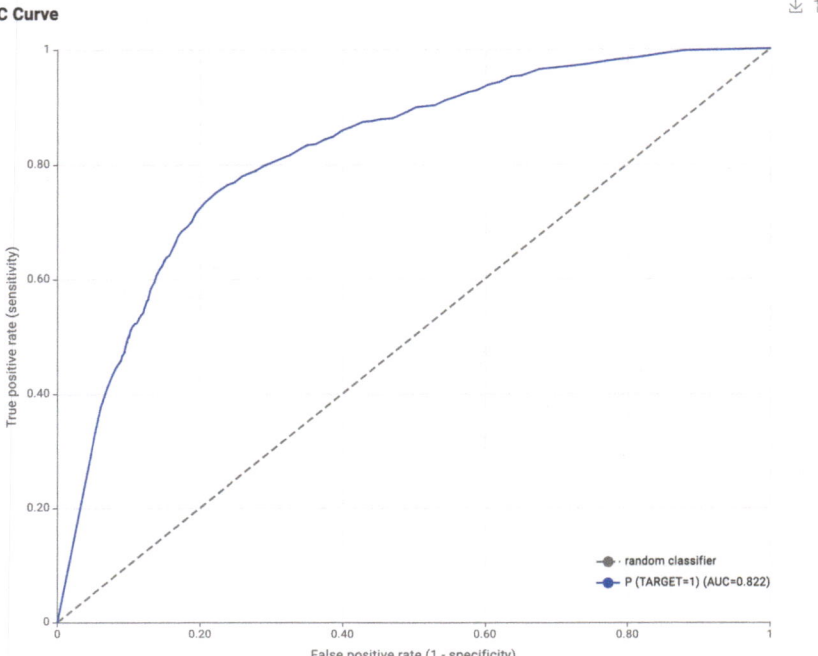

Fig. 8.19 ROC of minority data oversampling

File Hilite		
TARGET \ ...	0	1
0	12241	2339
1	218	406

Correct classified: 12,647	Wrong classified: 2,557
Accuracy: 83.182%	Error: 16.818%
Cohen's kappa (κ): 0.187%	

Fig. 8.20 Confusion matrix of minority data oversampling

In the data shown in Fig. 8.21, there are many more red data points than blue data points. If I only classify all points to be red, the accuracy will still be okay. But if our goal is to find all the blue data points, then we'll need to improve the classification.

The SMOTE algorithm will synthesize new minority points between existing data points. As shown in Fig. 8.22, assuming that we connect each blue point and artificially synthesize several data points on each line, this can greatly increase the number of blue points and achieve a balance between the two types of data points.

Is it realistic to connect each minority point with all other points when synthesizing new points using SMOTE? Clearly, this approach is not feasible. In real-world

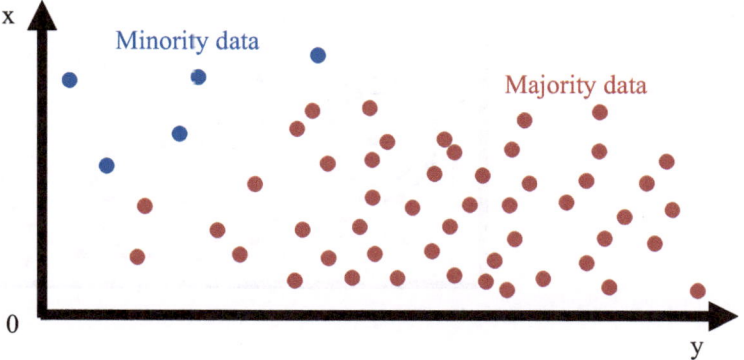

Fig. 8.21 Imbalanced data distribution

Fig. 8.22 Synthesize new
points

Fig. 8.23 $k = 1$

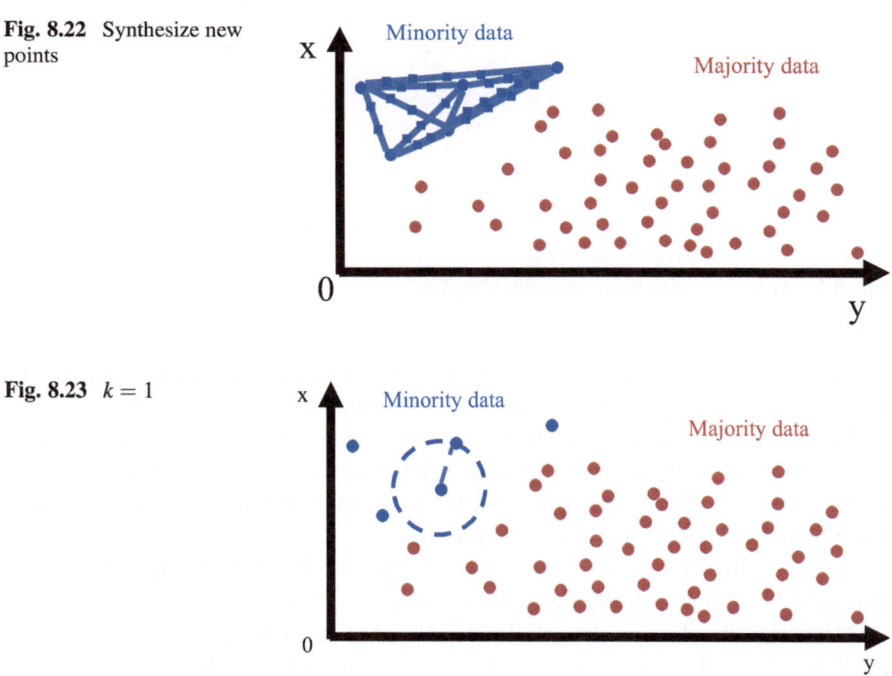

applications, we determine the connections through the parameter k. As illustrated in
Fig. 8.23, when $k = 1$, let's consider a specific point. We draw a circle with this point
as the center and gradually increase the radius of the circle until it touches another
point. We then connect these two points and randomly generate new points along the
line segment formed by the connection.

Similarly, when $k = 2$, as shown in Fig. 8.24, we draw a circle with this point as
the center and slowly increase the radius of the circle. Stop when the circle touches

Fig. 8.24 $k = 2$

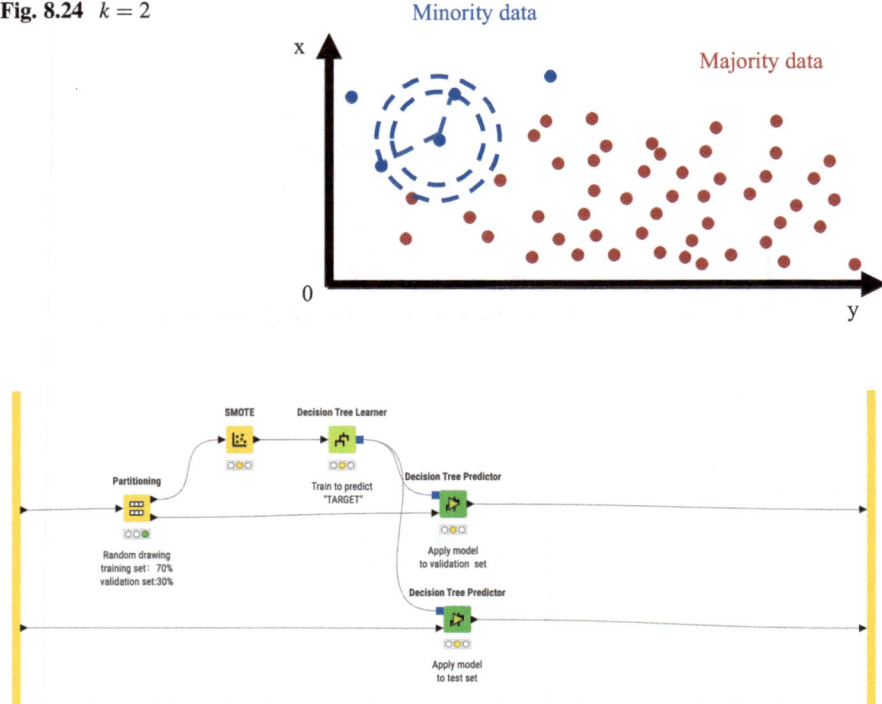

Fig. 8.25 Workflow of decision tree with SMOTE

two points. Then connect the point at the center with the two points touched, and randomly generate new points on the two line segments.

KNIME kindly provides us with a "SMOTE" node, which oversamples the input data (i.e., adds synthetic rows) to enrich the training data. Figure 8.25 shows its corresponding workflow in a decision tree. Note that "SMOTE" can only be used on the training set, never on the validation or test set.

However, this algorithm suffers from a significant drawback, namely, its computational efficiency drastically decreases when dealing with large datasets. Therefore, for scenarios involving substantial amounts of data, this algorithm may not be practical. Nevertheless, let's provide a brief overview of the settings for the KNIME "SMOTE" node, so that we can better use it in different scenarios accordingly. As depicted in Fig. 8.26, the "Nearest neighbor" parameter corresponds to the previously mentioned k-value, "Oversample by" determines the multiple of minority data to be added, and "Oversample minority classes" denotes the direct oversampling of minority data to match the quantity of majority data.

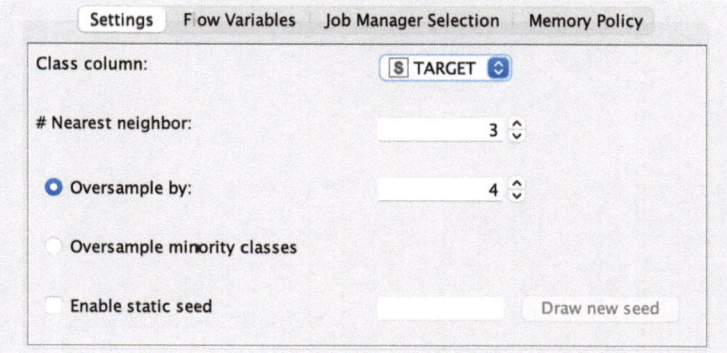

Fig. 8.26 Configure "SMOTE"

8.3 Practice

1. Have you heard of "heat death of the universe"? Please find out more about this concept.
2. What are the dangers of data imbalance?

References

https://en.wikipedia.org/wiki/Decision_tree_learning
https://en.wikipedia.org/wiki/Entropy_(information_theory)
https://machinelearningmastery.com/what-is-imbalanced-classification/
https://machinelearningmastery.com/tactics-to-combat-imbalanced-classes-in-your-machine-learning-dataset/

Chapter 9
Bayesian Analysis

When you have eliminated all which is impossible, then
whatever remains, however improbable, must be the truth.
—*Sherlock Holmes*

Abstract The chapter explores the fundamental concepts and applications of
Bayesian statistics. It begins with an introduction to basic terminologies like probability, sample space, and conditional probability, illustrating how probability quantifies the likelihood of events. The chapter delves into Bayes' theorem, explaining
how it updates prior beliefs with new evidence to form posterior beliefs. Concepts
like total probability theorem, sample spaces, and the importance of conditional
probability in Bayesian reasoning are discussed with practical examples. The chapter
also addresses the optimization of imbalanced data through techniques like
downsampling majority data and oversampling minority data. Additionally, it introduces the SMOTE algorithm for synthetic data generation. Practical applications of
Bayesian analysis in sentiment analysis and bank customer classification are
presented, highlighting the use of Naive Bayes learner in KNIME and other data
analysis tools. The chapter concludes with practical exercises to deepen the understanding of Bayesian principles in real-world scenarios.

Keywords Bayesian analysis · Sample space · Conditional probability · Bayes'
theorem · Total probability

This chapter covers

- Conditional probability.
- Sample space.
- Bayesian analysis and Bayes' theorem.

We often talk about probability, but what exactly is probability? What is the
probability of rain tomorrow? Does the probability of rain increase if you are in the
Sahara Desert? What about if you are in London? If you see a scene with sand swirling
all around, where do you think you are? The Sahara Desert or London? When thinking
over these questions, you are already employing Bayesian thinking to some extent.

© The Author(s), under exclusive license to Springer Nature Singapore Pte Ltd. 2024   255
Y. Geng et al., *Practical Machine Learning Illustrated with KNIME*,
https://doi.org/10.1007/978-981-97-3954-7_9

9.1 Bayes' Theorem

9.1.1 Basic Terminology

9.1.1.1 Probability

We often talk about probability, but what exactly is probability? Probability is commonly used to quantify how likely an uncertain event is to occur. This level of certainty can be represented by a numerical value between 0 and 1, which is called probability. The higher the probability of an event occurring, the more certain we are that the event may happen. For example, in the case of flipping a coin, the two possible outcomes of heads and tails are considered equally likely, with each having a probability of one-half. This means that the probability of getting heads and the probability of getting tails are both 50%.

The probability of an event E can be represented by $P(E)$.

9.1.1.2 Sample Space

In probability theory, the sample space of an experiment or random trial is the set of all possible outcomes or results of that experiment. And each possible outcome in the random trial is called a sample point, while the universal set (Ω) is the space that contains all the sample points. For example, in the case of flipping a coin, the sample space is the set {heads, tails}, which is also the universal set. If it is throwing a dice, the sample space will be {1, 2, 3, 4, 5, 6}, and this space is also the universal set.

9.1.2 Conditional Probability

$P(E)$ is actually the probability of event E occurring in the universal set, which can be written as $P(E|\Omega)$.

What would be the outcome when we already have a part of information? This situation involves the question of reducing the sample space. Taking the example of rolling a dice, the initial sample space A is {1, 2, 3, 4, 5, 6}. However, I have obtained information from a reliable source indicating that the dice has been manipulated, and the possible outcomes are now limited to {4, 5, 6}. Consequently, the sample space is no longer the entire universal set, but a smaller subset denoted as B, which consists of {4, 5, 6}. When calculating the probability $P(E)$, it becomes inadequate to express the narrowed-down sample space. To account for this modified sample space, we can denote the probability of event E occurring in the space B as $P(E|B)$ (Fig. 9.1).

Clearly, by narrowing down the sample space, we have become more certain about the occurrence of various values. If we can further narrow down the sample

Fig. 9.1 Reduced sample space

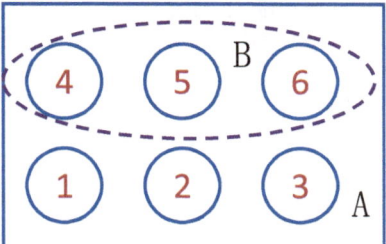

Fig. 9.2 Conditional probability

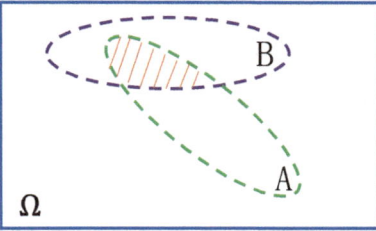

space, we can further determine the outcomes. On the other hand, if you roll a die and always get 4, 5, or 6 without getting 1, 2, and 3, what do you think is going on? Can you deduce anything from this?

9.1.2.1 Understanding the Sample Space

Suppose we now need to calculate $P(A|B)$ and we try to understand it using the concept of sample space. We want to know the probability of event A occurring in space B, as shown in Fig. 9.2. The original universal set space is reduced to the B, and now we need to calculate the probability of A occurring in the space of B.

We calculate in two steps:

The first step is to determine where exactly is $P(A|B)$ in the space?

As shown in Fig. 9.2, the probability of A occurring in B can correspond to the part "A and B both occurring" in the universal set space, i.e., $P(AB)$.

The second step is to enter the space B. We were viewing the universal set space just now, but since we are to calculate the probability in B, we need to enter the space of B.

The space B is smaller than the universal set space, and entering the space B is akin to a process of narrowing the field of view from a distance. Everything within the space B appears larger in comparison. In the example shown in Fig. 9.3, space A is represented by a dog, and space B is a specific part of the dog's body where there are parasites. The probability of space A having parasites may be very low, but the probability of space B having parasites is 100%.

As shown in Fig. 9.4, assuming there is another subspace C within space B, we have measured its volume relative to space B as c_b%.

Fig. 9.3 Larger field of
view in space B

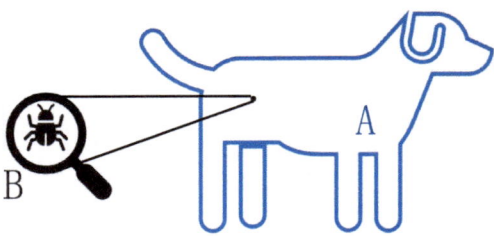

Fig. 9.4 Subspace

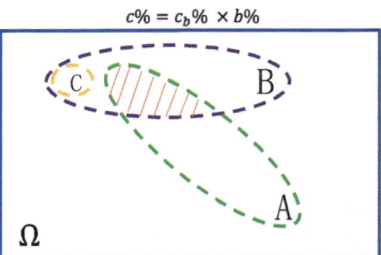

If the volume of space B occupies $b\%$ of the entire universal set space, the
proportion of volume that space C occupies in the universal set space can be
calculated as follows. We need to transition back to the entire universal set from
the B space.

$$c\% = c_b\% \times b\%$$

Conversely, if we are aware of the measure of subset C within the universal set,
and we want to determine the measure of C within the B space, we need to execute
the following operation, which involves transitioning from the universal set to the
B space:

$$c_b\% = \frac{c\%}{b\%}$$

Above, we have developed an intuitive understanding of the proportion of
volume occupied by a subspace within different parent spaces using the concept of
spatial volume. The subsequent step involves translating these intuitive perceptions
of volume into probabilities. Probability can be envisioned as a volume ratio, where
a larger volume occupied by an object within a space corresponds to a higher
likelihood of detection.

9.1.2.2 Calculating Using the Concept of Sample Space

We calculate $P(A|B)$ by following the above two steps. Based on the first step, we already know that $P(A|B)$ in the universal set can be seen as $P(AB)$. According to the analysis in the second step, we now need to get to space B. The probability of B occurring in the entire universal space is $P(B)$. It can be known:

$$P(A|B) = P(AB)/P(B)$$

The aforementioned formula can be interpreted as follows: the proportion of A within the B space is equivalent to the proportion of the intersection of A and B within the universal set, relative to the proportion within the B space. Developing a solid intuitive understanding of probability space at this stage will greatly facilitate the handling of total probability and Bayes' theorem in future applications.

9.1.3 Theorem of Total Probability

Please refer to Fig. 9.5 for better comprehension. The universal set is partitioned into subspaces $A_1, A_2, A_3, \ldots, A_n$, and our objective is to determine the probability (area) of B. If the area of B within each subspace can be readily calculated, then the total area of B can be expressed as the sum of B's areas within each subspace. This transformation yields a probability formula:

$$P(B) = P(A_1)\, P(B|A_1) + P(A_2)\, P(B|A_2) + P(A_3)\, P(B|A_3)$$

We can understand this strategy of seeking probabilities as a "divide and rule" strategy.

9.1.4 Bayes' Theorem (https://www.deeplearning.ai/courses/ machine-learning-specialization/)

With the foundational knowledge we have acquired, let's now delve into understanding the essence of Bayes' theorem. Bayes' theorem can be perceived as a

Fig. 9.5 Total probability theorem

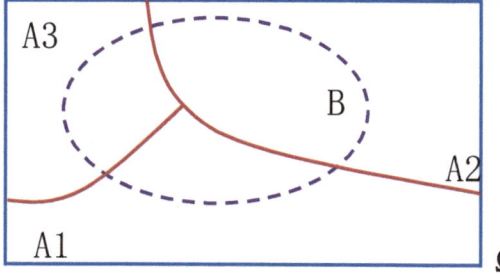

process of determining one's location. Simply put, we want to know the probability of being within the A_3 space within the entire space. The area of A_3 is represented as $P(A_3)$. To pinpoint our exact location, we require additional evidence. As more evidence accumulates, we find ourselves within the space B, with the probability of B occurring in the entire set denoted as $P(B)$. Therefore, the probability of A_3 needs to be reevaluated in light of new observational data, resulting in $P(A_3|B)$, signifying the probability of being within the space of A_3 given the occurrence of B. By employing the formula for conditional probability, we can derive the following calculation formula:

$$P(A_3|B) = \frac{P(A_3B)}{P(B)}$$

$$P(A_3|B) = \frac{P(B|A_3)P(A_3)}{P(B)}$$

This formula can be understood in combination with Fig. 9.6. Analyze the process above; initially, we have no information and can only make a judgment in the universal set space to obtain a prior probability: $P(A_3)$. Then, with the new information, we can narrow down the scope based on the new information and obtain a posterior probability: $P(A_3|B)$.

Let's consider a simple example to illustrate this concept. Suppose we want to determine whether a dice has been manipulated. We may roll the dice multiple times and analyze the outcomes. Have you ever wondered why this method works? You might argue that if the dice has been tampered with, the frequencies of certain numbers appearing will be different. But why does a significant deviation in number frequencies indicate a potential issue with the dice? Let's demystify it using Bayesian reasoning. We can define one space, denoted as X, to represent the scenario where the dice is fair and another space, denoted as Y, to represent the scenario where the dice is manipulated. Our objective is to determine which space we are in. Assuming we have prior knowledge (prior probability) based on past statistical experience,

$$P(X) = 0.8$$

$$P(Y) = 0.2$$

By rolling the dice multiple times, we have obtained new information. If this information is "the number of occurrences of each side of the dice varies

Fig. 9.6 Intersection of two sets

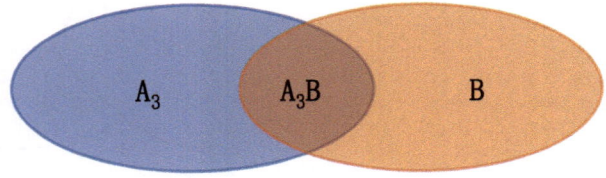

A$_3$ A$_3$B   B

significantly after multiple tosses," we would be more inclined to believe that our space is Y. Why? Let's denote this information as A. Based on experience, the probability of A occurring in X space is 0.1 and in Y space is 0.9:

$$P(A|X) = 0.1$$
$$P(A|Y) = 0.9$$

We want to determine whether X or Y is more likely to be the space we are in given the condition of knowing A:

$$P(X|A) = P(A|X)P(X)/P(A) \propto 0.1 \times 0.8 = 0.08$$
$$P(Y|A) = P(A|Y)P(Y)/P(A) \propto 0.9 \times 0.2 = 0.18$$

Comparing these two probabilities, we find that $P(Y|A)$ is larger, so we are more inclined to believe that knowing A has occurred, Y is more likely to be our space.

Do you find this process similar to solving a case? Let's imagine we are Sherlock Holmes and have identified a suspect in a murder investigation. The initial probabilities of the suspect being guilty or innocent are $P(\text{guilty}) = 0.5$ and $P(\text{not guilty}) = 0.5$. As the investigation progresses, we uncover evidence that the suspect has committed a robbery. Based on this new information, we can infer that $P(\text{guilty}|\text{robbery}) = 0.6$ and $P(\text{not guilty}|\text{robbery}) = 0.4$. With further observations, as more information becomes available, the probabilities evolve to $P(\text{guilty}|(\text{robbery} \cap \text{arson} \cap \text{trafficking})) = 0.99$ and $P(\text{not guilty}|(\text{robbery} \cap \text{arson} \cap \text{trafficking})) = 0.01$. At this stage, with substantial evidence, the probability of this individual being the murderer is already very high, providing strong grounds to believe that this individual is indeed the culprit. Bayesian theory allows us to infer the underlying causes based on knowledge of the outcome, recognizing that multiple factors can contribute to a particular result.

In standard terminology, Bayes' theorem provides a framework for calculating the conditional probability of an event or hypothesis by incorporating both evidence (our current location within a sample space) and prior beliefs (the universal set space or a broader space). The process begins with initial beliefs, which are then refined based on acquired data, resulting in the posterior probability (or posterior belief). As more data becomes available, the previous posterior probability becomes the new prior, and this iterative process continues.

Bayes' theorem can be written as

$$P(\theta|X) = \frac{P(X|\theta)P(\theta)}{P(X)}$$

X is the evidence. θ is the belief.

- $P(\theta)$. The prior probability represents our belief obtained through past experiences, such as in the above example where the probability of the suspect being the perpetrator is 0.5 without any additional evidence. However, based on the information "the suspect has committed a robbery," we can use the "P-(perpetrator|robbery) = 0.6" as a belief derived from past experiences and continue with subsequent calculations.

- $P(X)$. Probability of evidence appearing.
- $P(X|\theta)$. The likelihood is the possibility of certain evidence appearing under a certain belief condition. In the example above, it would be the probability of finding evidence of robbery if we believe the suspect is the murderer. Generally, the likelihood is denoted as $L(\theta|X)$.
- $P(\theta|X)$. Posterior probability refers to the belief updated with new evidence.

Because ultimately what we are comparing is which posterior belief is greater, we choose the one that is "believed" to be greater. And in this comparison process, $P(X)$ remains the same. So in general usage, we write the above equation as

$$P(\theta|X) \propto P(X|\theta)P(\theta)$$

Simply put, it is the update of old beliefs ($P(\theta)$) to new beliefs ($P(\theta)$) supported by new evidence ($P(\theta|X)$). If there are multiple pieces of evidence, multiply all the evidence together to obtain a new belief, i.e.,

$$P(\theta|X \cap Y \cap Z) \propto P(X|\theta)P(Y|\theta)P(Z|\theta)P(\theta)$$

In machine learning, data is the evidence and the results are the beliefs. Our goal is to update our beliefs using data.

9.1.5 Sentiment Analysis

Please analyze the individual's sentiment as positive or negative based on the comments in Fig. 9.7. We currently have a total of 100 comments, with 60 being positive and 40 being negative.

- **Task implementation.**
- Compute prior probability.
  According to the data, prior probability $P(\text{Positive}) = 0.6$, while $P(\text{Negative}) = 0.4$.
- Calculate posterior probability.

Next, we can start collecting evidence. The number of times "YES," "OK," and "GOOD" appear when emotions are positive is 10, 5, and 20, respectively. Therefore, in all the spaces where evidence exists, $P(\text{YES}|\text{Positive}) = \frac{10}{60} = 0.1667$, $P(\text{OK}|\text{Positive}) = \frac{5}{60} = 0.0833$, and $P(\text{GOOD}|\text{Positive}) = \frac{20}{60} = 0.3333$. By using Bayes' theorem, we can obtain $P(\text{Postive}|\text{YES} \cap \text{OK} \cap \text{GOOD}) \propto P(\text{YES}|\text{Positive})P(\text{OK}|\text{Positive})P(\text{GOOD}|$

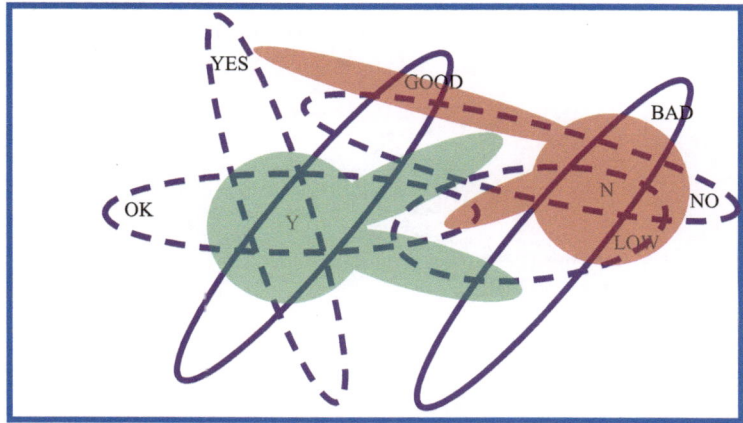

Fig. 9.7 Example of sentiment analysis

Positive)*P*(Positive). Next, we can calculate *P*(Negative| YES ∩ OK ∩ GOOD) using a similar method. Now we may compare the two posterior probabilities and make a decision based on which one is greater. This way, we can determine whether the sentiment is positive or negative under certain word combinations.

9.2 Address the Bank Customer Classification Case with Bayesian Algorithm

9.2.1 Workflow

We may continue with the case of bank customer classification discussed in the previous chapter. This time we will use the Bayesian algorithm to solve it. The workflow, as shown in Fig. 9.8, has not changed much overall. Only the model needs to be replaced. In this workflow, we use the "Equal Size Sampling" node for data balancing.

9.2.2 Bayes Learner

There isn't much to elaborate on regarding this workflow. Our focus here is to introduce the "Naive Bayes Learner" node, which is a crucial component of the Bayesian algorithm. As depicted in Fig. 9.9, the "Default probability" is typically set to a small value. Its purpose is to mitigate the issue of encountering new data with values that are absent in the training set, which would result in a probability of 0. By setting this value, any unseen data is assumed to have a probability of 0.0001.

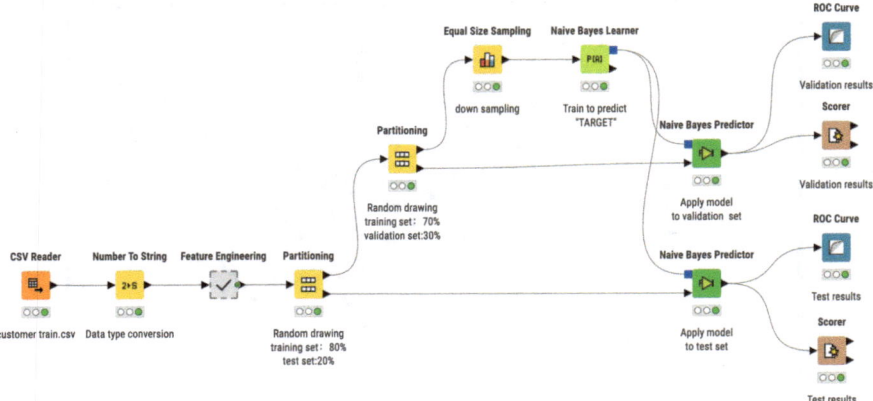

Fig. 9.8 Bayesian algorithm workflow

| Options | Flow Variables | Job Manager Selection | Memory Policy |

Classification Column: [S] TARGET

Default probability: 0.0001

Minimum standard deviation   0.0001

Threshold standard deviation   0.0

Maximum number of unique nominal values per attribute: 20

Ignore missing values   Create PMML 4.2 compatible model

Fig. 9.9 Configure the leaner

Another important setting is the "Maximum number of unique nominal values per attribute." This setting ensures that if the number of categories in a column exceeds 20, the training model will exclude that particular column. This precaution is taken because as the number of categories increases, the probability of each category occurring decreases. If this probability decreases significantly, it becomes less representative and can potentially lead to overfitting.

However, it can be observed from the test results that using the Bayesian algorithm in KNIME is not appropriate to solve this problem because, as seen in Fig. 9.10, the model considers almost all samples as 0 (not dissatisfied).

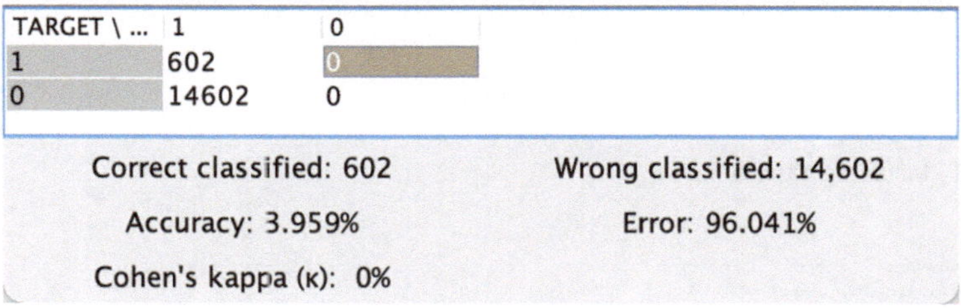

TARGET \ ...	1	0
1	602	0
0	14602	0

Correct classified: 602   Wrong classified: 14,602

Accuracy: 3.959%   Error: 96.041%

Cohen's kappa (κ): 0%

Fig. 9.10 Confusion matrix

9.3 Case of Sentiment Analysis

Below we will use KNIME to complete a sentiment analysis case in Kaggle competition (https://www.kaggle.com/crowdflower/twitter-airline-sentiment). To complete this task, first we need to install a KNIME extension and then modify a workflow from the KNIME open source community to achieve this functionality.

9.3.1 Install Extension

To install extensions in KNIME, ensure that your computer is connected to the Internet. Then, click on the icon [i] in the top right corner of the KNIME interface to access the "Go to info page" option, as illustrated in Fig. 9.11. From there, select the "Install Extensions" button, which will open the interface displayed in Fig. 9.12. In the input field, you can enter "text" to search for the desired extension. In this case, we will choose "KNIME Textprocessing" and click the "Next" button to proceed. KNIME will then analyze and download information about the extension from the Internet, as shown in Fig. 9.13. Click the "Next" button again. To initiate the installation, select "I accept the terms of the license agreement" as shown in Fig. 9.14. After the KNIME extension is installed, a restart is necessary for the changes to take effect. A restart window will appear, as shown in Fig. 9.15. Click on "Restart Now," and once the software restarts, the "KNIME Textprocessing" extension will be ready for use.

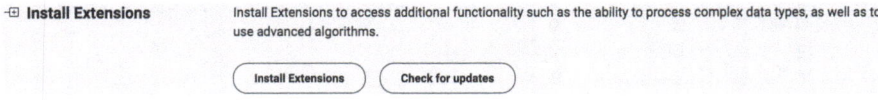

Fig. 9.11 Installing extensions

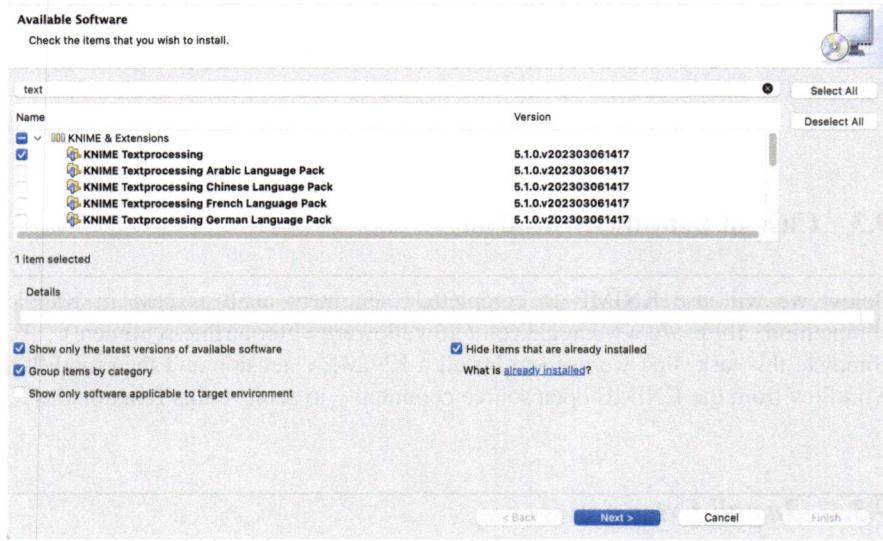

Fig. 9.12 Select the extension to install

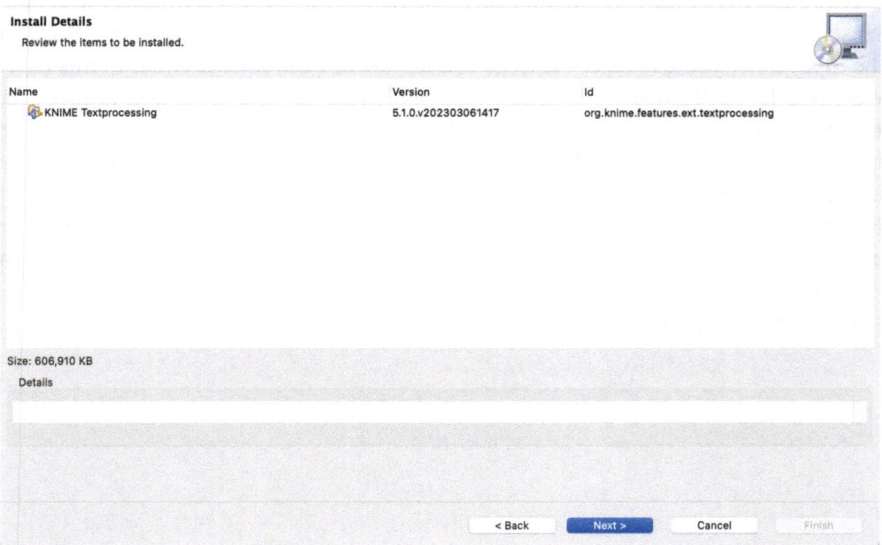

Fig. 9.13 Click "Next" to continue

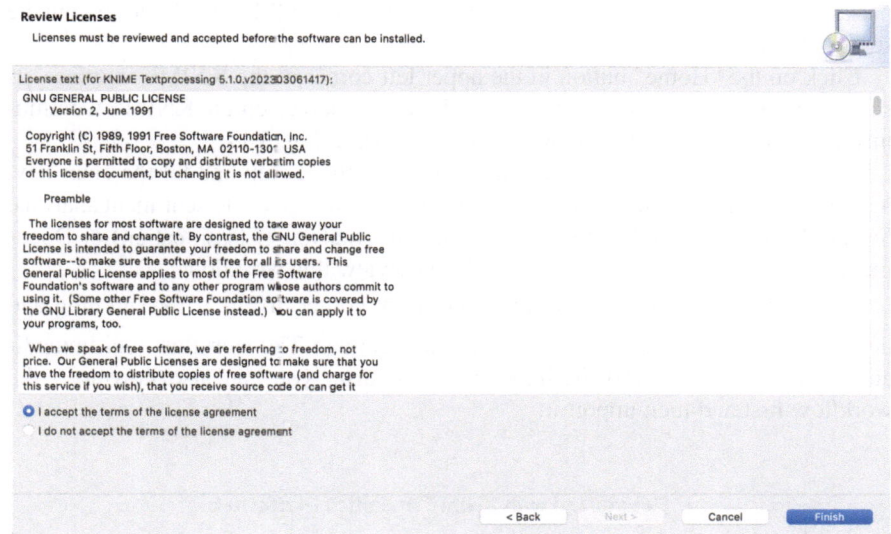

Fig. 9.14 Agree to install

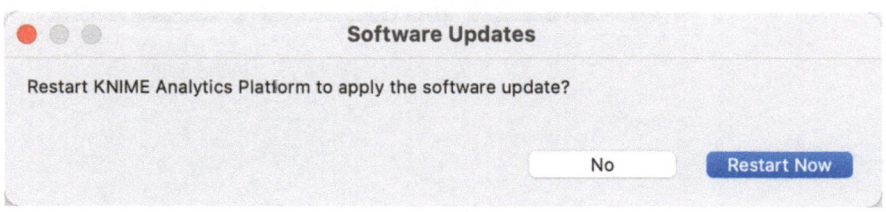

Fig. 9.15 Restart

9.3.2 Build a Workflow

9.3.2.1 Origin of Workflow

KNIME provides two types of workflow examples for us to download and readily use. One is the KNIME Hub, an open source community, and the other is KNIME EXAMPLES.

KNIME Hub

The KNIME open source community provides us with a large number of nodes, components, and workflows. We can search for relevant workflows in the community and load them onto our local computers for use. You can download the "Sentiment Analysis (Classification) of Documents" workflow in the workflow package of this book, or you can download it from the KNIME Hub by following the steps below:

Click on the "Home" button in the upper left corner of the KNIME interface, as shown in Fig. 9.16, and click on the "Find more resources on the KNIME Community Hub" link to enter the KNIME Hub as shown in Fig. 9.17.

Enter "sentiment" in the search field on the KNIME Hub and hit the enter key. You will find many nodes, components, and workflows related to sentiment analysis. Select "Workflows" and choose the "Sentiment Analysis (Classification) of Documents" workflow, as shown in Fig. 9.18, to preview this workflow.

Hold down the button [⌗] in the upper right corner of Fig. 9.18, and drag and drop this workflow into the KNIME workflow canvas. The workflow will immediately appear in your KNIME interface. Of course, you can also download the workflow first and then import it.

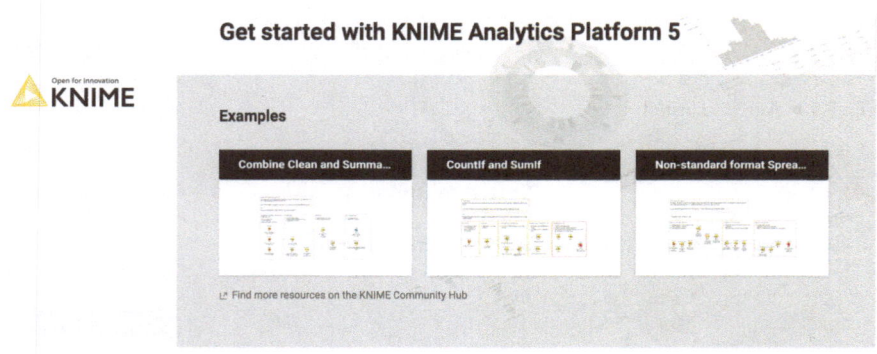

Fig. 9.16 New home interface

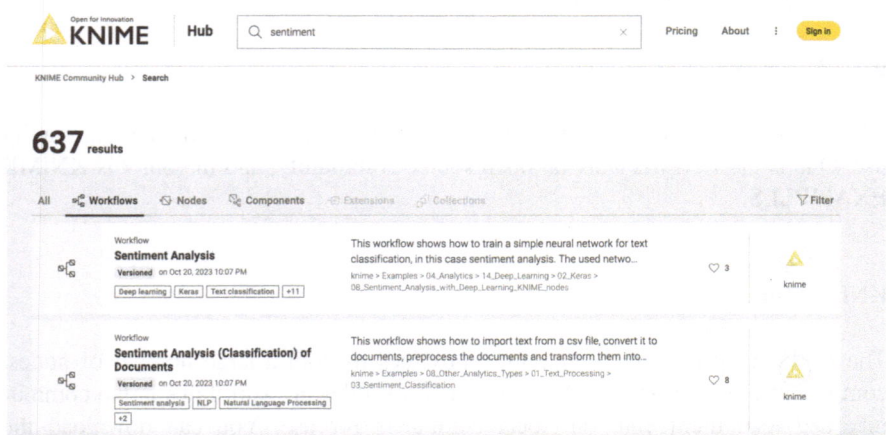

Fig. 9.17 KNIME Hub interface

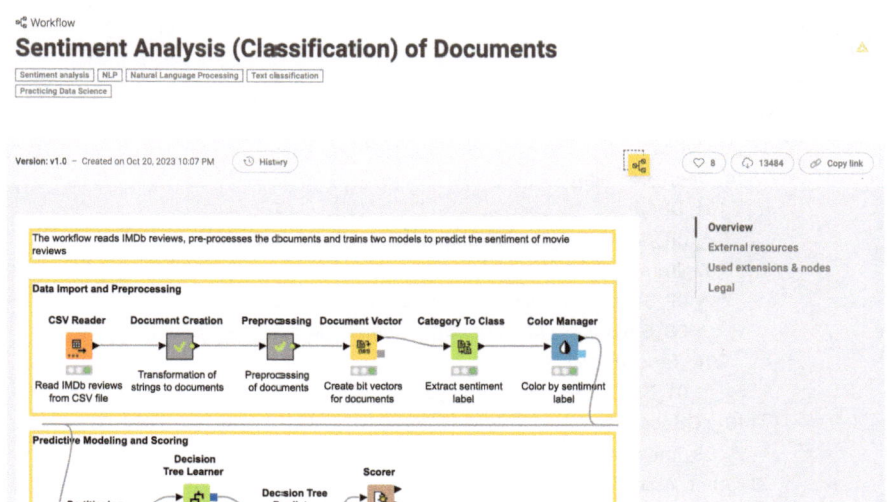

Fig. 9.18 Workflows in the KNIME Hub

KNIME Examples

In addition to searching for workflows in the KNIME Hub, you can also use the built-in workflows in the classic interface of KNIME. Follow Fig. 9.19, double-click on "EXAMPLES" in the workflow area of the classic interface, and search for "sentiment." You will find many sentiment analysis workflows in the "EXAM-PLES." Here, we can choose a relatively simple one, "03_Sentiment_Classification." The workflow demonstrated below is a modification of "03_Sentiment_Classification."

Since we want to try the Naive Bayes algorithm, we need to modify the decision tree algorithm in "03_Sentiment_Classification" to a Naive Bayes algorithm node. As shown in Fig. 9.20, this workflow first reads the data and then builds a document using our data, followed by preprocessing the document for vector conversion and finally feeding it to the model for training.

This process highlights the importance of exploring the KNIME Hub or utilizing the built-in workflows when encountering unfamiliar problems that require solutions. While the performance of the Bayesian algorithm may not always surpass that of other algorithms, it is worth noting that the results of this particular example may not meet expectations. However, substituting the Bayesian-related nodes with nodes from alternative algorithms can potentially yield significantly improved outcomes.

Let's take a rough look at what needs to be done in each step. The principles of word processing will be discussed in the next chapter, and we don't need to grasp more technical details at the beginner stage.

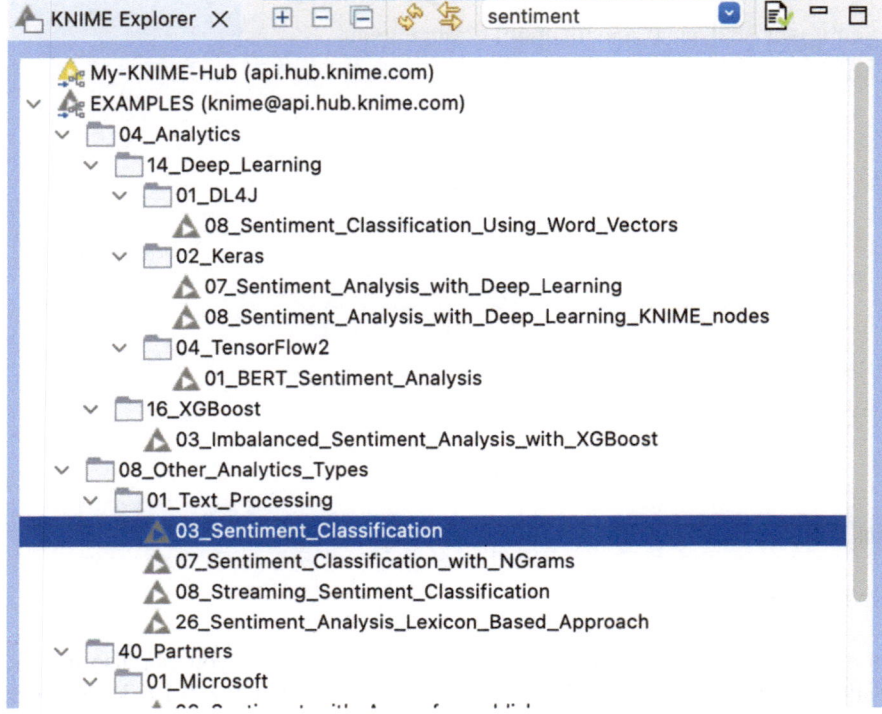

Fig. 9.19 Search for workflows in EXAMPLES

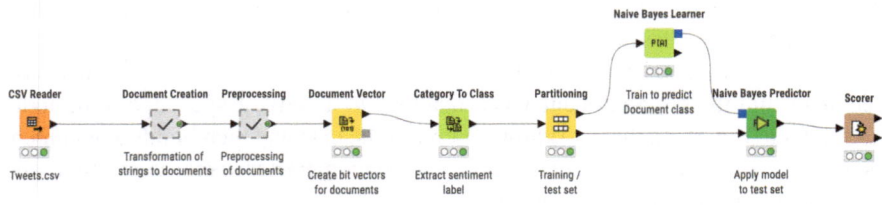

Fig. 9.20 Sentiment analysis workflow using Naive Bayes

9.3.2.2 Document Creation

The content of document creation is in "Document Creation," as shown in Fig. 9.21. First, we can use the "Strings To Document" node to convert the sentences to be analyzed into documents and then use the "Column Filter" to only keep the converted documents.

Now let's take a look at the settings for "Strings To Document," as shown in Fig. 9.22. The most important part of this node is the "Text." Here, we can choose "Full text." Then set the training objective to "Use categories from column," and

Fig. 9.21 Document creation workflow

Options	Flow Variables	Job Manager Selection	Memory Policy

Title

○ Column   ○ Row ID   ● Empty string

Title column   S user_timezone

Text

Full text   S text

Meta Information

Document source

☐ Use sources from column   Document source column   S user_timezone

Document category

☑ Use categories from column   Document category column   S airline_sentiment

☑ Use authors from column   Authors column   S name

Author names separator   ,

Default author first name   –   Default author last name   –

Type and Date

Document type   UNKNOWN

Date: 2017–12–13

☐ Use publication date from column   Publication date column

Column

Document column: Document

Processes

Number of maximal parallel processes   4

Tokenization

Word tokenizer   OpenNLP English WordTokenizer

Fig. 9.22 Configure the "Strings To Document" node

select "Document category column" as "airline_sentiment," which is the training objective. Other options can be ignored; just click confirm.

The next step is to configure the "Column Filter" node, as shown in Fig. 9.23, to only keep the "Document" column created earlier.

9.3.2.3   Preprocessing

The preprocessing section does not require any additional setup; it can be directly executed as is. As illustrated in Fig. 9.24, the "Punctuation Erasure" node is responsible for removing punctuation marks, followed by the filtering out of numbers and end characters and the conversion of all letters to lowercase. Subsequently, all the words undergo processing using the "Snowball Stemmer" node. Other nodes

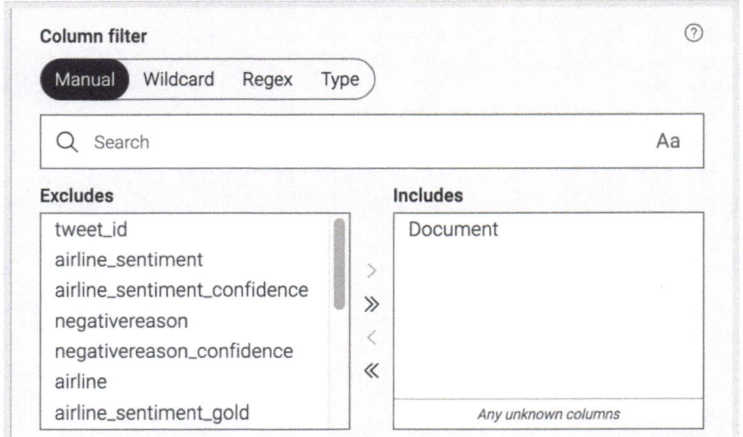

Fig. 9.23 Only keep "Document"

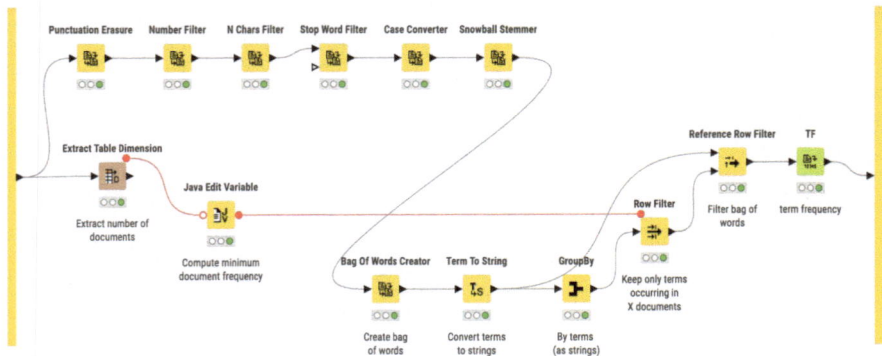

Fig. 9.24 Preprocessing workflow

in this section involve specific algorithms, which will not be described here. These nodes can be easily copied for immediate use.

9.3.2.4 Vectorization

In order to use machine learning algorithms, we need to process all words into numerical representations that are easier for machines to understand. This is where "Document Vector" is used to vectorize them.

9.3.2.5 Classification

"Category To Class" node assigns class labels to the dataset based on the provided categorization information.

We are familiar with all subsequent parts and will not introduce them here.

9.4 Practice

1. What is a sample space?
2. Please raise an example of conditional probability in your daily life.

Reference

https://www.deeplearning.ai/courses/machine-learning-specialization/

Chapter 10
Computer Vision and Natural Language Processing

More is Different.
—*Philips Anderson*

Abstract The chapter provides a brief exploration of deep learning applications in computer vision and natural language processing (NLP). It begins with an introduction to deep learning, its key elements, and the significance of neural networks, activation functions, and hyperparameters. The chapter then delves into the realm of computer vision, discussing the evolution and contributions of prominent convolutional neural network (CNN) models like LeNet-5, AlexNet, VGGNet, GoogLeNet, and ResNet. The practical implementation of CNN using KNIME, a data analytics platform, is also covered. Furthermore, the chapter addresses NLP, discussing the challenges of digitizing natural language and various approaches like one-hot encoding, bag-of-words, neural network language models, and feature extraction techniques like LSTM and Transformer. The chapter concludes by introducing large language models (LLMs) such as ChatGPT, Google's Bard, and Baidu's Wenxin Yiyan and their implementation in KNIME through AI assistant and AI extension tools. Practical exercises are provided to reinforce understanding and application of these concepts.

Keywords Computer vision · Natural language processing · Deep learning · Neural networks · Convolutional neural networks · Bag-of-words · LSTM · Transformer · Word embedding · Large language models · ChatGPT

This chapter covers

- **Neural network.**
- **Computer vision.**
- **Common models of computer vision.**
- **Natural language processing.**
- **Large language models.**

With the rapid advancement of artificial intelligence technology, machines are now capable of performing tasks that were once considered impossible. These tasks include facial recognition, intelligent customer service, and more. In order to delve into the field of deep learning, let's begin by exploring computer vision and natural language processing.

10.1   Introduction to Deep Learning

Deep learning is an important branch of machine learning. As shown in Fig. 10.1, it is an algorithm that attempts to use multiple processing layers with complex structures or multiple nonlinear transformations to achieve high-level abstraction of data. You may not know what I am talking about, but don't worry, let's just keep on going.

Figure 10.1 is actually a relatively simple neural network. It can be seen that this network consists of several layers, each layer having several nodes, and each node is connected to nodes in the adjacent layers. Since the emergence of deep learning, it

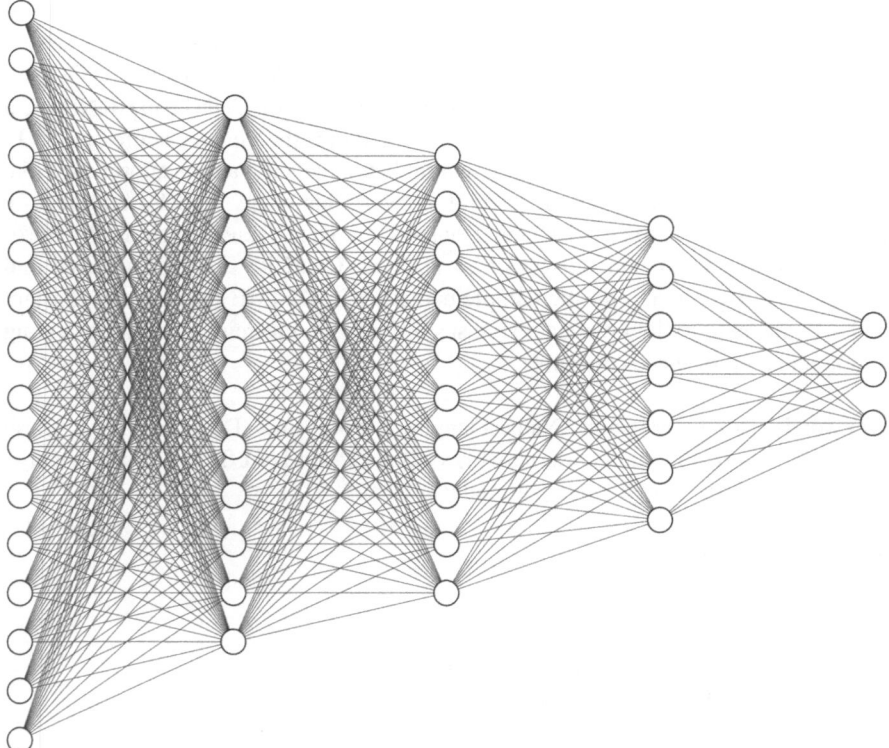

Fig. 10.1 Deep learning network

has become an indispensable part in many fields, especially in computer vision and natural language processing (NLP).

Training deep learning models is resource-intensive and requires tuning a significant number of hyperparameters.

10.1.1   Key Elements to Deep Learning

Deep learning has three key elements: suitable algorithms, powerful hardware, and large amounts of data (Fig. 10.2).

The emergence of deep learning did not happen recently; however, it gained significant popularity around a decade ago only because of the absence of the three key elements.

In 1989, Yann LeCun and his colleagues applied the standard backpropagation algorithm, initially proposed in 1974, to deep neural networks for the purpose of handwritten postal code recognition. Although the algorithm demonstrated success, it came with a high computational cost and limited availability of hardware resources. The training process for the neural network alone took up to 3 days, rendering it impractical for real-world deployment. However, with continuous advancements in algorithms and hardware, the execution time of machine learning algorithms has significantly decreased. Furthermore, the continuous progress in big data technology has provided deep learning with access to vast amounts of data, enabling it to effectively leverage its inherent advantages.

Fig. 10.2 Key elements of deep learning

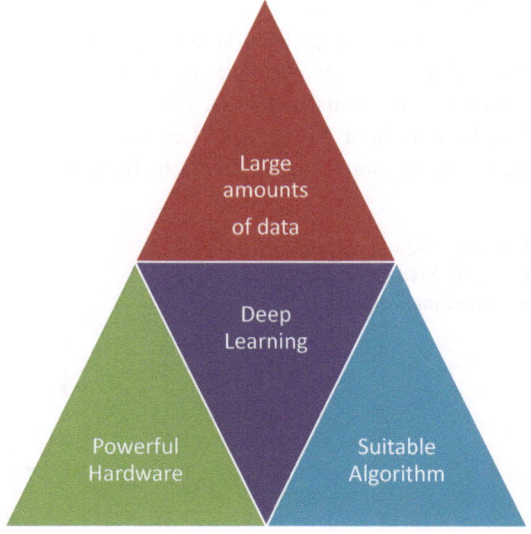

10.1.2   Our Goal

At the beginner stage, it is not necessary to delve into extensive knowledge about deep learning. When presented with a task, we can have a general understanding of which deep learning model to employ and simply utilize its API (application programming interface), without the need to train the model or comprehend the intricate workings of each individual model.

Generally speaking, for image-related tasks, convolutional neural networks (CNNs) are typically used. If it is a text-based task, generally recurrent neural networks (RNNs) or models based on Transformers are used.

10.1.3   Overview of Deep Learning

10.1.3.1   Neural Networks

As depicted in Fig. 10.3, the schematic diagram of a neural network receives input data from the input layer, undergoes transformation through a sequence of hidden layers, and ultimately produces output data from the output layer. Each layer consists of a collection of neurons, with every neuron in a given layer being fully connected to all neurons in the preceding layer. The final layer represents the prediction stage.

Each circle represents a neuron, and the layers between the input and output layers are the hidden layers.

10.1.3.2   Activation

Nonlinearity is a crucial aspect of neural networks, as it enables them to capture complex relationships. Neural networks achieve their output by passing the weighted sum of their inputs through an activation function. There are various types of activation functions available, but as beginners, it is not necessary to delve too deeply into their intricacies. Let me introduce one commonly used activation function, the sigmoid (σ) function. Its graph is depicted in Fig. 10.4. It is worth

Fig. 10.3 Neural network (https://cs231n.github.io/convolutional-networks/)

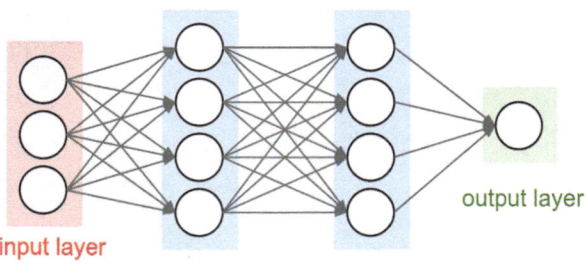

input layer

hidden layer 1   hidden layer 2

output layer

Fig. 10.4 Activation
function sigmoid

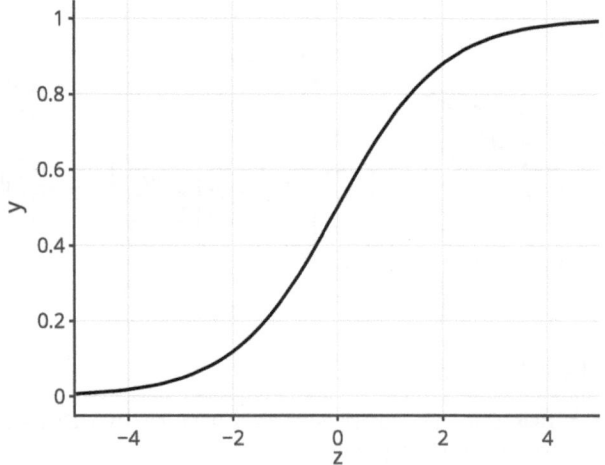

Fig. 10.5 Linear function
and activation function

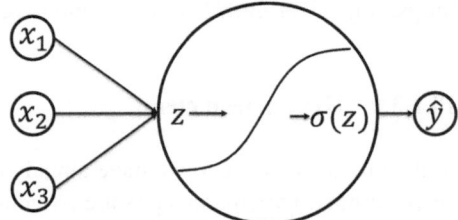

noting that any linear relationship, when multiplied by an activation function, becomes nonlinear.

Based on Fig. 10.3, we can now observe Fig. 10.5. We may notice that linear functions produce nonlinear functions when the neurons pass through an activation function. These nonlinear functions are then weighted and linearly combined, followed by activation through the neurons in the subsequent layer. This iterative process continues until reaching the final layer.

Therefore, the values in the final feature maps are not mere sums but rather the result of applying activation functions to them. It is the presence of activation functions that empowers neural networks with remarkable capabilities.

10.1.3.3   Convolutional Neural Network (CNN)

If we organize each layer of a neural network based on the dimensions of width, height, and depth, as illustrated in Fig. 10.6, where neurons in one layer are exclusively connected to a localized region in the subsequent layer, the final output will be reduced to a probability value vector along the depth dimension. This type of

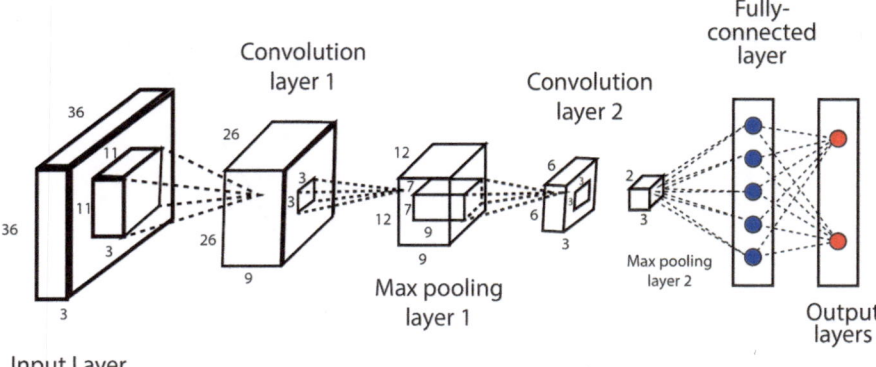

Fig. 10.6 Convolutional neural network (https://www.analyticssteps.com/blogs/common-architectures-convolution-neural-networks)

neural network is commonly referred to as a convolutional neural network (CNN) (https://ujjwalkarn.me/2016/08/11/intuitive-explanation-convnets/).

10.1.3.4   Hyperparameters

In the previous chapters, we have already discussed hyperparameters. However, in deep learning, hyperparameters are particularly important and play a crucial role in model training. But at the initial stage of learning, we don't consider these, as we simply utilize the model without training it.

10.2   CNN in Computer Vision

Computer vision enables machines to perceive the world like humans do with their eyes. Here we introduce several commonly used convolutional neural network models that can be used in our own image recognition projects. Please check more about these models on the web yourself.

10.2.1   LeNet-5

LeNet-5 is a convolutional neural network developed by Yann LeCun in 1998 specifically for handwritten digit recognition. It gained significant popularity during that time and was extensively utilized by numerous American banks for recognizing handwritten digits on checks. LeNet-5 stands as one of the most influential experimental systems from the early stages of convolutional neural networks.

Comprising a total of seven layers (excluding the input layer), each layer of LeNet-5 encompasses varying numbers of trainable parameters.

However, due to the lack of large-scale training data and limited computing power at that time, and the relatively simple network structure, LeNet-5 did not yield satisfactory results in tackling complex problems.

10.2.2   AlexNet

AlexNet, introduced in 2012 by Alex Krizhevsky, Ilya Sutskever, and Geoffrey Hinton, made a groundbreaking impact by achieving remarkable performance in the 2012 ILSVRC (ImageNet Large-Scale Visual Recognition Challenge). This marked a significant milestone as it was the first instance where a CNN outperformed other methods, notably surpassing the second-place support vector machine by a considerable margin. This achievement sent shockwaves throughout the academic community, leading to widespread recognition and awareness of CNN networks among the general public.

AlexNet demonstrated the effectiveness of CNNs in complex models and proved that using GPUs for training can yield results within an acceptable timeframe.

10.2.3   VggNet

VGGNet, proposed by Karen Simonyan and Andrew Zisserman from the University of Oxford in 2014, stands out for its remarkable "depth and simplicity." The depth refers to VGGNet's architecture, which consists of 19 layers, surpassing its predecessors by a significant margin. The simplicity lies in its uniform structure, where 3×3 filters with a stride of 1 and 2×2 max pooling with a stride of 2 are consistently employed.

VGGNet encompasses six distinct network architectures, each comprising five sets of convolutions. Each set of convolutions employs a 3×3 convolutional kernel, followed by a 2×2 max pooling operation. Subsequently, three fully connected layers are employed. To expedite the convergence of higher-level networks, it is possible to initialize them using the weights obtained from training lower-level networks.

10.2.4   GoogLeNet

GoogLeNet, also known as Inception, is an innovative deep learning architecture introduced by Christian Szegedy in 2014. Prior to its development, models like AlexNet and VGGNet achieved improved training performance by increasing the depth of the network while presenting challenges such as overfitting, vanishing gradients, and exploding gradients. But GoogLeNet aimed to address these issues by taking a different approach. It focuses on efficiently utilizing computational resources and extracting more features with the same computational capacity, thereby enhancing training outcomes.

10.2.5   ResNet

ResNet, proposed by researchers from Microsoft Research Asia in 2015, addresses a common challenge faced by convolutional neural networks as their depth increases. This challenge often manifests as a bottleneck or even a decrease in effectiveness, primarily due to the presence of vanishing or exploding gradients. To overcome this issue and enable the training of deeper networks, ResNet introduces a novel concept called the residual block.

10.3   Implementing CNN with KNIME

10.3.1   Environment Construction

To use KNIME for deep learning, we also need some third-party tools. The third-party tools used include the following:

- TensorFlow: a deep learning framework by Google.
- Keras: a more user-friendly framework based on several deep learning frameworks including TensorFlow.
- Python programing environment.
- Extensions for KNIME integrating Python, Keras, and TensorFlow.

10.3.2   Install the Necessary Tools

10.3.2.1   Install Miniconda

Miniconda is a lightweight Python environment management tool. Enter https://conda.io/miniconda.html in the browser address bar to visit the website, choose your operating system, and download and install miniconda.

10.3.2.2 Create Conda Keras Environment

The initial step involves installing the CPU version of the conda keras environment within Anaconda. To accomplish this, we will need to utilize the command line interface of your operating system. In Windows, we can use Anaconda Prompt, while in MacOS, we can use Terminal. Let's proceed by creating a virtual environment named "py36_knime." Please enter the following command in the command line tool:

```
conda create -y -n py36_knime python=3.6 keras=2.1.6 pandas
```

Here, we have installed Python 3.6, Keras 2.1.6, and the pandas packages. Miniconda will automatically install the appropriate version of TensorFlow for us.

Next, we may activate and use the "py36_knime" environment.

Windows/Mac command line

```
conda activate py36_knime
```

After entering the "py36_knime" environment, we sequentially input "python" and "import keras" in the console. When the console displays both Python version information and "Using TensorFlow backend," respectively, it indicates that our environment is successfully configured, as shown in Fig. 10.7.

Finally, we may install the KNIME extensions as follows (refer to the previous chapter for the installation guide):

- KNIME Deep Learning—Keras Integration.
- KNIME Deep Learning—TensorFlow Integration.
- KNIME Image Processing.

10.3.2.3 Setting Up the Environment

After installing the conda environment and KNIME extensions, we need to start and use the conda "py36_knime" environment in KNIME. Specific setups are as follows:

First, set the location of conda installation. Sequentially click on the settings (preferences) ⚙ in the upper right corner of the software → KNIME → Conda, as shown in Fig. 10.8, and select the installation directory of Miniconda in "Path to the

```
       ~ %conda activate py36_knime
(py36_knime)       ~ %python
Python 3.6.13 |Anaconda, Inc.| (default, Feb 23 2021, 12:58:59)
[GCC Clang 10.0.0 ] on darwin
Type "help", "copyright", "credits" or "license" for more information.
>>> import keras
Using TensorFlow backend.
>>> []
```

Fig. 10.7 Successful configuration of conda environment

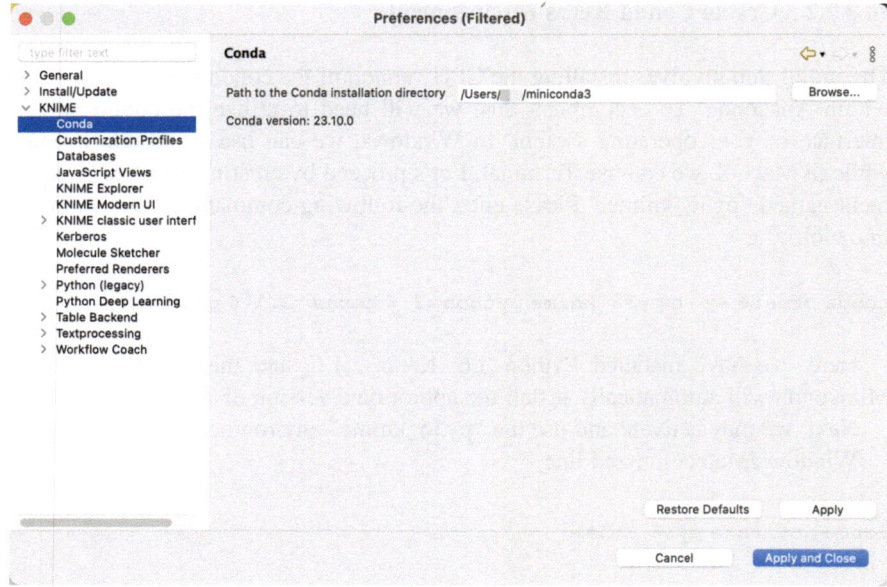

Fig. 10.8 Directory setup

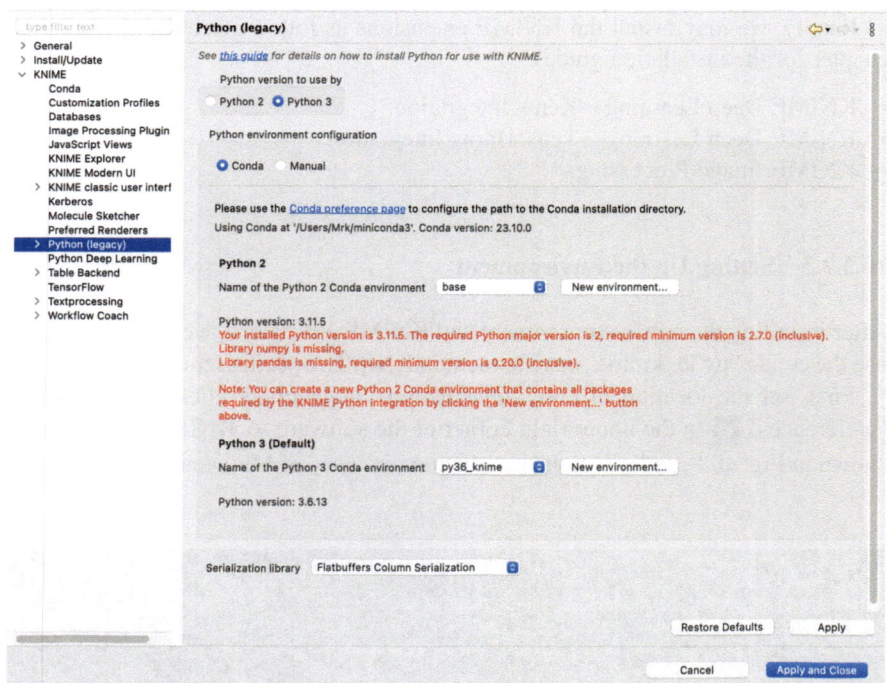

Fig. 10.9 Python environment setup

Conda installation directory." KNIME will automatically detect the version of Conda, and, finally, click "Apply" to finish the configuration.

Subsequently, we can proceed with configuring the Python environment. Choose the "Python(leagacy)" option within the current settings interface, as depicted in Fig. 10.9. Then, designate the "Python version to use by default" as "Python3," set the "Python environment configuration" to "Conda," select the "py36_knime" environment within "Python3 (Default)," and conclude by clicking "Apply and Close."

Here, we can set up the Python deep learning environment following these steps: select the "Python Deep Learning" option under "Python (legacy)" as shown in Fig. 10.10. In the "Library used for 'DL Python'" tab, choose the "Keras" framework. In the "Deep Learning Python environment configuration" tab, select the "Conda." In the "Keras" tab, choose the "py36_knime" environment that we just created. Finally, click on "Apply and Close" to apply the changes. Please note that the following workflows only support Keras, so we won't configure TensorFlow 2 for now.

For more detailed information, please visit the official website: https://docs.knime.com/latest/python_installation_guide/index.html#_introduction.

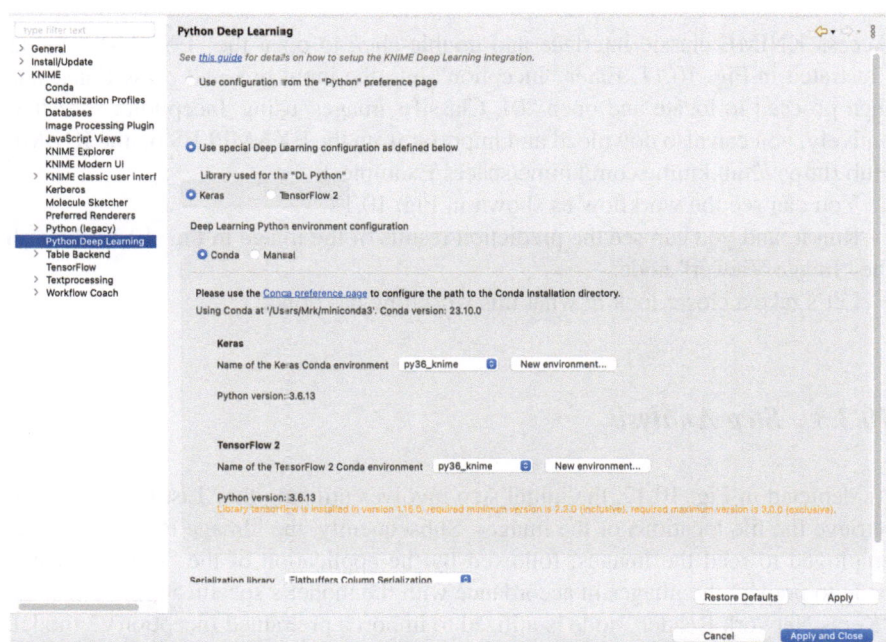

Fig. 10.10 Deep learning environment setup

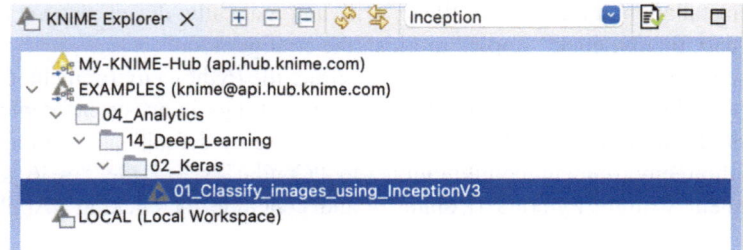

Fig. 10.11   01_Classify_images_using_InceptionV3

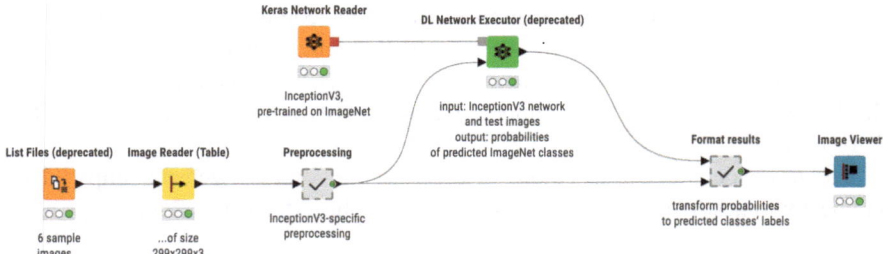

Fig. 10.12   Inception network workflow

10.3.2.4   Preliminary Trial

Access KNIME classic interface and double-click to open the "EXAMPLES," as illustrated in Fig. 10.11. Enter "Inception" into the input box and press Enter, and then proceed to locate and open "01_Classify_images_using_InceptionV3." Alternatively, you can also download and import it from the EXAMPLES on the KNIME Hub (https://hub.knime.com/knime/spaces/Examples).

You can see the workflow as shown in Fig. 10.12.

Run it, and you can see the prediction results of the image in Fig. 10.13 through the "Image Viewer" node.

Let's take a closer look at what this workflow has done.

10.3.3   Step Analysis

As depicted in Fig. 10.12, the initial step involves utilizing the "List Files" node to retrieve the file locations of the images. Subsequently, the "Image Reader" node is employed to read the images, followed by the application of the "Preprocessing" node to prepare the images in accordance with the model's specifications. Then, the "Keras Network Reader" node is utilized to import a pretrained InceptionV3 model. The subsequent procedure entails feeding the model and preprocessed image data into the "DL Network Executor" node for computation, undergoing additional data

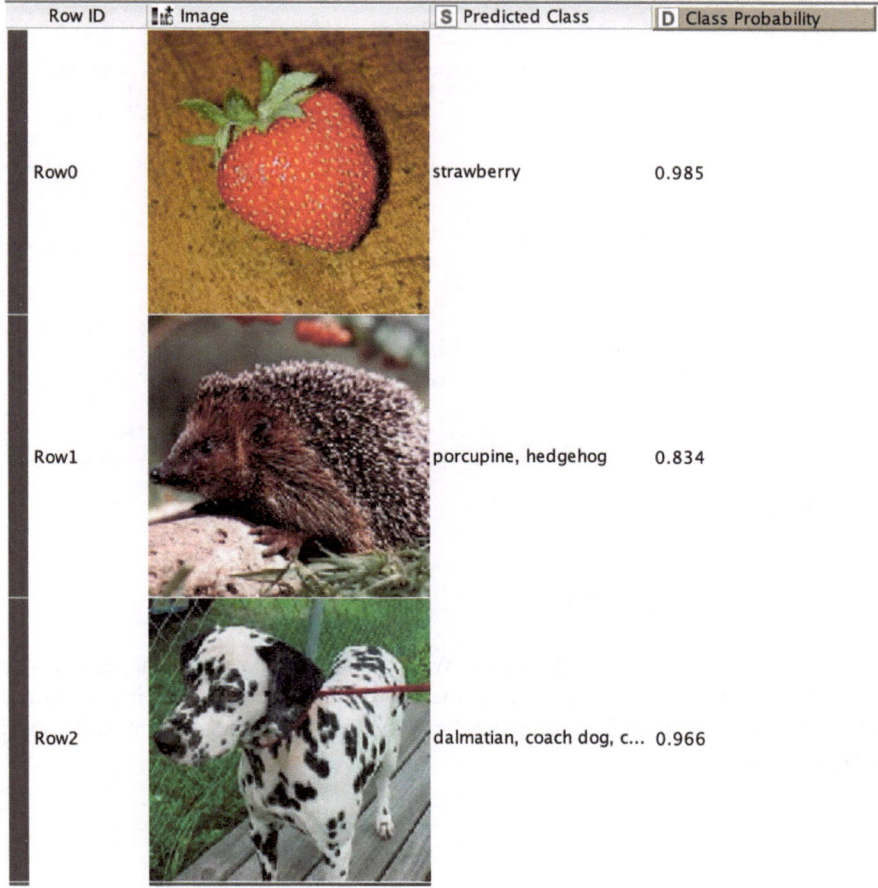

Row ID	Image	S Predicted Class	D Class Probability
Row0		strawberry	0.985
Row1		porcupine, hedgehog	0.834
Row2		dalmatian, coach dog, c...	0.966

Fig. 10.13 Image viewer

processing, and ultimately visualizing the image prediction results using the "Image Viewer." It is evident that this process is solely about the utilization of a model, without any training involved.

10.3.3.1 Preprocessing

We can briefly look at what preprocessing has done. Expand the "Preprocessing" node and find "Image Calculator." As shown in Fig. 10.14, you may notice that its function is to store the pixels of the original image as integers ranging from 0 to 255 before transforming the original image to integers ranging from −1 to 1.

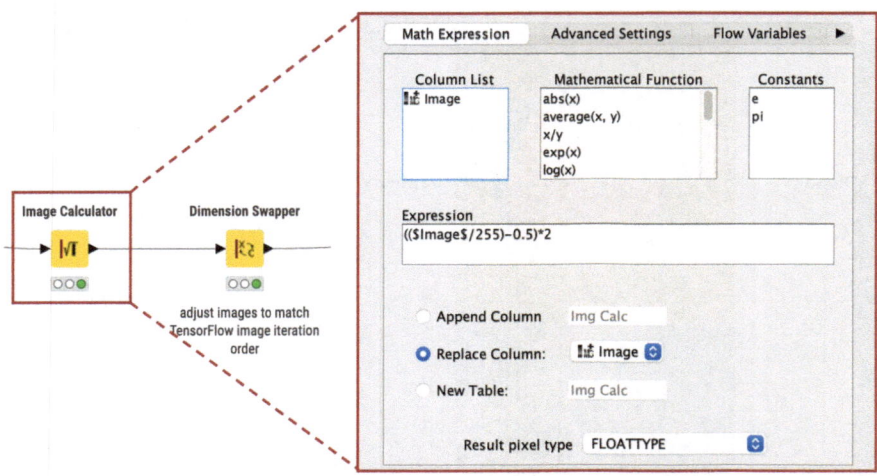

Fig. 10.14 Preprocessing

10.3.3.2   Image Preparation

What about predicting our own images? We need to first resize our images to a size compatible with the model input. That means we should change the image size to 299×299 as per the model requirements, specify the file storage location, and import using KNIME.

10.3.3.3   Training

At the beginner stage of deep learning, there is no need to worry about the model training; simply utilize the pretrained model.

10.4   Natural Language Processing

Natural language processing facilitates seamless communication between humans and computers through the use of natural language. This interdisciplinary scientific field integrates linguistics, computer science, and mathematics. Research within this domain focuses on the study of natural language, which encompasses the language utilized in everyday human communication. While closely related to the field of linguistics, it also presents distinct characteristics. Natural language processing includes various components, including semantic analysis, information extraction, and machine translation. Key challenges involve delineating word boundaries, disambiguating word meanings, addressing syntactic ambiguity, and managing imperfect or nonstandardized input, among other considerations.

10.4.1 How to Digitalize Natural Language?

For computers to effectively process natural language, it is imperative that they possess the capability to comprehend it. In the realm of image processing, images are digitally represented by pixels, with each pixel's numerical value carrying specific significance. For instance, the pixel value [0, 0, 0] denotes black, while [255, 255, 255] signifies white. However, the digitalization of language presents unique challenges. While some may propose the utilization of Unicode, language inherently embodies meaning and encompasses phenomena such as synonyms. Addressing these complexities within a digital framework raises questions about representing natural language as digital signals and extracting linguistic features, including semantic relations, from these digital signals. Can we really manage to do that?

One approach is word embedding, which maps words to a vector space, embedding words into another space that is easier for computation. As shown in Fig. 10.15, we map some words in three-dimensional space. Now the question is: how can we embed the words?

10.4.1.1 One-Hot Encoding

Initially, individuals employed a rudimentary approach to map each common word to a vector using one-hot encoding as shown in Fig. 10.16. One-hot encoding involves utilizing a binary vector with as many bits as there are words, where only one bit is set to 1 and all others are set to 0. For instance, the genders "male" and "female" can be encoded as 01 and 10, respectively. Following this encoding, each word represents a dimension, resulting in a large vector with many zeros (i.e., a sparse matrix). However, a significant issue arises from the fact that each word is orthogonal to every other word, signifying that each word occupies a distinct dimension. For example, in the scenario depicted in Fig. 10.17, let's consider the

Fig. 10.15 Word embedding

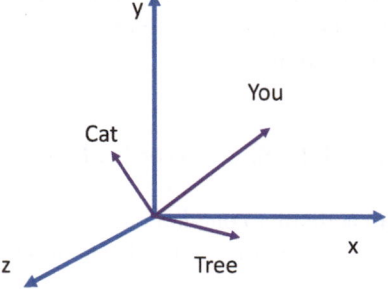

Fig. 10.16 One-hot encoded words

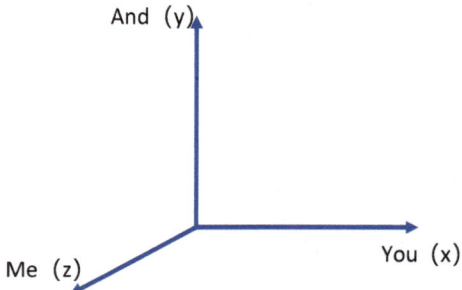

a	abbreviation		zoology	zoom
1	0		0	0
0	1		0	0
0	0		0	0
0	0		0	0
0	0		0	0
.	.	...	.	.
.	.		.	.
.	.		.	.
0	0		0	0
0	0		1	0
0	0		0	1

Fig. 10.17 Three-dimensional coordinates with examples of three words

And (y)

Me (z)

You (x)

phrase "you and me." If we aim to represent the three words in an n-dimensional space (assuming someone provides us with an $(n-1)$-dimensional paper), we can assign the words "you," "and," and "me" to the x, y, and z axes, respectively. Consequently, the word "you" can be vectorized as [1, 0, 0], "and" as [0, 1, 0], and "me" as [0, 0, 1]. These three words are orthogonal, as indicated by the dot product of their vectors being 0, suggesting no relationship between the words. However, based on our linguistic understanding, "you" and "me" should still exhibit some connection, given that they are both personal pronouns. Evidently, the current one-hot encoding method is inadequate in addressing the issue of word meaning correlation.

10.4.1.2   Bag-of-Words Model

In addition to one-hot encoding, another simple and straightforward method known as the "bag-of-words" (BoW) model has been developed. The bag-of-words model

Fig. 10.18 Bag-of-words model

conceptualizes sentences or texts as a bag containing various words, where the order of the words in the sentence is disregarded as shown in Fig. 10.18. The bag is represented in the form of {word: frequency of word occurrence}. For instance, the sentence "Learning methods are important for learning" can be transformed into {learning: 2, methods: 1, are: 1, important: 1, for: 1,}. If we are analyzing a world consisting solely of these few words, we can depict this bag as a vector: [2, 1, 1, 1, 1].

After implementing the bag-of-words model, the relationship between sentences can be assessed by computing the distance between them. This distance can be evaluated using measures such as cosine distance, Euclidean distance, and so on. For instance, if we have another sentence, "Learning methods are important" can be vectorized as [1, 1, 1, 1, 0]. It is evident from the vector perspective that these two sentences exhibit a relationship, and the comparison of different sentences can be facilitated by evaluating their distances to measure their similarity.

10.4.1.3 Neural Network Language Model

Whether it's one-hot encoding or the bag-of-words model, each word corresponds to one dimension. However, do we really need such a multitude of dimensions? Is it truly essential to represent "you" and "me" as separate dimensions, as demonstrated in the previous example? Can we categorize words manually along these lines of reasoning? For instance, "male," "female," "he," "she," etc. could be grouped under the dimension of "gender," while "me," "you," "it," etc. could be grouped under the dimension of "person." Is this approach overly laborious? Can the machine autonomously generate these dimensions for us? This is where a neural network language model (NNLM) comes into play. It not only predicts a specific word based on the context but also generates a matrix of word embeddings. Building upon this principle, methods such as Word2Vec are employed to compute word embeddings. These methods capture the nuances of natural language far more effectively than previous approaches. For instance, they can accurately identify the similarities between "Paris–France" and "Beijing–China" due to their word vector representations. From this perspective, word vectors encapsulate semantic meaning (Fig. 10.19).

Fig. 10.19 Word vector
relationships

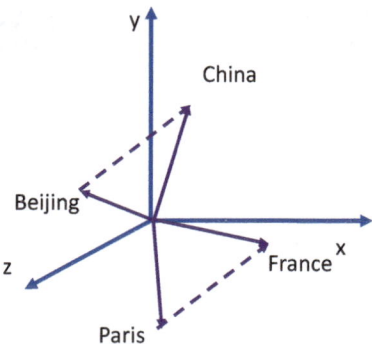

10.4.1.4   Polysemy

While methods like Word2Vec may initially seem promising, they often fall short in practical applications. The primary challenge lies in polysemy.

In order to solve this problem, we arrive at ELMo, which stands for Embeddings from Language Models. ELMo has the capability to dynamically adjust the vector representation (i.e., semantics) of words. For instance, to achieve this dynamic semantic adaptation, ELMo can train three types of word embeddings: word features, syntactic features, and semantic features. In practical application, these three-word embeddings are combined in a specific proportion based on the context.

ELMo uses LSTM to extract features. So, what is feature extraction, and what is LSTM?

10.4.2   Feature Extraction

How should language vectorization be utilized? Are all vectors input as features to a model? Similar to image recognition, directly manipulating thousands of vectors within the model can be challenging. Instead, the objective is to extract features and employ these features as inputs to subsequent models. The next question is how to extract features from words. The bottom layer of this network comprises individual words, representing the most fundamental level of features, while the layers above encompass higher-level syntactic and semantic features associated with context (https://colah.github.io/posts/2015-08-Understanding-LSTMs/).

An essential architecture for extracting word vector features is the recurrent neural network (RNN). As illustrated in Fig. 10.20, it is evident that each output is not only linked to the current input but also influenced by the preceding outputs.

However, the meaning of a word is not solely dependent on the preceding word but also on other words, whether they are in close proximity or distant. This necessitates the network to remember not only the nearby words but also those that are distant. Consequently, a specialized type of RNN network known as long

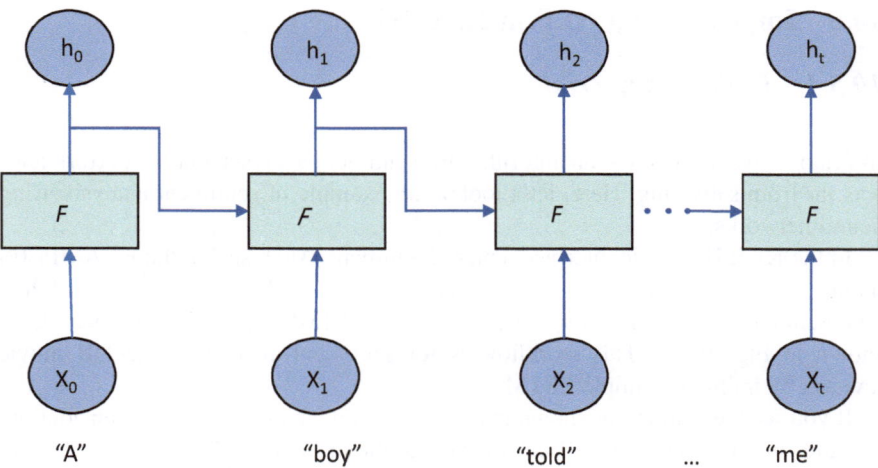

Fig. 10.20   Recurrent neural network (x is the input, h is the output, F is a unit of RNN)

short-term memory (LSTM) networks was devised. When integrated with multiple layers of LSTM, its capabilities will be further enhanced.

However, LSTM is not ideal for handling long sentences and has limited parallel computing capabilities.

10.4.2.1   Transformer

Given the issues with LSTM, is there a solution to address them? The Transformer, introduced by Google in the paper "Attention Is All You Need" for machine translation tasks in 2017, offers a potential solution (Ashish Vaswani et al. 2017). This approach incorporates the "Attention" mechanism, which mimics human attention by assigning varying attention weights to different words within each sentence and executing them in parallel. When humans read a paragraph, it is impractical to assign equal importance to every word, as some words hold greater significance and command more attention. The attention mechanism effectively addresses these challenges encountered with LSTM. Furthermore, BERT (Bidirectional Encoder Representations from Transformers), introduced in 2018, further enhances the performance of NLP across diverse tasks, building upon ELMo and Transformer (https://www.turing.com/kb/brief-introduction-to-transformers-and-their-power).

Note: CNN can also be used for feature extraction, which means CNN can also be applied in natural language processing tasks.

10.5   Implementing NLP in KNIME

10.5.1   Preliminary Trial

In Chap. 9, we used Bayesian algorithm for sentiment analysis, but the performance was far from satisfying. Here, let's look at an example of sentiment analysis using neural networks.

In the KNIME classic interface, enter "Sentiment_Analysis" in the EXAMPLES input field. Find the workflow "08_Sentiment_Analysis_with_Deep_Learning_KNIME_nodes," and open it, as shown in Fig. 10.21. This workflow is for sentiment analysis on IMDB movie reviews by training a simple LSTM.

If you see the warning as shown in Fig. 10.22, indicating that some extensions are missing, simply click on the "Yes" to agree to the installation. Then we just need to proceed with the next steps as instructed. After restarting the software, the changes will take effect. The process of installation is quite straightforward.

Run the workflow and output the predicted result ("prediction") of "sentiment" in the "Extract Prediction" node, as depicted in Fig. 10.23. Subsequently, assess the

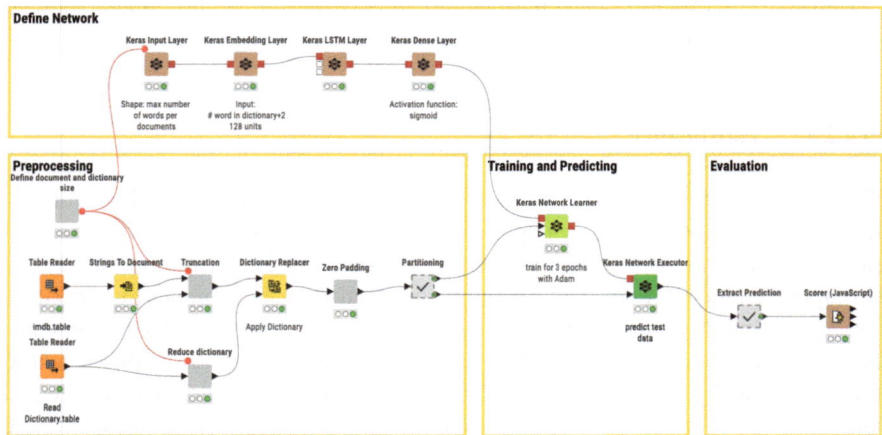

Fig. 10.21 "08_Sentiment_Analysis_with_Deep_Learning_KNIME_nodes" workflow

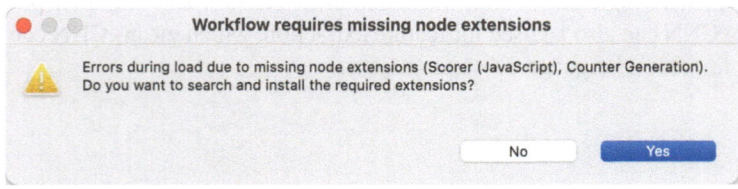

Fig. 10.22 Prompt indicating missing extensions

Rows: 25000	Columns: 4			
#	Row...	sentiment	AggregatedValues	dense_1/Sigmoid:0_0	prediction
		String	List	Number (double)	String
1	Row...	0.0	[20001,20001,20001,...]	0.947	1.0
2	Row...	1.0	[20001,484,21,...]	0.934	1.0
3	Row...	1.0	[1194,17932,22,...]	0.014	0.0
4	Row...	0.0	[542,1096,3,...]	0.052	0.0
5	Row...	1.0	[1407,20001,14,...]	0.117	0.0
6	Row...	0.0	[52,22,8,...]	0.281	0.0
7	Row...	1.0	[20001,400,33,...]	0.107	0.0
8	Row...	1.0	[722,2,407,...]	0.635	1.0
9	Row...	1.0	[1784,4164,46,...]	0.923	1.0
10	Row...	0.0	[20001,20001,6004,...]	0.031	0.0
11	Row...	0.0	[342,72,51,...]	0.038	0.0
12	Row...	0.0	[20001,1,870,...]	0.021	0.0

Fig. 10.23 Output results of "Extract Prediction"

Scorer View
Confusion Matrix

	0.0 (Predicted)	1.0 (Predicted)	
0.0 (Actual)	10512	1988	84.10%
1.0 (Actual)	3493	9007	72.06%
	75.06%	81.92%	

Overall Statistics

Overall Accuracy	Overall Error	Cohen's kappa (κ)	Correctly Classified	Incorrectly Classified
78.08%	21.92%	0.562	19519	5481

Fig. 10.24 Prediction accuracy

accuracy of the output prediction using the "Scorer (JavaScript)" node, as shown in Fig. 10.24, where the overall accuracy stands at 78.08%, indicating an overall satisfactory performance.

Now let's see how this workflow performs sentiment analysis.

10.5.2 Step Analysis

In Fig. 10.21, the first two "Table Reader" nodes read "imdb.table" and "Dictionary. table," respectively. Subsequently, "imdb.table" is digitized and partitioned into training and test sets using "Dictionary.table." Concurrently, an LSTM neural network is configured using four Keras nodes. The training set is then trained using the "Keras Network Learner" for neural network training, while the test set is evaluated using the "Keras Network Executor." The resulting test outcomes are further processed, culminating in the calculation of the prediction score through the "Scorer (JavaScript)" node.

10.5.2.1   Data Input

The first "Table Reader" node inputs 50,000 lines of IMDB movie reviews, as shown in Fig. 10.25; the data includes text description "text" and sentiment label "sentiment": 0 for negative and 1 for positive. The second "Table Reader" node reads the English vocabulary dictionary, as shown in Fig. 10.26. This table maps each English word to unique indices "Trunaction" and "WordAsinteger." The digitization of the text is achieved through the vocabulary dictionary "Dictionary.table."

10.5.2.2   Text Processing

The text processing involved is quite complex, and we only need a brief understanding of its functionality. Upon expanding the "Truncation" component, it is evident that the "Dictionary Replacer" node replaces complete terms matching the specified dictionary "Dictionary Replacer" terms ("Term as String") in the input document "imdb.table" with index values ("Truncation") while utilizing the "OpenNLP English WordTokenizer" as the WordTokenizer, as depicted in Fig. 10.27.

Then, through the "Zero Padding" component, as shown in Fig. 10.28, the "text" is processed into a numerical vector "AggregateValues" as the input of the model.

Rows: 50000	Columns: 2		Table Statistics
#	Row...	text *String*	sentiment *Number (integer)*
1	Row0	Story of a man who has unnatural feelings for a pig. Starts out with a opening scene that is a terrific exa...	0
2	Row1	Airport '77 starts as a brand new luxury 747 plane is loaded up with valuable paintings & such belonging...	0
3	Row2	This film lacked something I couldn't put my finger on at first: charisma on the part of the leading actres...	0
4	Row3	Sorry everyone,,, I know this is supposed to be an art film,, but wow, they should have handed out guns a...	0
5	Row4	When I was little my parents took me along to the theater to see Interiors. It was one of many movies I ...	0
6	Row5	It appears that many critics find the idea of a Woody Allen drama unpalatable. And for good reason: the...	0
7	Row6	The second attempt by a New York intellectual in less than 10 years to make a Swedish film - the first be...	0
8	Row7	I don't know who to blame, the timid writers or the clueless director. It seemed to be one of those movie...	0
9	Row8	This film is mediocre at best. Angie Harmon is as funny as a bag of hammers. Her bitchy demeanor fro...	0
10	Row9	The film is bad. There is no other way to say it. The story is weak and outdated, especially for this countr...	0
11	Row10	This film is one giant pant load. Paul Schrader is utterly lost in his own bad screenplay. And his directing...	0

Fig. 10.25   IMDB movie review data

Rows: 211065	Columns: 3		Table Statistics		C
#	Row...	Term as String *String*	Trunaction *String*	WordAsInteger *String*	
2	Row1	the	10198095	2	
3	Row2	a	10108235	3	
4	Row3	and	10111291	4	
5	Row4	.	10006966	5	
6	Row5	of	10168880	6	
7	Row6	to	10199625	7	
8	Row7	is	10154264	8	
9	Row8	in	10152022	9	
10	Row9	this	10198645	10	
11	Row10	it	10154375	11	

Fig. 10.26   English dictionary data

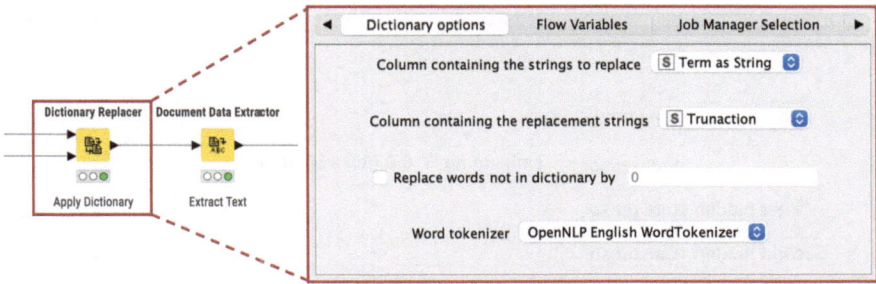

Fig. 10.27 Text processing

Rows: 50000	Columns: 2		Table Statistics	
#	Row...	sentiment Number (integer)	AggregatedValues List	
1	Row0	0	[1898,6,3,...]	
2	Row1	0	[12160,20001,543,...]	
3	Row2	0	[52,29,3057,...]	
4	Row3	0	[2876,328,5,...]	
5	Row4	0	[266,13,17,...]	
6	Row5	0	[43,706,12,...]	
7	Row6	0	[18,388,585,...]	
8	Row7	0	[13,51,23,...]	
9	Row8	0	[52,29,8,...]	
10	Row9	0	[18,29,8,...]	
11	Row10	0	[52,29,8,...]	
12	Row11	0	[18,122,15,...]	

Fig. 10.28 Text to vectors

10.5.2.3   Network Construction

In the workflow depicted in Fig. 10.21, the input layer, embedding layer, LSTM layer, and dense layer are sequentially employed to customize the LSTM neural network. Specific parameters are configured for each type of neural network. Using the LSTM layer as an example for better comprehension, upon accessing the configuration interface of the "Keras LSTM Layer" node in Fig. 10.29, the input tensor dimension is set to be 80 × 128, consistent with the preceding layer's embedding layer. The activation function is Tanh, while the recurrent activation function is a piecewise linear approximation of sigmoid known as "Hard sigmoid." Both the dropout and recurrent_dropout rates are set at 0.2, and the implementation mode is specified as 1.

10.5.2.4   Network Training and Testing

Open the "Keras Network Learner" node, and you can see the following settings for the neural network:

- Input data. As shown in Fig. 10.30, in the input data options, the input columns will include the "AggregateValues" after text digitization. Since it is an integer array, the "Conversion" is set to "From Collection of Number (integer)."

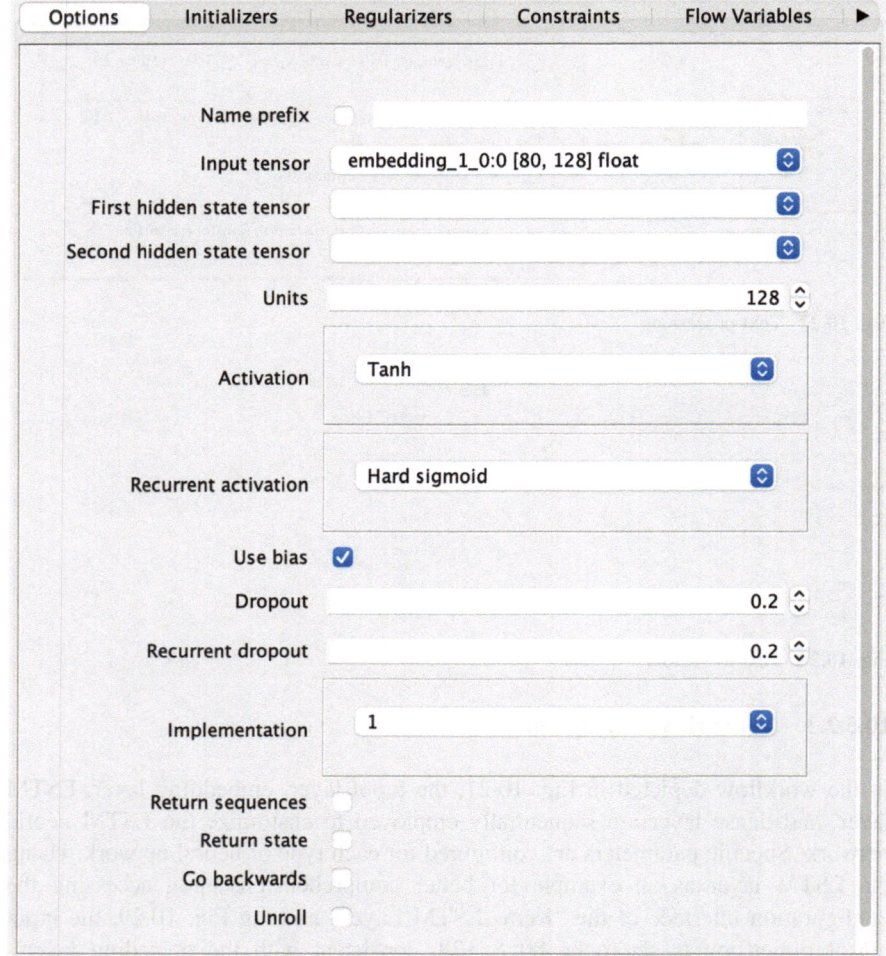

Fig. 10.29 Configure LSTM layer network

- Training target. As shown in Fig. 10.31, in the "Target Data" option, the output target (target columns) is selected as "sentiment." Since it is an integer ranging from 0 to 1, the "Conversion" is set to "From Number (integer)." Standard loss function is set to "Binary cross entropy."
- Hyperparameter. As shown in Fig. 10.32, in the "Options," you can see the settings of neural network hyperparameters.

Open the "Keras Network Executor" node and you will see the settings shown in Fig. 10.33.

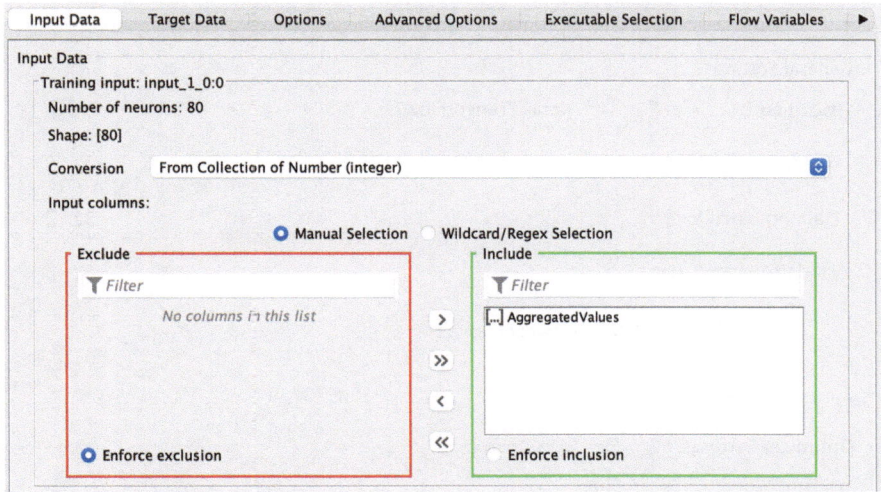

Fig. 10.30 Neural network training—input data

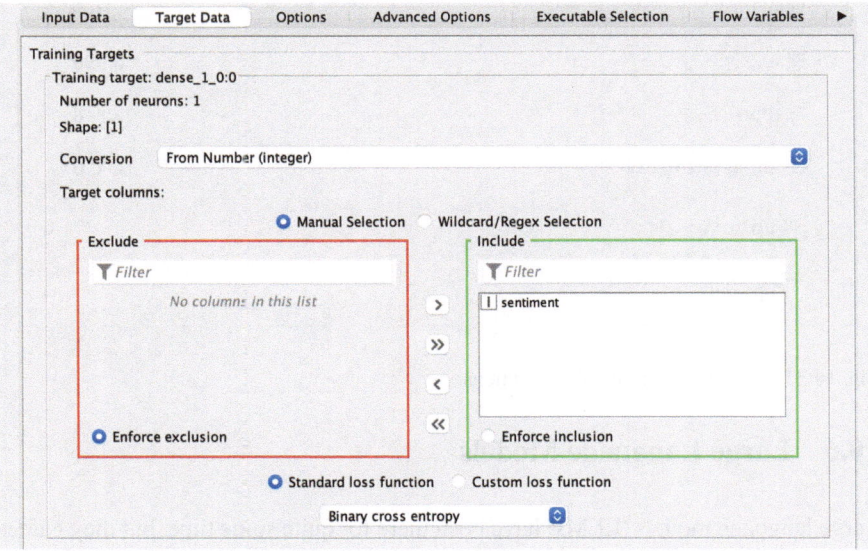

Fig. 10.31 Neural network training—target data

- "Back end" is set to "Keras (TensorFlow)."
- Inputs. The input column is "AggregateValues," and the "Conversion" selects "From Collection of Number (integer)."
- Outputs. Since its output are numbers, "Conversion" is set to "To Number" and "Output columns prefix" is set to the prefix "dense_1/Sigmoid:0_."

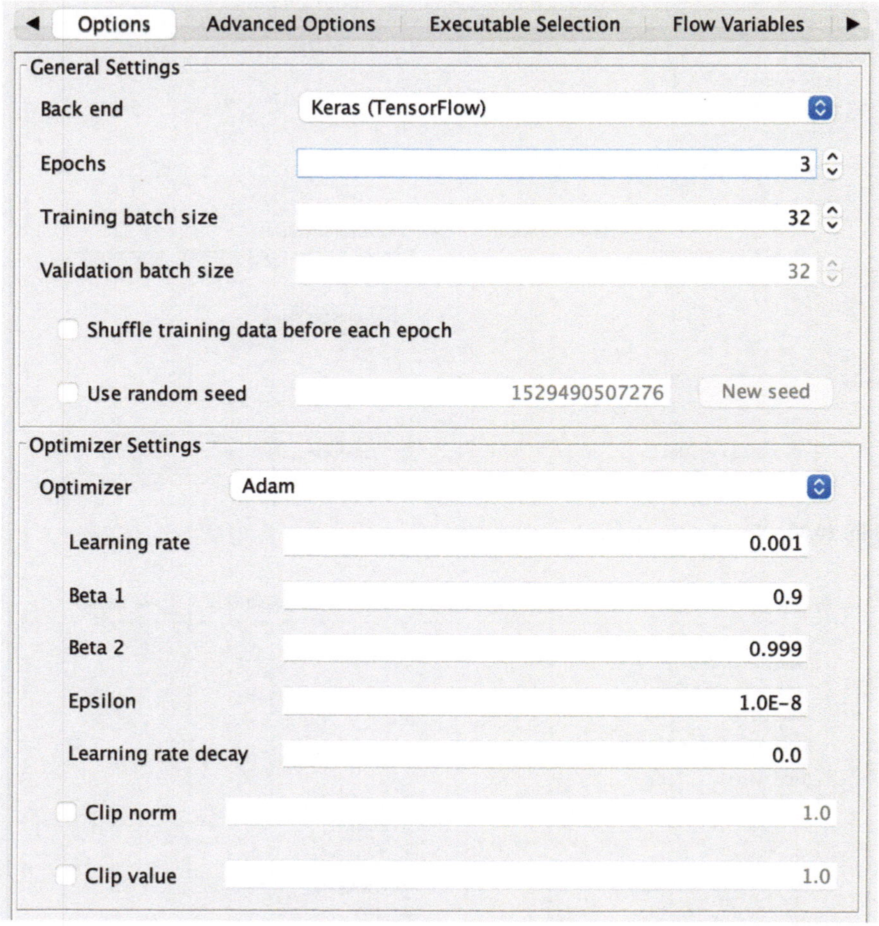

| ◄ | Options | Advanced Options | Executable Selection | Flow Variables | ► |

General Settings

Back end	Keras (TensorFlow)	⬍
Epochs		3 ⬍
Training batch size		32 ⬍
Validation batch size		32 ⬍

☐ Shuffle training data before each epoch

☐ Use random seed   1529490507276   New seed

Optimizer Settings

| Optimizer | Adam | ⬍ |

Learning rate	0.001
Beta 1	0.9
Beta 2	0.999
Epsilon	1.0E–8
Learning rate decay	0.0
☐ Clip norm	1.0
☐ Clip value	1.0

Fig. 10.32 Neural network training—options

10.6   Large Language Models

Large language models (LLMs) have been there for quite some time, but they gained widespread attention when OpenAI introduced the ChatGPT model in 2022 for natural language text generation. Therefore, many individuals are only familiar with ChatGPT and may not be aware of the broader concept of large language models, often using ChatGPT as a generic term for large language models in their daily life. In reality, there are more LLMs other than OpenAI. For instance, in February 2023, Google unveiled the chatbot Bard, powered by Google's extensive language model LaMDA. Additionally, in March 2023, Baidu in China officially launched Wenxin Yiyan, underpinned by the Wenxin LLM as its technical foundation.

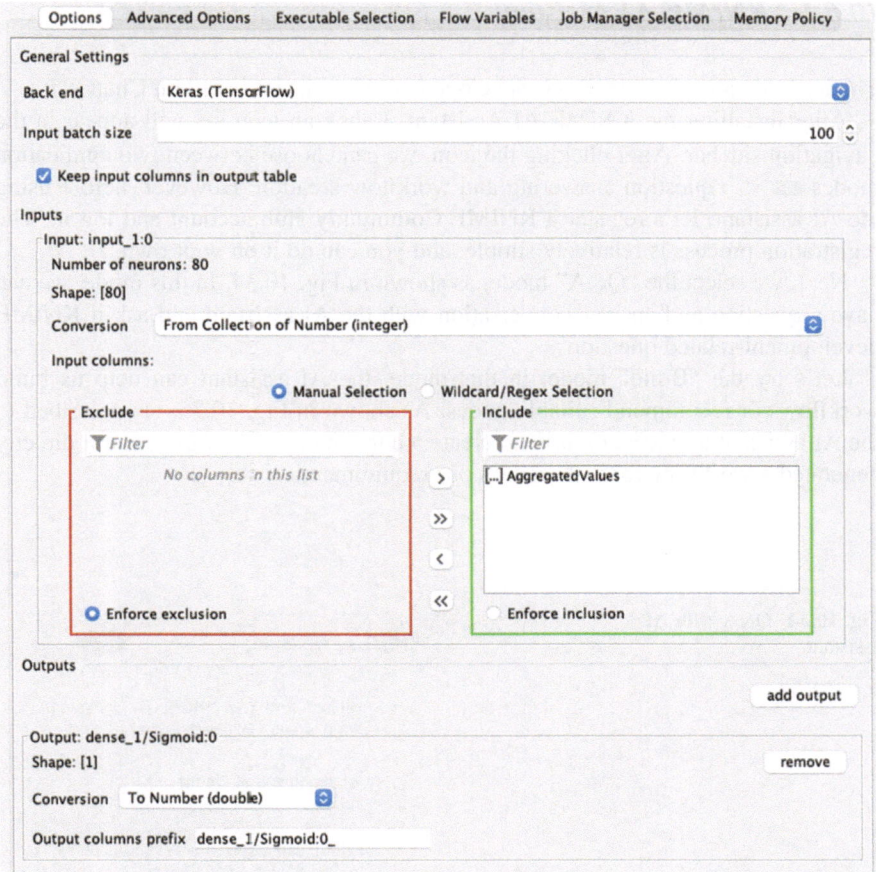

Fig. 10.33 Neural network testing

The characteristic of a LLM is that it is large, which means it has many parameters. The numerous parameters help the model learn the complex patterns inherent in the language and generate text that can be indistinguishable from real text.

Now, the new version of KNIME provides us with an AI assistant and AI extension based on LLM. Let's give it a try. First, install the following two extensions (you may refer to the previous chapter for detailed installation guide):

- KNIME AI Assistant (Labs).
- KNIME AI Extension (Labs).

We can try them out, respectively, in the following sections.

10.6.1 KNIME AI Assistant

The new version of KNIME has developed an AI assistant based on ChatGPT.

After installing the KNIME AI Assistant (Labs), an icon will appear in the navigation sidebar. After clicking the icon, we can choose between two application modes : question answering and workflow creation. However, before using the AI assistant, let's register a KNIME Community Hub account and log in. The registration process is relatively simple, and you can do it on your own.

Next, we select the "Q&A" mode, as shown in Fig. 10.34. In this mode, we can have a question-and-answer conversation with the AI assistant and ask it KNIME development-related questions.

Let's try the "Build" mode; in this mode, the AI assistant can help us build workflows or recommend suitable nodes. As shown in Fig. 10.35, we described to the AI assistant a request on how to create a char chart and box plot, and it directly generated a workflow for us based on our requirements.

Fig. 10.34 Q&A with AI assistant

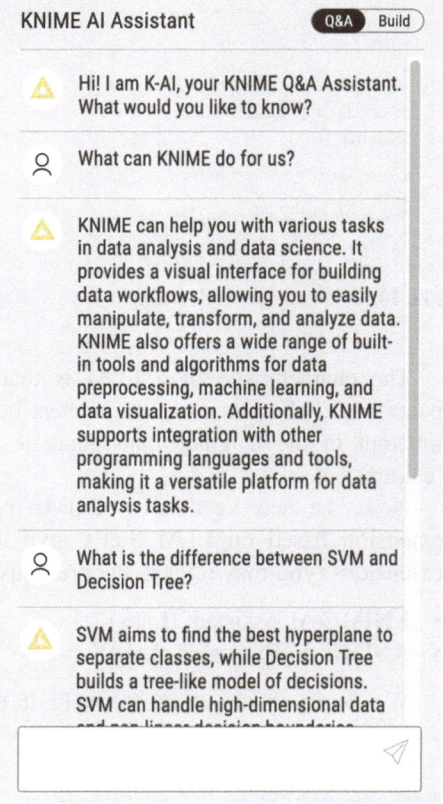

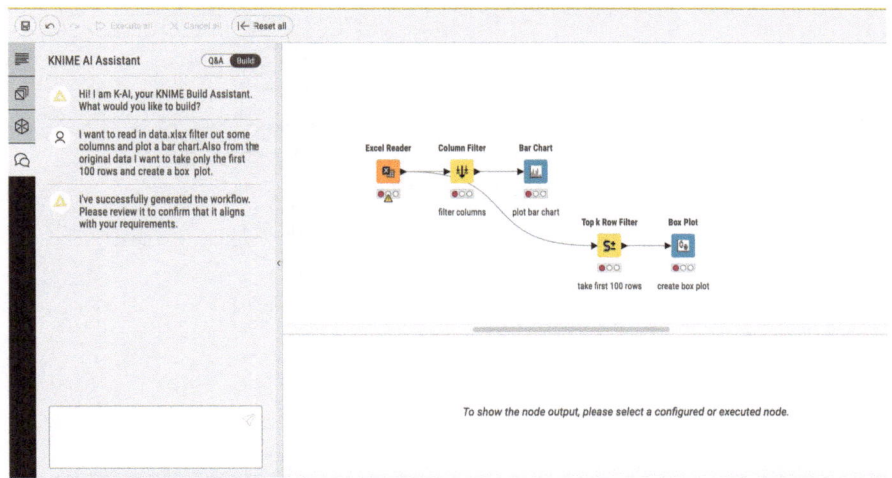

Fig. 10.35 Build a workflow using AI assistant

10.6.2 KNIME AI Extension

The KNIME AI extension provides nodes for building LLM-powered applications (such as chatbots) through KNIME workflows. It supports both OpenAI models and the open-source LLM from the Hugging Face Hub.

Next, we will use a workflow provided by "OpenAI Agent" on the KNIME Hub to experience how to build LLM-driven applications through KNIME workflows, as shown in Fig. 10.36. This workflow can also achieve human-machine interactive question-answering functionality.

However, this approach usually requires an OpenAI key and is not very convenient to use. If you are interested, you can download it and try it out.

For more AI extension example workflows, please refer to the official tutorial at https://hub.knime.com/knime/spaces/AI%20Extension%20Example%20Workflows/~CzxTTU_Gi4moY5Gz/.

10.7 Practice

1. What model can be used to solve image recognition problems?
2. With knowledge in this chapter, we can use KNIME to do some deep learning-related tasks. Open "03_Train_MNIST_classifier" and give it a try.
3. In the "08_Sentiment_Analysis_with_Deep_Learning_KNIME_nodes" workflow, how should text processing be carried out if it is switched to your native language?

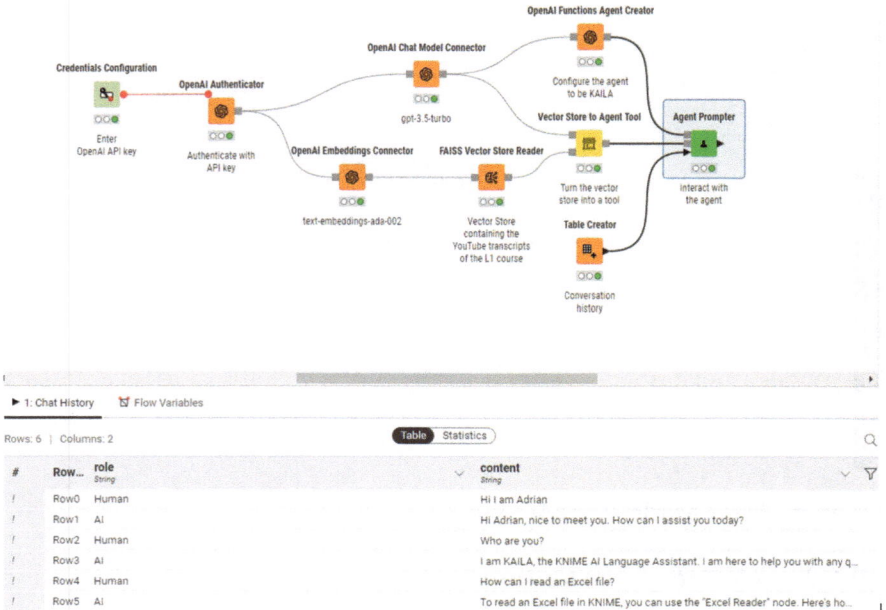

Fig. 10.36 OpenAI agent

References

Ashish Vaswani, Noam Shazeer, Niki Parmar, Jakob Uszkoreit, Llion Jones, Aidan N. Gomez, Łukasz Kaiser, and Illia Polosukhin. 2017. Attention is all you need. In Proceedings of the 31st International Conference on Neural Information Processing Systems (NIPS'17). Curran Associates Inc., Red Hook, NY, USA, 6000–6010.

https://ujjwalkarn.me/2016/08/11/intuitive-explanation-convnets/

https://www.turing.com/kb/brief-introduction-to-transformers-and-their-power

https://colah.github.io/posts/2015-08-Understanding-LSTMs/